ASTRONOESIS

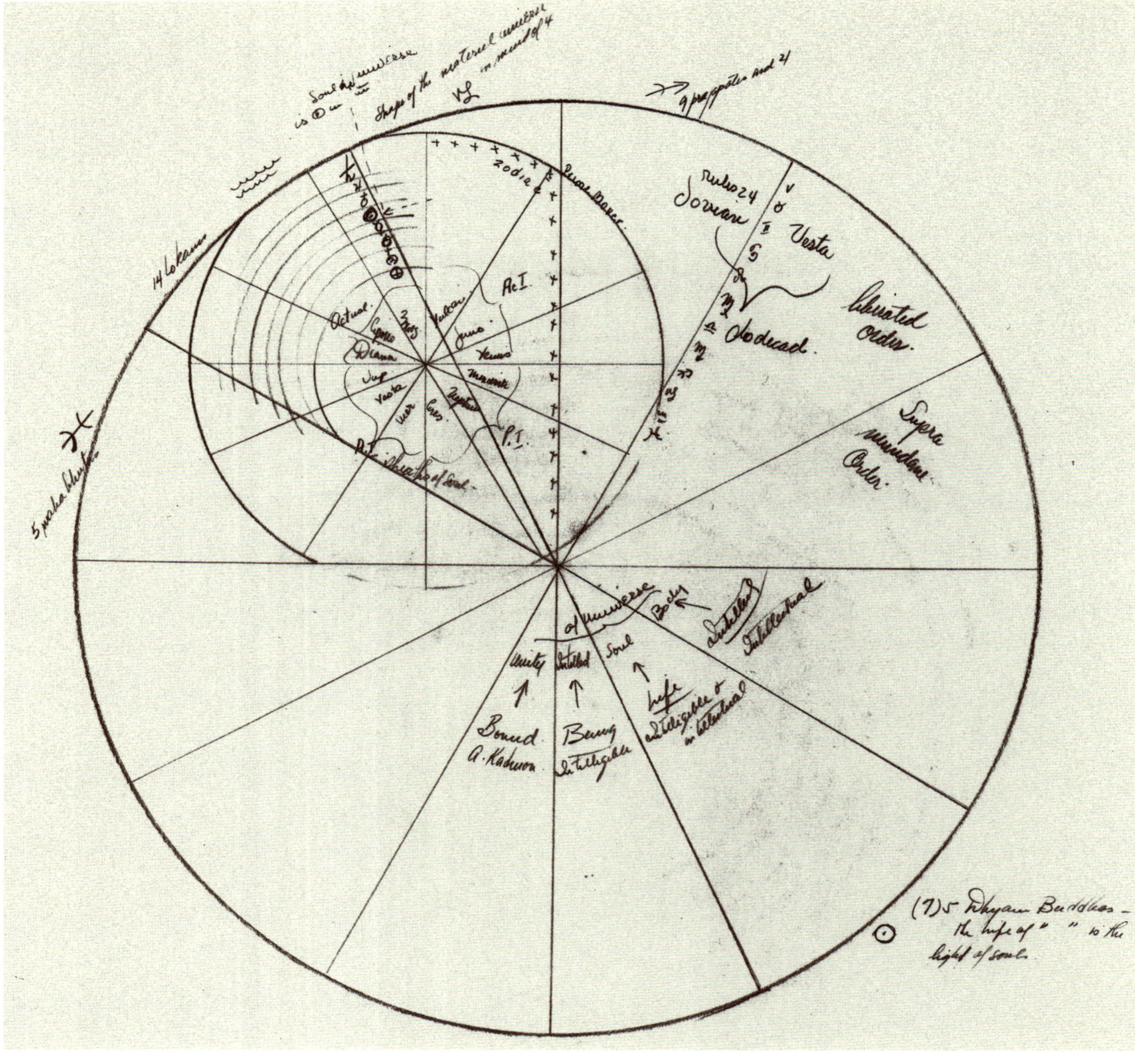

14 lokams
zodiac
Vesta
Dodecad
Elevated order.
Supra mundane Order.
Ac. I.
Vulcan
Juno
Venus
Mercury
Neptune
Diana
Jup
Vesta
Mar
Ceres
P. I.
of universe
Body
Intellect
Intellectual
Unity
Intellect
Soul
Bound
Being
Intelligible
Life
(7) 5 Dhyani Buddhas –
light of soul.

ASTRONOESIS

Philosophy's Empirical Context ◆ Astrology's Transcendental Ground

Anthony J. Damiani

Published for Wisdom's Goldenrod, Ltd. by Larson Publications

Publisher and editors gratefully acknowledge the following individuals and institutions for permission to reprint material under copyright: p. viii photograph of Anthony Damiani by Jan Hallén © 1983; pp. ix and x text courtesy of Kathleen Damiani; p. 5 by permission of the British Library, (item 30.G.9); pp. 13 and 257 drawings by Rosemarie Quebral; p. 316 photograph of Kalachakra mandala © 1988 Namgyal Monastery; p. 316 Kalachakra deity, drawing by Sidney Piburn © 1986; pp. 289–299 Sabian symbols © 1993, by permission of Aurora Press.

International Standard Book Number: 0-943914-00-0

Library of Congress Control Number: 00-133496

Published for Wisdom's Goldenrod, Ltd. by
Larson Publications
4936 NYS Route 414
Burdett, New York 14818 USA
a division of the Paul Brunton Philosophic Foundation

07 06 05 04 03 02 01 00
10 9 8 7 6 5 4 3 2 1

The dedication is eternal and to the Eternal.

Anthony Damiani

august 24, 1922–october 2, 1984

Contents

CHAPTER THREE ◆ SOUL

CHAPTER FOUR ◆ THE SYSTEM OF NATURE

Anthony Damiani in 1983

Foreword: An American Philosopher

His love for wisdom and meaning in all its historic diversity may only be compared to his love of humanity and its higher potentials. It was an occasion to find a thinker or philosopher with whom he wasn't acquainted. For philosophy to him meant comparative philosophy, involving dialogue, debate, discussion, and exchange of ideas. In this—the realm of thought—he was most American, accepting all philosophies from around the world as valid perspectives but demanding an open exchange through rational discussion. And if he demonstrated a preference, it was for the spirit of Greek philosophy where the spirit and principles of democracy itself were born some 2500 years ago.

He is an American philosopher in the tradition of Ralph Waldo Emerson, wherein he placed upon himself the demand to acknowledge and investigate man's striving for a nobler life as bodied forth in its diverse cultures, religions, and philosophies. He studied under Christian Jesuits, Hindu pundits, Zen Buddhists, N.Y.U. professors. Yet, like Emerson, he had to go his own way in the face of opposition from all quarters—friends, foes, family, and students; not in the spirit of criticism and rejection but in the need for universalism of approach and understanding in our times which flowed from his vision.

Like other great leaders he had an ennobling vision, in which different thinkers, religions, philosophers could think together, contemplate divinity together, laugh together, without giving up their own heritage, could side by side enrich life through their different approaches.

For all his erudition he was no scholar or academician, but more like an artist inspiring those around him to explore the mysteries of the universe, soul, and God. Always egging one on, ensnaring one into *thinking*, and in the act of thinking he would point to those diviner implications in Man, the light of his soul, its ordering presence, and the facticity of each person's higher potentials.

Philosophy for Anthony was this inalienable aspect of Man. No school, religion, or group has a monopoly on it, for it flows and follows along the spirit of wonder and enquiry itself. For, as he once expressed, philosophy aims to bring about Good Citizenship with the whole of mankind and the world.

—composed for Anthony's memorial service, October 7, 1984, by his son, Paul Damiani

Editors' Introduction

ANTHONY DAMIANI WAS AN INFORMED VISIONARY. He learned all he could from giants who came before him. Then—as carriers of a new impulse so often do—he stood on the shoulders of those he thought tallest, and saw more. What he saw inspired him, and in turn inspired others who experienced the spiritual passion and human warmth it sustained in him.

Much of Anthony's creative work was done in the context of preparing and leading classes at Wisdom's Goldenrod Center for Philosophic Studies, which came into being under his directorship in 1972. In the mid-seventies, he started to express his key ideas in writing and enlisted a few students to help write papers that would form the basis of a book. Under the working title of *Astronoesis* (derived from Greek words that mean Star-Wisdom), the writing evolved in the spirit of a suggestive enquiry rather than with the goal of producing a polished, final statement.

Astronoesis is a maiden voyage into an unusually fertile way of thinking. As Anthony became increasingly adept at it, he evolved ways of expressing his vision more vividly. In the process, he discovered and developed a language increasingly more adequate to the details of that vision—one that sustains deep thought on the nature of reality and the complex interrelationships of its metaphysical, cosmological, psychological, and physical levels. His method often makes "thinkable" things we could not think before for lack of adequate language.

Anthony was not the type of thinker who was meticulous about using primary sources, in dotting all his philological i's, or crossing all his historical t's. His strength was in penetrating deeply into big ideas and seeing how they live together as parts of an evolving organic whole.

Many of us who embark from this seminal work onto journeys of our own will respond at two levels to Anthony's inspiration. Both responses will draw power from the Vision we encounter through reflecting deeply on this pioneering work.

At one level, we will try to express in more familiar terms what catches our breath in Anthony's writing. We will try to translate his multivalent intuitions into the kind of thinking that our conventional dualistic, school-fed intellect is more familiar with (and that our separative ego is more comfortable with). To some extent we will succeed. But on the whole this kind of thinking will give photographs, or at best video clips, of a majestic bird in flight. It won't deliver the experience of flying *as* that bird. And it won't domesticate the raw, challenging power of the text that urges us to gather our energies, soar to its height, and See.

As we rise more often to this challenge, we will awaken in jolts and flashes at a higher level and become part of a motley crew: sometimes seeing, sometimes blind; sometimes hearing, sometimes deaf; sometimes flying, sometimes losing even the memory of wings. Painfully honest with ourselves. Clumsily creative. Plugged in to the Vision one moment, aware of having lost it the next. Always seeking more of it. Often experiencing what Anthony felt when he could find no good way to express what he had seen, or to express it even as it shone before his rapt eyes.

* * * *

Anthony invoked the third-century sage Plotinus, founder-genius of the Neoplatonic tradition, as his guide for writing *Astronoesis.* Through many years of extensive study and rigorous comparing of the major spiritual traditions, his admiration of Plotinus had increased steadily. He was awed and inspired by Plotinus' unusual combination of skills: the ability for profound mystical introversion and the intellectual development to articulate his immediate mystical experience in precise, thrilling detail.

When it came to understanding, developing, and deepening his own vision, Anthony's repeated readings of Plotinus' *Enneads* offered the most fertile context for his own insights to emerge. He felt that he owed much of his understanding to these intense readings, and that he could best develop his own original contribution with ideas and terms acquired through them. The Plotinian framework was sufficiently deep and inclusive for Anthony to juxtapose and integrate *detailed* intuitions that he had earned through personal experience and broad study of many sacred teachings.

Astronoesis accordingly has four major chapters. Each focuses on one facet of the four-fold nature of reality as Plotinus saw it: the One, the Intellectual-Principle (Nous), Soul, and the system of Nature. Each of these fundamental features of reality—as one, as intelligent, as alive, and as manifesting—yields a unique perspective on the whole. Each contributes a different richness to the mandala in which Anthony planted and nurtured the seed ideas of his relentlessly creative work.

* * * *

But *Astronoesis* is not direct from Anthony's pen. His life ended before his work on it was complete. Shortly before his death in October, 1984, he confided to his wife that the manuscript included all of the ideas he planned to discuss, but much work remained undone: The chapter on the One was "finished"; the Intellectual-Principle was "almost finished"; Soul was still "a series of papers"; and Nature "needed work," including more on astrology.

Anthony's teacher Paul Brunton had told him in 1981 that he didn't have to finish the book, just get the intuitions onto paper well enough that they wouldn't be lost; and in the process he should do his best to see that others around him could complete the book in the event that he could not. He had done the first. It remained to be seen how well he had succeeded with the second.

Several of us who had studied, meditated, and worked on this text with Anthony at Wisdom's Goldenrod eventually found out. It took years of exploring options to reach consensus on how best to complete and present the material Anthony left. Finally, thanks to the patient, persevering, and often inspired leadership of Avery Solomon (who worked most closely and continuously with Anthony on the manuscript), we are ready for publication.

The main text is Anthony's work, essentially as he left it. Sentences and paragraphs often have been moved within or between sections for better continuity and/or clarity. Words have been changed only when the improvement is significant and unquestionably clarifies the meaning. We have put interpretive additions that may influence meaning in [square brackets].

We have provided several levels of commentary:

Editors' Introductions and Editors' Extensions. These papers, which the editors (primarily Avery Solomon) have based on Anthony's ideas and notes, appear in some sections. We feel that the information in them is critical to understanding the sections they accompany.

Margin Notes. These appear on many pages to enhance or bring necessary clarification or development to the text they accompany. ▼

▼ Notes: Two facing pages may have up to five notes, correlating to points in the text, each successive note marked by one of the following glyphs: ▼ ■ ● ▲ ◆

Epilogue. This important section gives a brief overview of how Anthony combined the principles in *Astronoesis* with traditional astrology to appreciate the flow of meaning in an individual life. It also brings a "roundness" to the material in The System of Nature that we were unable to develop to our satisfaction within that chapter itself. Detailed applications of this method will appear in a series of books to be published through the next several years.

Supplementary Notes. The superscripts that appear within the flow of the text refer to numbered notes in the back matter. These consist of scholarly references, technical information, and short essays (in some cases Anthony's own) that expand the text. Consult them only for supplemental information.

Glossary and Index. We have also developed a glossary to familiarize readers with Anthony's sometimes unique use of terms. Consult this first when uncertain of terms that appear in the flow of reading. If that doesn't provide enough help, then locate the term in the index, where asterisks indicate pages on which the term is explained in the text.

Anthony customarily illustrated and developed his ideas with diagrams and pictures. Facsimiles of some of his hand-drawn diagrams provide an aesthetic ambiance to the beginning of each of the major sections; they are not meant to illustrate or accompany any particular section of the text. The majority of the diagrams and figures that do illustrate text have been computer generated; most of them are reproductions of diagrams Anthony used and indicated for use in the text; the remainder are based on diagrams he used in classes. A few in-text diagrams are facsimiles of Anthony's originals.

Quotations from Plotinus' *Enneads,* unless otherwise noted, are from the fourth edition of Stephen MacKenna's translation, which Anthony preferred. The definitive edition of that text, published by Larson Publications in 1992, has an appendix that gives a concise overview of key ideas in *Astronoesis.*

Quotations from Plato, unless otherwise noted, are from *Plato: The Collected Dialogues,* edited by Edith Hamilton and Huntington Cairns. Anthony also prized the Thomas Taylor and F. Sydenham translation of Plato, primarily because of Taylor's valuable notes.

* * * *

We express deep gratitude to all past and current members and associates of Wisdom's Goldenrod who contributed skills, financial support, and many other resources to this project through the years. Completion of this work has truly been a community effort. May we all be worthy of the community it has helped foster.

ASTRONOESIS

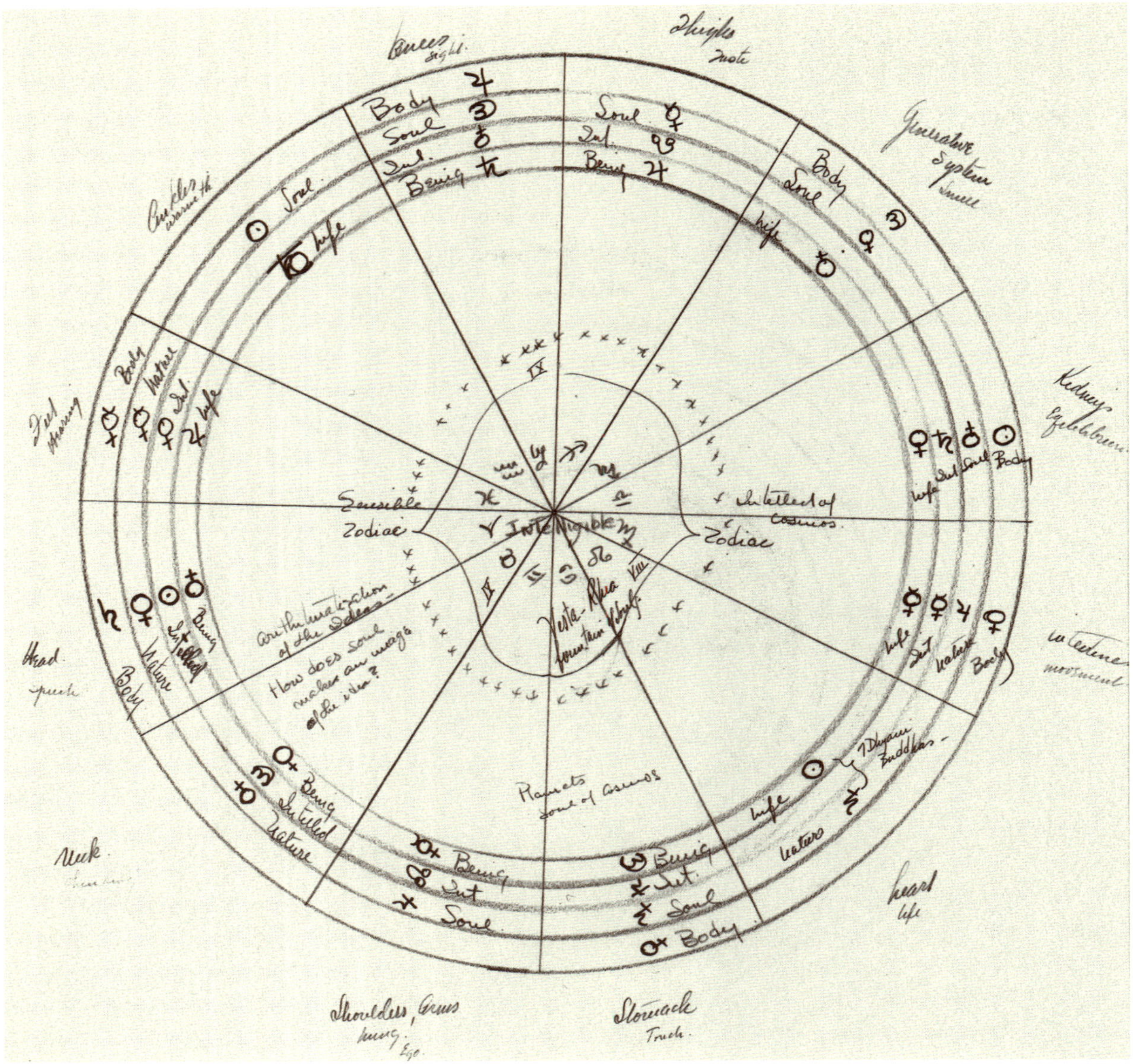
Knees
Thighs
Generative System
Body
Soul
Int.
Being
Life
Nature
Kidneys
Intestines
movement
Head
Neck
Feet
hearing
Shoulders, arms
Stomach
Touch
heart
life
Sensible
Zodiac
Intellect of cosmos
Zodiac
How does soul makes an image of the idea?
Planets
soul of cosmos
7 Dhyani Buddhas

Invocatory Prelude

The sublime majesty of Truth's presence in the sacred sermons of so many wise ones, those of good will, ultimately guides us to the ways of peace and wisdom, the lovely, the holy. And so our troubled hearts are reassured, and we can read and reread about God's word in many scriptures.

However incomprehensible, a faith settles into our hearts that awakens the desire to understand Infinity and Forever. The dedication is eternal and to the Eternal. For the Word becomes flesh—and this is an eternal utterance, an act that includes the human heart mounted on a towering aspiration. Yet even this pyramiding love is without effect unless winged with His Grace. So we shall let fly our supplications, enfolded in holy silence on wings of veneration, to the throne of our most Sovereign Good with the request to guide the spirit of our understanding, wherein we stand within His Light, in order to see the pouring out of the Sacred Word made manifest.

We read, in Eckhart's seventeenth sermon:

> *In principio erat verbum.* (John 1:1) Theologians talk of the eternal Word. God never spoke but one word, and that is still unspoken. The explanation is this. The eternal Word is the logos of the Father which is his only begotten Son our Lord Jesus Christ. In him he pronounces all creatures without beginning and without end. This accounts for the Word remaining unborn, for it never came out of the Father. (p. 57)

Deepening the same idea, and perhaps stating it more profoundly, in sermon 35:

> *Dominus dicit: sta in porta domus domini et praedica verbum istud.* (Jer. 7:2) The Lord says 'Stand in the gate of God's house and proclaim his word, extol his word.' The heavenly Father speaks one Word and that he speaks eternally and in this Word expends he all his might: his

> entire God-nature he utters in this Word and the whole of creatures. This Word lies hidden in the soul unnoticed and beyond our ken, and were it not for rumours in the ground of hearing we should never heed it; but all sounds and voices have to cease and silence, perfect stillness, reign. This is a meaning I will not pursue. (p. 95)

In this relatively short compass, Eckhart indicates an overwhelmingly vast revelation which has preoccupied the human intellect, despite all the historical differences in the spirit of any age—one to which any number of sacred discourses can testify. Among its many formulations we find Plato's "Idea of the Good," John's "the Word made flesh," "the conceptions of the Father" in the Chaldean sacred oracles, and the sacred syllable *Pranava Aum* of the Hindus. [1, p. 302]

We intend to pursue meanings that Eckhart did not elaborate. In doing so, we pursue a vision, for in that Vision all of it lies. What is this vision? Is it granted to all? Yes, but the degree of penetration into it varies. The heavens are spread out before us in this vast placelessness and are directly perceived. Within it lies the key to the infinite treasure, the holiest temple of wisdom and abode of our Sun of Glory, manifesting the unfathomable Word of God, the primal scripture, expertly revealed by the nineteenth Psalm:

> The heavens tell out the glory of God, the vault of heaven reveals his handiwork.
> One day speaks to another, night with night shares its knowledge,
> and this without speech or language or sound of any voice.
>
> Their music goes out through all the earth, their words reach to the end of the world.
> In them a tent is fixed for the sun, who comes out like a bridegroom from his wedding canopy,
> rejoicing like a strong man to run his race. His rising is at one end of the heavens,
> his circuit touches their farthest ends; and nothing is hidden from his heat.
> (New English Bible)

To explore this vision, this contemplation, the utmost stillness of the mind and all its faculties is required. Too few of us today have even any idea of what such a requirement means. For the extreme difficulty of attaining the stage of introversion that is beyond and more fundamental than even the archetypal imaginal reality has not yet dawned on the horizon of our contemporary mentality. Only a sage—one who has achieved such a state and can abide in or re-enter it at will—can guide us. For even when we get a glimpse, our murky reflection of consciousness—infested as it is with the potentialities and traces of other experiences—reasserts itself and distorts our understanding of what we in that moment produced. [2, p. 302]

So difficult is this knowledge that Hegel was forced to say:

> Knowledge of spirit is knowledge of the most concrete and consequently of the sublimest and most difficult kind. Know thyself, this absolute commandment, is not concerned with a mere self-knowledge, with the particular abilities, character, inclinations and foibles of the individual, but in its intrinsic import, as in the historical contexts in which it has been formulated, it is concerned with cognition of human truth, with that which is true in and for itself, —with essence itself as spirit. . . . cognition of that which is universal . . . (p. 3)

The sage to whom we turn our sextant minds as to a radiant star in navigating through these troubled seas of mental imperfection is Plotinus, mystical philosopher *par excellence.* Despite the lack of recognition of his deserved status by this irreverent and secular age, we see the uncanny precision and skillful dialectic of his creative exposition of the perennial doctrine as a result of the final stages of ultimate human intelligence. We have gathered, and will comment upon, a garland of quotations from his work—a garland whose unifying thread bespeaks that breathless intimacy with the grace of divine understanding which any encompassing philosophy of truth requires. We incline in the main toward MacKenna's quasi-poetical and inspiring translation of the *Enneads* and will, when necessary, supplement it with other translations. Throughout, however, we are face to face with an extreme difficulty—for Plotinus is poised as a knower within the Divine Mind, and we are but humble students.

There is a contemporary need to organize our understanding of fundamental metaphysical principles by the use of a symbolism grounded in sensible appearances. The astrological mandala is an open-ended schema which not only exposes our understanding and organizes it, but also reciprocally guides and directs it. Without such a schema, the web of interrelationships would overwhelm us.

Not to initially restrict ourselves in this manner is to risk having our understanding relativized *prematurely* by the immensities of universal manifestation. So in the beginning we restrict ourselves to the scale of perception provided by this sensible universe. More precisely, we restrict ourselves to the particular world revealed by our egos. For the transcendental understanding we seek has its humble origins in this empirical world. Here we find the only phenomenon adequate to serve us in the task our aspiration has been assigned. This symbolism is vast enough to provide the polyvalent systems of conception and logic that will be necessary to synthesize the many-faceted aspects of humanity's perpetual quest, and to reconcile the seeming contradictions among the various philosophic and religious traditions. The sensible world is sufficiently determined in its parts both to keep our understanding under rein and to spur it beyond its limits.

This, the sensible universe, is the primordial scripture, the Vision previously referred to. As such, it embodies the wisdom of the primal principles which are beyond it. Who has not thrilled to the excitement of the starry skies, to the silent unanswered questions?

Grounding ourselves, therefore, in the experience that we human beings on this Earth have of the unique integrity of our own world—that is, our solar system—we take as our model the geocentric world-view that has been made familiar through the diagrams of Ptolemaic astronomy. [*fig. 1*] ▼ We will tentatively accept and inquire into the nature and structure of the cosmos as it was conceived by the ancient cosmologists and astrologers who claimed that its principles are gods whose activity is conascent with the being, knowledge, and experience of this very world. Our hope is that we will be led to a superior understanding, or rather, to a qualitatively different kind of understanding of our being here—an understanding that does not sidestep or deny the anomaly of sense perception but, instead, reveals something of the meaning implicit in the orderly and evolving transformation that our minds and bodies undergo. In short, we are going to enter sympathetically into an analysis of experience as conceived by *some* of the ancient philosophers—to

▼ Editors: By placing the Earth at the center, Ptolemy indicates that the individual is the center of his or her own universe. As science shows, every description of motion must have a reference frame. For example, with respect to a stationary solar reference, we are traveling at more than 60,000 miles per hour through space; but the Sun is moving through the galaxy as well. Experientially, every one of us is the still center of our own frame of reference. The geocentric view is not only psychologically valid, but it has theophanic implications as well: For Plotinus the Earth is a god, and we most immediately participate in the universe as given to us through its soul or mind. Anthony expands on this outlook a few pages later.

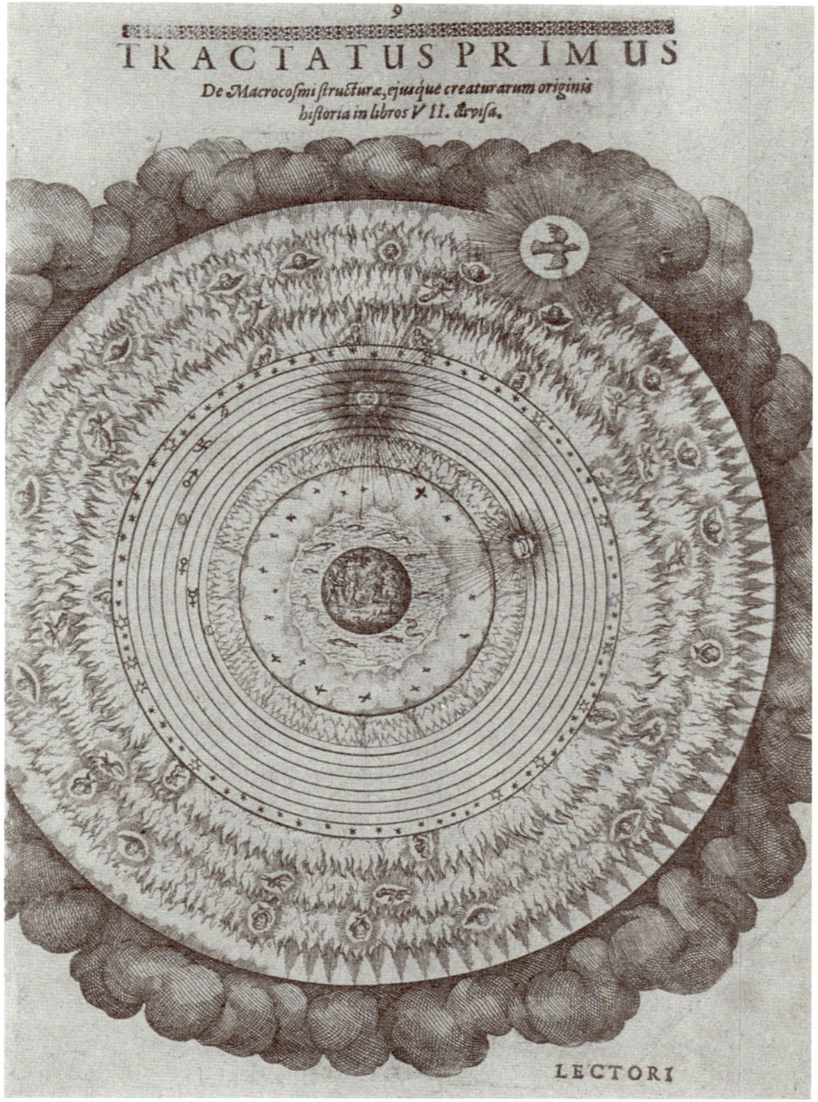

FIG. 1: ROBERT FLUDD'S GEOCENTRIC WORLD-VIEW

ADMIRING THE WORLD OF SENSE AS we look out upon its vastness and beauty and the order of its eternal march, thinking of the gods within it, seen and hidden, and the celestial spirits and all the life of animal and plant, let us mount to its archetype, to the yet more authentic sphere: there we are to contemplate all things as members of the Intellectual—eternal in their own right, vested with a self-springing consciousness and life—and, presiding over all these, the unsoiled Intelligence and the unapproachable wisdom.

PLOTINUS, V.1.4

try to understand and appreciate the philosophic insight they summarized in the dictum: The world is full of gods.

The astrological mandala provides us with the multiple levels of symbolism that will help to make explicit many meanings that are hidden in the Neoplatonic tradition as a whole, most specifically in the *Enneads* of Plotinus. This diagram will allow us to juxtapose philosophy and astrology in such a way as to reveal that the sensible universe with its starry host *is* the primordial living scripture that embodies the wisdom of primal principles—principles that are as far beyond it as the heavens are beyond the Earth—and that the basis of astrology is an acquaintance with the wisdom-knowledge that sustains the cosmos. [*fig. 2*]

In the course of this juxtaposition, we will distinguish philosophy and astrology from one another, define the terms of each, and illustrate how they combine in manifestation as mutually implicating elements of a single whole. This method allows us to develop a sharp contrast, to conceptually bifurcate the Indivisible in such a way that the understanding (in a broad sense of meaning) which can operate only in the realm of dualities, may see how it itself is manifested within the cosmic harmony that is evolving it.

We will speak, on the one hand, of immaterial knowledge and being: the pure hypostases in themselves. These principles or wisdom-emanations constitute a spiritual hierarchy that is broader and more inclusive than can even be adequately *suggested* by any verbal schema that may be employed to represent or define them. Their self-sufficing and eternal pure natures and the grandeur of their authenticity are not available to our undisciplined, undeveloped, and unillumined intellects. Their own proper subsistence constitutes a realm that is beyond the reach of *any* imaginative activity, heights that are *literally* unimaginable.

Nonetheless, these principles are precisely what constitute the content of philosophic wisdom. Though an encounter with these supernals may occur to someone who is not yet prepared to comprehend the experience, we are concerned primarily with that which illumines and permanently enlightens the proficient, the adept, who understands these

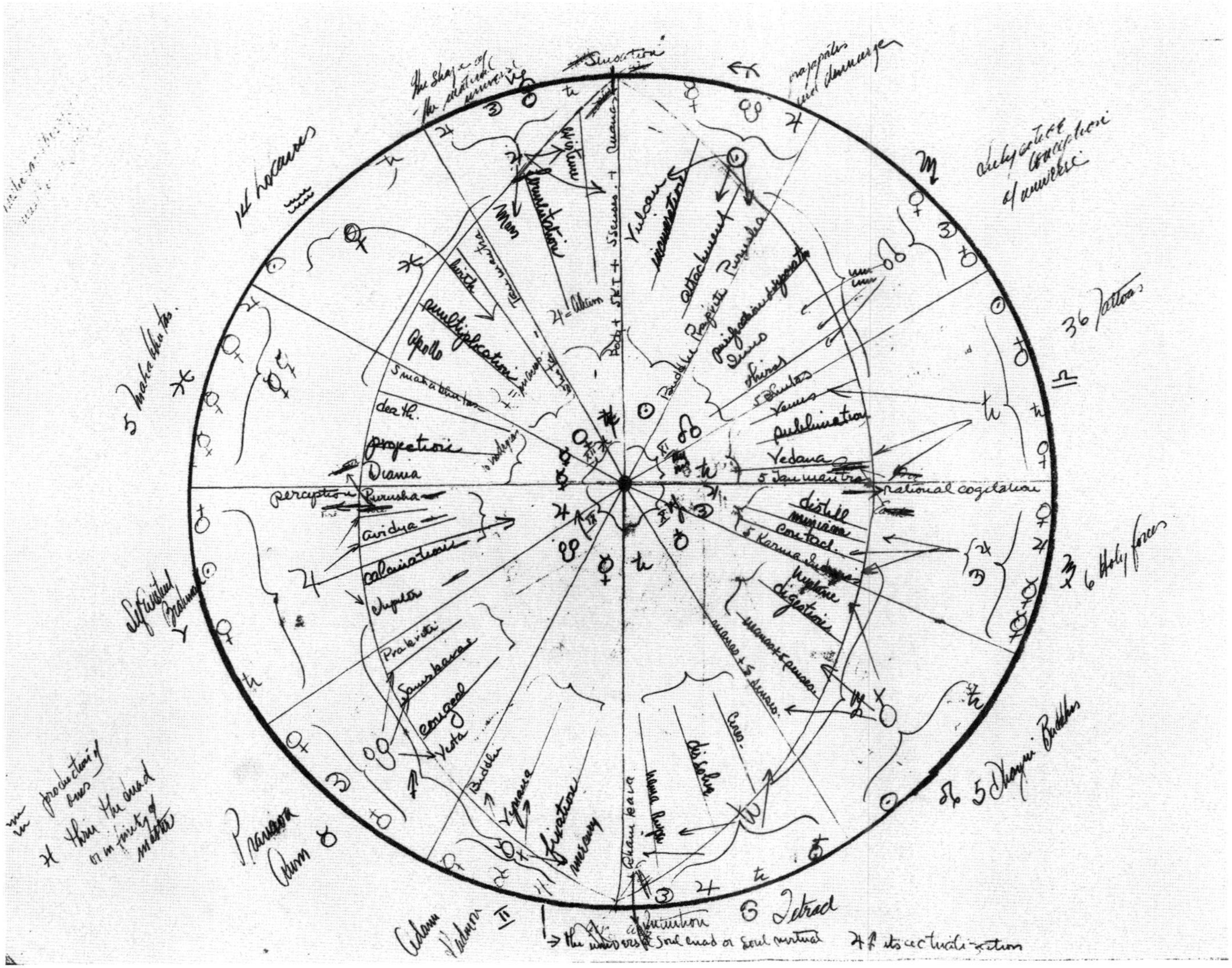

Fig. 2: Anthony Damiani's drawing of one version of the metaphysical chart

principles. The experiences that occur among the unprepared must be discussed separately and must not be confused with the achievements of the philosophic sage.

It would be premature, at this point, to introduce much detail concerning the levels of reality and knowledge discussed by the Platonists. But it can be confidently asserted that these philosophers did not restrict reality to the imagination—not even to imagination in its most comprehensive sense as the Earth's memories.

Concomitantly with the analysis of these principles, we will look to their immanence in the realm of manifestation—as deciphered by means of the astrological symbolism that we employ. The use of this symbolism allows us to considerably amplify and objectify our internal and abstract verbal understanding. It also allows us to bring that understanding to a vital and spiritual fulfillment in the natal chart, the horoscope of an individual. For the elements of this symbolic language participate in deific power, as Plotinus seems to suggest in v. 8.6:

> Similarly, as it seems to me, the wise of Egypt—whether in precise knowledge or by native intuition—indicated the truth where, in their effort towards philosophic statement, they left aside the writing-forms that take in the detail of words and sentences—those characters that represent sounds and convey the propositions of reasoning—and drew pictures instead, engraving in the temple-inscriptions a separate image for every separate item: thus they exhibited the absence of discursiveness in the Intellectual Realm.
>
> For each manifestation of knowledge and wisdom is a distinct image, an object in itself, an immediate unity, not an aggregate of discursive reasoning and detailed willing.

So, also among the disciplines that were to be included in the education of a sage-king was the study of astrology. An abundance of evidence to be found throughout the writings of the Neoplatonists indicates that the study of astrology is a task of the first magnitude. Two of the many primary sources of evidence are *On the Theology of Plato,* by Proclus, and *On the Mysteries,* by Iamblichus. [3, p. 302]

Plato himself, though through a veil, reveals something about the mystery of astrology in books VI, VII, and IX of the *Republic.* For example:

> We establish then, said I, astronomy as the fourth discipline. . . . if any one contemplates the various bodies in the firmament, and, by earnestly looking up, apprehends every thing, you think that he has intelligence of these things; and does not merely see them with his eyes; and perhaps you judge right, and I foolishly. For I, on the other hand, am not able to conceive, that any other discipline can make the soul look upwards, but that which respects being, and that which is invisible; and if a man undertakes to learn any thing of sensible objects, whether he gape upwards or below downwards, never shall I say that he learns; for I aver he had no science of these things; nor shall I say that his soul looks upwards, but downwards, even though he should learn lying on his back, either at land or at sea. . . . But what was the proper way, said you, of learning astronomy . . . ? In this manner, said I, that these variegated bodies in the heavens, as they are varied in a visible subject, be deemed the most beautiful and most accurate of the kind, but far inferior to real beings. . . . Which things truly are to be comprehended by reason and the dianoetic power, but not by sight . . . that variety in the heavens [is] to be made use of as a paradigm for learning those real things. . . . (Thomas Taylor trans., v. 1, pp. 373–374)

Of course there existed then, as now, those for whom there simply are no mysteries, and Plato refers to them sarcastically in this quote. Whether lying on their backs on land or at sea (and, we would add, with or without telescopes), they could have no access to astrology in its truest sense, for their souls were bound to Earth-images. His remarks apply not only to modern astrologers preoccupied with the flow of sensible events and materialistic concerns (those with a "cookbook" approach to deriving the meaning of an individual's horoscope) but also to those other investigators of reality who fail to perceive the intelligible significance of the appearances that they are trying to explain. We see a situation today where the *intelligibility* of scientific laws is subsumed under and reduced to the same level as the material phenomena that those laws are meant to explain. Astrologers and astronomers of today—both alike, Plato would say—fail to use their understanding as a springboard into the higher realms of being.

Much of the Platonic opus remains a closed book to anyone who insists on ignoring evidence that indicates that Plato had a thorough comprehension of astrological first principles.

His reference to the sensible appearance of the heavens themselves as the model on which our learning is based is a direct reference to astrology. We intend to demonstrate that the cosmic soul in the *Timaeus* and the "divided line" in the *Republic,* [4, p. 302] for example, cannot be thoroughly appreciated without explicit recognition of the type of astrology that will be explored in the course of this exposition.

The perennially living philosophy of Plato and the Neoplatonists has a vital meaning for us that can be unfolded and applied to the contemporary situation. These teachings are not merely relics from the past, fit only for respectable entombment in libraries; they are the sculptured breath of living Wisdom. They teach us how to live life, here and now. While we should not let the past mesmerize us, we also should not hesitate to further its findings. We should amplify the meanings in the writings of those whom the Earth and heavens have made wise before us, and we should put our understanding of their teachings to the test of living and understanding life itself.

So let us reconsider our attitude toward astrology, clearly one of the means that the ancients employed in their investigation of reality. The modern superficial acquaintance with astrology, however, lacks an understanding of the comprehensive philosophic principles that underlie it. Such superficiality prevents us from reaching the core of meaning buried in the individual's horoscope. It reveals neither the framework of principles presupposed nor the context in which those principles are significant.

From the point of view of the ancient, philosophically oriented cosmologist, the astrological mandala is a wisdom-example of the intellectual imagination—a symbolic reflex portraying how the Gods or Principles are involved in the fabrication of the vehicles that the individual soul employs for its manifestation. The mandala exemplifies and embodies the knowledge of superior beings—seers—capable of functioning with intuition and transmitting their knowledge of how these fundamental principles are operative in the individual and can be understood from one's natal chart. [*fig. 3*] The chart makes available, through its deific imagery, a formulation *by* reason-principles of a symbolic method through which those very same reason-principles are being manifested: It can reveal, through sensible becoming, the topography of the subtle world.

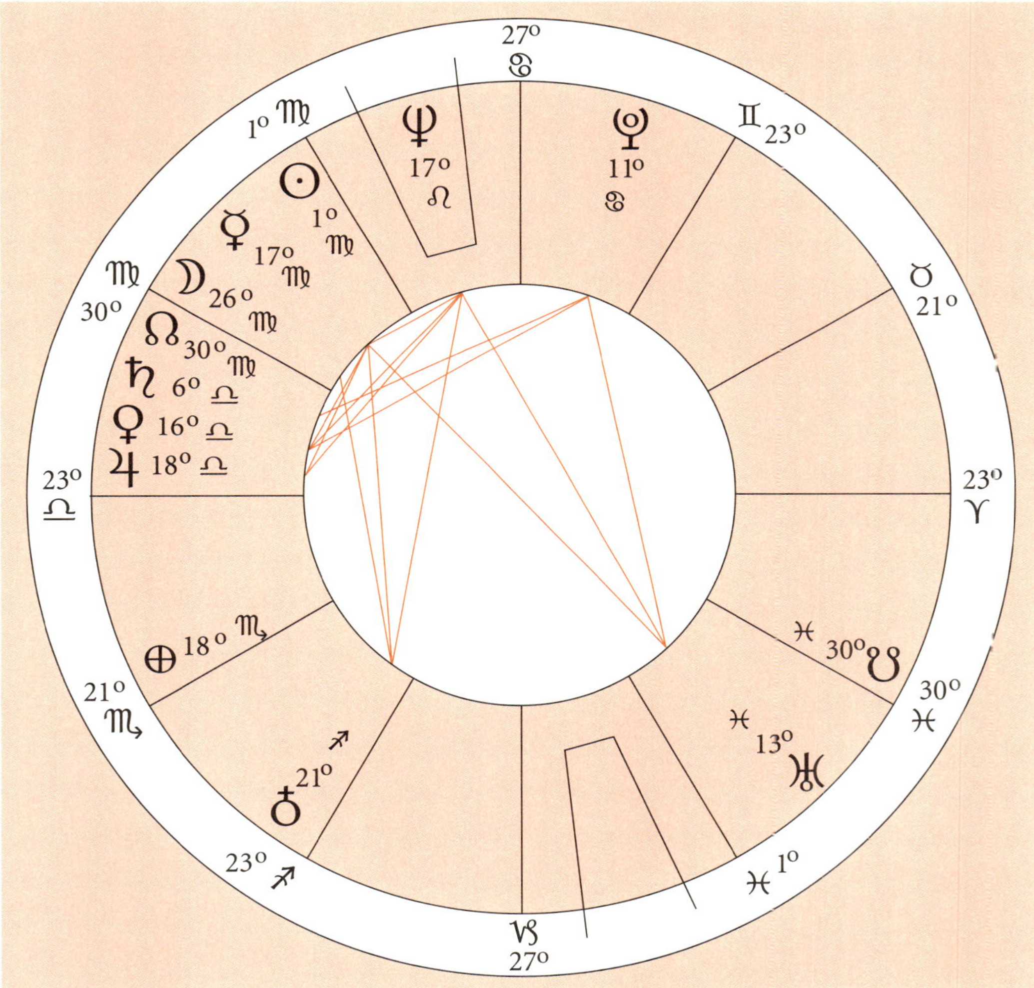

Fig. 3: Anthony Damiani, natal chart ◆ August 24, 1922, 10:43 AM EST ◆ Brooklyn, NY, USA

Latitude 40:45 N, longitude 73:57 W ◆ Placidus houses

The ancients spoke of the realm beyond the constellations as the World of Ideas. ▼ These Ideas or universal substances are not in any place, but were mythologically represented by the constellations. To try to pictorialize this realm we can imagine the stars as statues of the gods, [5, p. 302] or the Ideas, radiating a guiding influence throughout boundless space. The energy of their intelligence laid down the basic pattern within which our world, among others, would develop and unfold. The Idea of the cosmos—the World-Idea—and its inhabitants, evolving and manifesting, can be traced ultimately to this circle of Ideas.

One of the most original contributions to the meaning of the Ideas was made by T. Subba Row in a short, obscure essay ["The Twelve Signs of the Zodiac"] included in his book, *Esoteric Writings.* We strongly recommend to our readers an acquaintance with this essay. Some indication of the meaning of the mythological zodiac emerges in his attempt to interpret the Idea underlying each of the signs. ⊞ Through a semantic analysis of their Sanskrit names and by applying a prescribed procedure similar to the one used by the Kabbalists, Subba Row intuited the meaning of the Idea embodied in each zodiacal sign. The presupposition here is that the mystical experiences of certain exalted beings were deposited in these myths.

If we conceive of the universe as an Idea within the Self, we may be able to comprehend how various portions of that Idea were explored and described by these ancient seers, who then associated specific types of primal intelligence with the zodiacal signs. Each type of intelligence was pictorialized as an animal—as intelligence personified—which would ultimately be traced back to an Idea. [*fig. 4*] Each of these intelligences is a universal substance correlated to a sign. These reason-principles—the signs—are the paradigms of all living forms or species symbolized by the archetypal zodiac or inerratic sphere, which represents the immutability of the Ideas. In this sense the stars are a theophany of radiating intelligences, distributing their presence and informing the universe through patterns of intellectual energy. ⊕ So the stars, which are the bodies of the gods, sensibly represent Ideas that are being transmitted from the Universal Soul to the cosmos. It is not possible for the profane human mind to acquiesce to these truths, but knowledge

▼ Plotinus, v.1.4: That archetypal world is the true Golden Age, age of Kronos, whose very name suggests (in Greek) Abundance (*koros*) and Intellect (*nous*). For here is contained all that is immortal: nothing here but is Divine Mind; all is God; this is the place of every soul. . . . this is pure being in eternal actuality; nowhere is there any future, for every then is a now; nor is there any past, for nothing there has ever ceased to be; everything has taken its stand for ever, an identity well pleased, we might say, to be as it is; and everything, in that entire content, is Intellectual-Principle and Authentic-Existence; and the total of all is Intellectual-Principle entire and Being entire.

⊞ Editors: This fragment relates the Ideas to the astrological chart.

> Our attempt to indicate something of the meaning of each and every element within the mandala will force us to search out sources and hunt down clues scattered in many different texts. In the final analysis, it will be possible to demonstrate that the hidden meanings that went into the formation of the mandala are available in a body or as a total in Plotinus and are the underlie, the philosophy of astrology. . . . That the mythology attached to the zodiacal belt can

ROSEMARIE QUEBRAL

Fig. 4: Zodiac

> be traced back ultimately to the Logos, the sacred wisdom, has been hinted at in many and varied religions. That there could be a correspondence between the various "Ideas" and the sections of the chart may seem far-fetched and absurd to our critics. Nonetheless, we intend to show that it is applicable, with certain qualifications, even to an adventitious unity such as we find in the sensible world, and that the meaning of the houses—the soul's self-portrait—or our perceived world, is ultimately derived from the Ideas.
> —Anthony Damiani, Notes and Papers, p. 640

"Adventitious unity" means a temporary and compound entity, such as the personal ego.

Plotinus, iv. 4.8: If the stars live by a blessed life in their vision of the life inherent in their souls, and if, by force of their souls' tendency to become one, and by the light they cast from themselves upon the entire heavens, they are like the strings of a lyre which, being struck in tune, sing a melody in some natural scale: if this is the way the heavens, as one, are moved, and the component parts in their relation to the whole—the sidereal system moving as one, and each part in its own way, to the same purpose, though each too holds its own place—then our doctrine is all the more surely established.

and devotion may bring a person to that realization wherein the rising and setting of worlds is a terrestrial incident in Eternal Consciousness.

The world or cosmos that evolves with the creatures in it is therefore a manifestation of this Nous in the heavens as the ancients called it. ▽ Our sensible Sun took its place within this order and through its powers, the cosmocrators, articulated and manifested the predetermined order contained in this incomprehensible intelligence, which was symbolized by the various animals. An individual soul in its own right, our Sun comprehends the nature of these intelligences through its intellect and proceeds to unfold this prodigious intelligence. Each part of the whole (the solar system)—that is, each planet or rational power—may be considered in its own right, or it may be considered as part of the whole. When conceived of as part of the whole, each cosmocrator is a functioning power of the rational soul of the sensible Sun. ⊞

> Every sun has a solar spirit.... Every planetary orb of a terrestrial sort [has] a terrestrial spirit. The spirit of every orb is diffused into, and lives throughout, its mineral, vegetable, animal and human kingdoms, and is an immortal entity, a living, indivisible and instinctively conscious existence, but is without human personality, and so abstractly conscious of pleasure or pain, but without power to determine its own sensations.
>
> Insphered in every solar, aromal or terrestrial planetary world, is an appropriate world-soul, living in the life of all its distinct creations, and permeating alike its atmospheres, its waters, its material crust and its electro-igneous centre. These world-souls comprise the first family of God, and their number is as that of the stars.... They are absorptive organs for the Divine Spirit; and, pervading each its own world, and living in all its parts, they distribute throughout matter the Divine vitality. (T. Lake Harris, p. 100)

Our Earth is included within the functioning of the rational soul of the Sun, and the geocentric view comes into being when we localize or identify consciousness within one or all of the creatures inhabiting this planet. [*fig.* 5] For the Earth is also a recipient of the Ideas, and through the intermediary of the rational powers of the Sun gradually there is organized and manifested on Earth a variety of species that will partially and imperfectly

▽ Editors: For Anthony, the world is a manifestation of the zodiacal intelligence, and the latter is the "Nous in the heavens," or presence of the Nous (Intellectual-Principle). "Nous" means Divine Intelligence. It is both the realm of Ideas and the power of Intellection which is authentic knowing.

⊞ Editors: Each star is a sun, the body of a divine soul, radiating its intelligence into the universe. Furthermore, the planetary souls (cosmocrators) that function to manifest the wisdom of a given sun are divinities in their own right:

> We must conclude, then, that every part and member of the earth carries its vestige of this principle of growth, an under-phase of that entire principle which belongs not to this or that member but to the earth as a whole: next in order is the nature (the soul-phase) concerned with sensation, this not interfused (like the vegetal principle) but in contact from above: then the higher soul and the Intellectual-Principle, constituting together the being known as Hestia (Earth-Mind) and Demeter (Earth-Soul). —Plotinus, IV.4.27

The subsequent chapters on Soul and Nature unfold this view of cosmology. Appendix I to the Larson edition of the *Enneads* offers an overview of how Anthony applied it in his work.

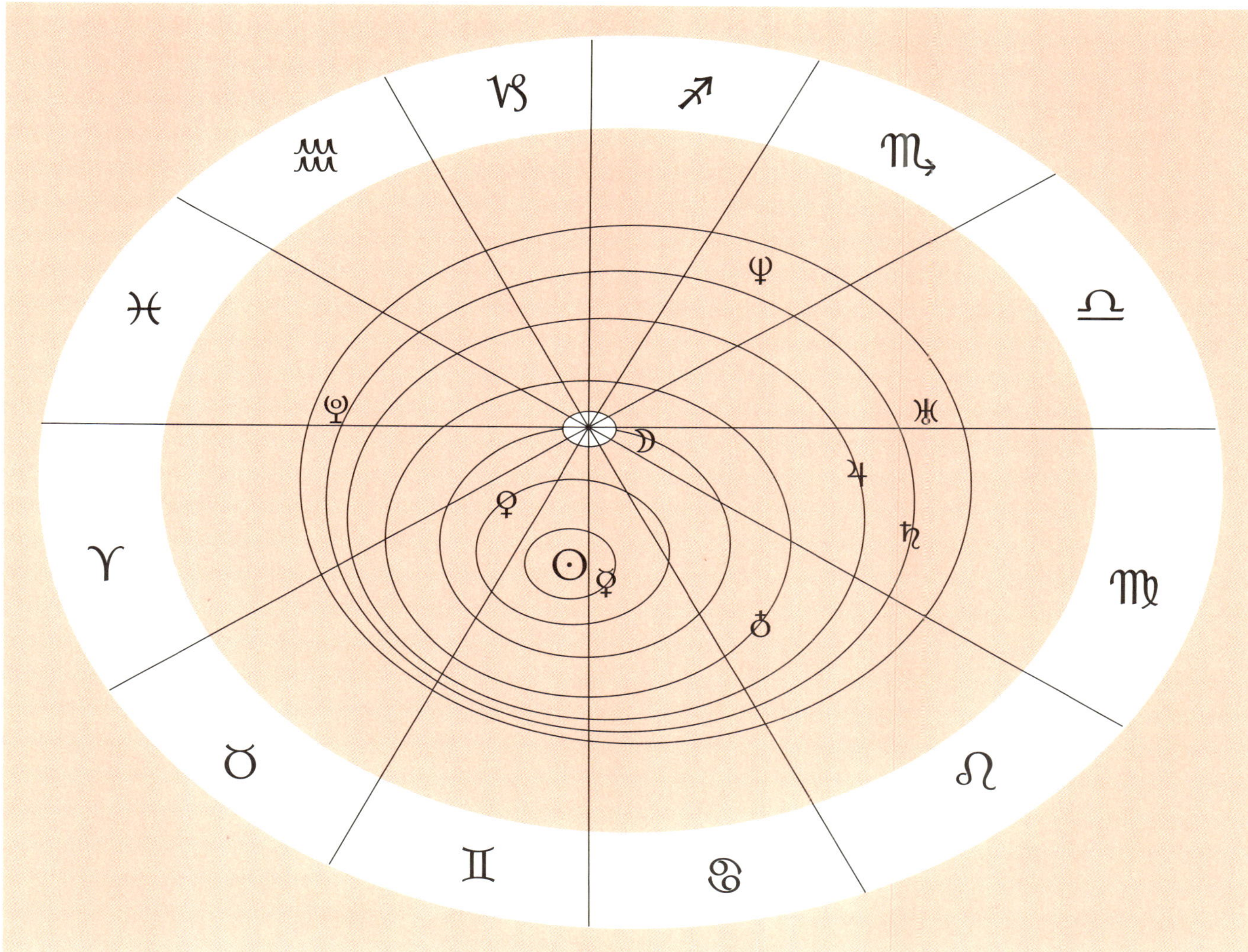

Fig. 5: Heliocentric and geocentric views

▼ PLOTINUS, III.4.6: Emancipated souls, for the whole period of their sojourn there above, have transcended the Spirit-nature and the entire fatality of birth and all that belongs to this visible world, for they have taken up with them that phase of the Soul's existence in which the desire of earthly life is vested. This phase may be described as 'that which becomes divisible among bodies', for it is what enters bodily forms and multiplies itself by this division among them.

reflect the Ideas or animal intelligences. The traces of life left behind by these creatures are retained by the soul of our Earth as the content of its memory, and this is often referred to in the literature of the ancients as the "Dragon." (This Dragon must be understood as distinct from the Celestial Dragon, the constellations previously referred to as the encircling Ideas represented by the stars.) The Dragon is the end-product of all the life lived on the Earth, the storehouse of all the images and the intelligence behind the functioning of animal bodies. This is what the degree symbolism of the chart refers to.

The organization of these tropes into animal bodies is [what constitutes] our inherited modes of functioning. The instant an individual is born, the distribution of planets in signs and houses shows a selection from within the Earth's memory of the tendencies that individual previously identified with most closely. This predetermination on the part of the soul as to which elements will constitute its psychosomatic organism will shape the postnatal history or *karma* of the individual. In essence it will be these *vasanas* or memory traces that will blossom forth as the individual's life. Psychologists refer to this intelligence as the subconscious—an intelligence so vast that it dwarfs the conscious mind. Consequently, these inherited modes of functioning constitute a system of beliefs for the soul that inhabits a given, specific body. We conceive of this as the psyche of the individual and as the basic presuppositions or attitudes that organize the person's perceptions and understanding of the world in which he or she lives. The incarnation of these tropes as an organized psychosomatic organism is the vehicle through which the soul lives and learns or, when its learning is completed, from which it liberates itself by transcending it. ▼

To summarize: The method here followed is the juxtaposition of philosophy and astrology—to define them and distinguish each from the other in order to illustrate how they reciprocally implicate one another into a whole and are combined in manifestation.

Therefore we shall speak on the one hand of immaterial knowledge or pure hypostases in themselves: the spiritual reality, the hierarchy of wisdom-emanations existing in a distinctionless mode within this reality. Their self-sufficing and eternally august nature, the grandeur of authenticity is an inscrutable abyss, a blinding darkness to our intellect, a radiance and joy that can suffocate the unprepared but illuminate and enlighten permanently the proficient, the adept. These principles are beyond the imagination—literally.

Concomitantly and parallel to the above discussion, these principles are applied throughout the realm of manifestation. They are portrayed by the use of the astrological symbolism [applied] to our world, so that our internal and abstract understanding can be further objectified and applied to the level of analytical and empirical psychology, epistemology, and to the problems of everyday living.

To communicate something about the untouchable marvel of the reason-principle directly without the aid of the imagination, to express in word and metaphor the mystery that silently embraces our everyday existence, to single out the intelligence that is the underlying ground of the experience of cosmic consciousness is beyond our power. Nonetheless the illuminaries of humankind—whose noble generosity and compassion have guided us so far—will not fail us in our present need to sacralize the world and ourselves, and to restore unity to thought and becoming.

The conscious "reinstatement" of purposive evolution into the divine scheme must be available to us at the most practical level whether through the philosophically oriented astrologer, psychologist, educator, or whomever. Our intention is to help reveal the verities we each embody, and that actualize life with the divine fragrance. A philosophy that enhances existence with wonder and meaning may easily encompass an indefinite number of schools of thought that supposedly contradict one another. Such a philosophy may make available to the subtle thinker unlimited vistas of conception regarding the depth of human being.

We repeat: There is a need for a thorough exposition and understanding of the fundamental metaphysical principles that underlie a spiritual knowledge of astrology. That astrology can be redirected or reoriented to its spiritual significance is a prerequisite to its transvaluation. All this groundwork, however, is only a preliminary way by which we can know ourselves. It will result in the perception that each individual's being, or rather each individual's manifestation, is rooted in the justice of the cosmos.▽

▽ Editors: In this work, Anthony is neither recreating nor limiting himself to the astrology that was available to Plato or Plotinus. Rather, he applies his understanding of their work to astrology as he studied it through personal practice and a broad variety of contemporary and classical resources.

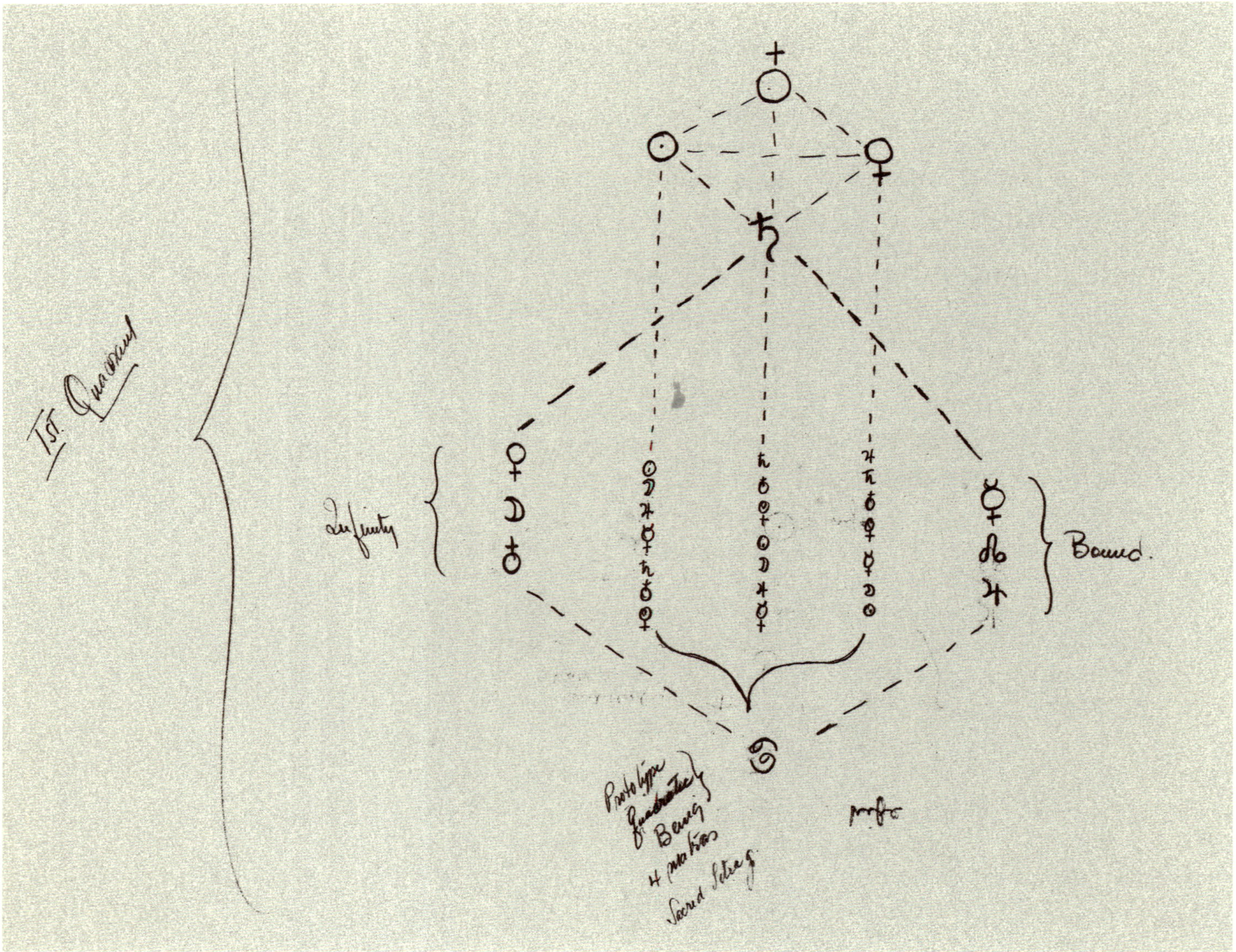
1st. Quadrant
Infinity
Bound.

Chapter One

THE ONE

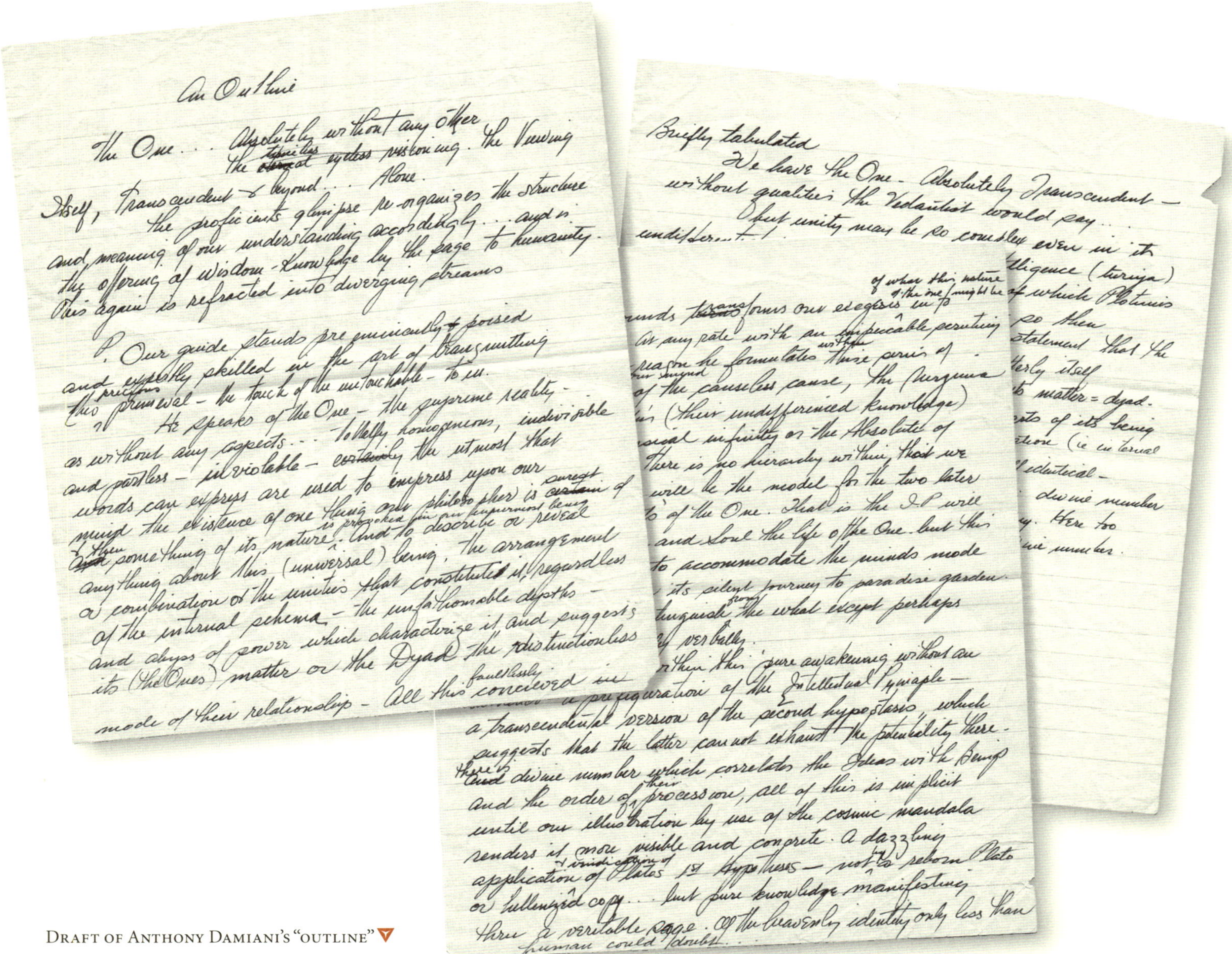
An Outline

The One . . . Absolutely without any other
The limitless eyeless visioning. The Viewing
Itself, Transcendent & beyond . . . Alone.
The proficient's glimpse re-organizes the structure
and meaning of our understanding accordingly . . . and is
thy offering of wisdom-knowledge by the sage to humanity.
This again is refracted into diverging streams

P. Our guide stands pre-eminently poised
and expertly skilled in the art of transmitting
this precious primeval - the touch of the untouchable - to us.
He speaks of the One - the supreme reality -
as without any aspects . . . totally homogeneous, indivisible
and partless - inviolable - certainly the utmost that
words can express are used to impress upon our
mind the existence of one thing our philosopher is except of
& then something of its nature. And to describe or reveal
anything about this (universal) being, the arrangement
or combination of the unities that constituted it, regardless
of the internal schema - the unfathomable depths -
and abyss of power which characterize it, and suggests
its (the One's) matter or the Dyad the distinctionless
mode of their relationship - All this faultlessly conceived in

Briefly tabulated
We have the One - Absolutely Transcendent -
without qualities the Vedantist would say . . .
but unity may be so considered even in its
undifferent . . .

. . . this pure awakening without an
. . . a prefiguration of the Intellectual Principle -
a transcendental version of the second hypostasis, which
suggests that the latter cannot exhaust the potentiality there.
There is divine number which correlates the Ideas with Being
and the order of their procession, all of this is implicit
until our illustration by use of the cosmic mandala
renders it more visible and concrete. A dazzling
application & vindication of Plato's 1st Hypotheses - not by a reborn Plato
or hellenized copy . . . but pure knowledge manifesting
thru a veritable sage. Of the heavenly identity only less than
human could doubt.

Draft of Anthony Damiani's "outline"

An Outline

THE ONE . . . ABSOLUTELY WITHOUT ANY OTHER. THE VIEWING ITSELF, TRANSCENDENT AND BEYOND . . . ALONE.

The proficient's glimpse re-organizes the structure and meaning of our understanding accordingly . . . and is the offering of Wisdom-Knowledge by the sage to humanity. This again is refracted into diverging streams of philosophy.

Our guide P[lotinus] stands pre-eminently poised and expertly skilled in the art of transmitting this precious primeval—the touch of the untouchable—to us.

He speaks of the One—the supreme reality—as without any aspects . . . totally homogeneous, indivisible and partless—inviolate—the utmost that words can express are used to impress upon our mind the existence of one thing our philosopher is surest of, and then something of its nature is provoked in our innermost being. To describe or reveal anything about this (universal) being, the arrangement or combination of these Transcendent unities that 'constitute' it, regardless of the internal schema—the unfathomable depths—and abyss of power which characterize it and suggest its completeness, the distinctionless mode of their relationship— All this faultlessly conceived and executed by sounds, transforms our exegesis of what this nature of the One might be into a hymn of adoration. At any rate, with an impeccable scrutiny of the range of human reason, he formulates within these series of quotes a paradigm for our mind of the causeless cause, the Nirguna Brahman of the Hindus (their undifferenced knowledge), the idea of Metaphysical infinity or the Absolute of certain Westerners. There is no hierarchy within, that we can point to, but/and it will be the model for the two later hypostases as the 'aspects' of the One. That

▽ EDITORS: The differences between Anthony's handwritten outline (*opposite*) and this typeset version reflect revisions made or approved by him, with these exceptions: We have introduced material in square brackets [ ], made minor corrections, and added a few helpful commas. Ellipses here (. . .) indicate Anthony's deliberate pauses, *not* omitted material.

is, the I.P. [Intellectual-Principle] will be teleology for the One and Soul its life, but this differentiation will be to accommodate the mind's mode of ascension ... in its silent journey to paradise garden. Once there, who can distinguish the what from the who except the godly ... once here again, only verbally.

He includes within this 'pure awakening without an awakener' a prefiguration of the Intellectual-Principle—a transcendental version of the second hypostasis, which suggests that the latter cannot exhaust the potentiality there. There is divine number which correlates the Ideas with Beings and the order of their procession; all of this is implicit until our illustration by use of the cosmic mandala renders it more visible and concrete. A dazzling application and vindication of Plato's 8 Hypotheses—not by a reborn Plato or Hellenized copy ... but pure knowledge manifesting through a veritable sage. Of the heavenly identity only less than human could doubt ...

Briefly tabulated:

We have the One—Absolutely Transcendent—without qualities, the Vedantist would say ...

> but unity may be so complex even in its undifferentiated simplicity or as pure intelligence (*turiya*) as to suggest immense mysteries . . . tracings in the dark—of which Plotinus delineates—so then we have first this very specific statement that the One or unity is absolutely transcendent, utterly itself.
>
> then power, intrinsic and non-different, characterizes it, is its matter = dyad—preontological henads discern [*sic*] as partless parts of its being regardless of their arrangement and/or combination (i.e., internal schema) no heterogeneity arises but all is self-identical—
>
> included is the prefiguration of the Intellectual-Principle; divine number and the unities that includes [*sic*] One-many. Here too the many essential numbers are derived from divine number.

The One: Exposition

THE SPECIFIC SEQUENCE OF QUOTES recited in the course of this discussion is meant to accomplish a dual purpose. On the one hand, it should reveal something about the mysterious and infinite complexity of the One.▽ Simultaneously, it should organize and order for our understanding the formless grades within the unshaped, the unfigured, which we will refer to as its indivisible aspects. The inquiry aims to reveal both the meta-ontological source of Plotinus' revelations—the Soul which is rooted in the One—and the reasoning that is based on and flows from the Soul as an authentic existent in the Intellectual-Principle. This is inspired Truth.

▽ ANTHONY DAMIANI, NOTES AND PAPERS, P. 681: In our discussion of Unity, the One, we must create for ourselves a way of seeing, a thinking organized. . . . By recruiting from the Master-Architect Plotinus the sequence of reasoned revelations, it may be possible for us to reconstruct conceptually his vision, and that would be tantamount to a penetrating into the secret adytum of his thought through which his revelations from the One become available to us.

The first priority is to arrive at some understanding of the complete perfection of the One itself. Secondarily and concomitantly will be the discovery of a paradigmatic thought-model that will indicate how any "one"—whether Idea, God, Number, or Being—is necessarily constituted. This developing insight may then become a key that will unlock some of the mysteries in Plotinus' teachings and disclose for the spiritual aspirant the inner significance of the theory of participation.

It is important to bear in mind continually that our thought must always keep the One intact while we develop our understanding of it as a paradigm for other referents or principles. Thus, though we necessarily employ an incorrect mode of thinking in the effort to glimpse realities that are beyond thought, we will never fragment the unity of the One. When we speak, for example, of the "partless aspects" of the indivisible Unity, we cannot help but recognize and try to account for the fact that we are employing and addressing an intellect that operates in dualities. We cannot do otherwise. But this very problem, strangely enough, precisely indicates that we are not to abolish the plurality of frameworks available to us and applicable to the universal relativity that is both suspended from the

One and included within it. We must explain how universal relativity is included within the One in a special way.▼ This interpretation will attempt to portray and to account for that possibility and will permit us to expound the two views of truth: the ultimate truth that is the substratum of the appearance, and the truths within the appearance—truths that are not reduced to or resolved into that Ultimate.

▼ Plotinus, v.3.15: All that is not One is conserved by virtue of the One, and from the One derives its characteristic nature: if it had not attained such unity as is consistent with being made up of multiplicity we could not affirm its existence: if we are able to affirm the nature of single things, this is in virtue of the unity, the identity even, which each of them possesses. But the all-transcendent, utterly void of multiplicity, has no mere unity of participation but is unity's self, independent of all else, as being that from which, by whatever means, all the rest take their degree of unity in their standing, near or far, towards it.

The Only Perfection

The philosopher-sage enunciates the Logos (the Word), and the knowledge he communicates is primarily Wisdom. ⊞ He is required to be precise within the given structures, the cultural mores, and the needs of his era, yet his stance is also outside of and beyond these forms. Hence it is only natural that first in the order of importance should be statements about the existence of God, the One-Only—for, following upon this certitude that is the very core of all that he will say, he also transmits for our benefit a wisdom-knowledge or *prima philosophia* conascent with the foundations of our world and our being.

⊞ Plotinus, iii.8.6: The Sage, then, has gone through a process of reasoning when he expounds his act to others; but in relation to himself he is Vision . . .

The One is absolute positivity: pure universal existence inbound with the totality of possibilities and infinity of power. Perhaps it might be more meaningful and provocative to say that it is intelligence—or, better still, pure intelligence. Sufficient and complete in itself, *the only Perfection* is without peer and is beyond any limitation, dependence, need, or multiplicity of any kind. This affirmation of its transcendence does not imply that the One is a blank homogeneity or a zero, but, rather, that it is unlimited fullness. To aim the understanding "beyond being" does not mean that it must at some point fall into an abyss of nothingness. ⊕ △ [6d, p. 304] On the contrary, whatever aseity characterizes any principle or idea is removed by this method, and the infinity of the transcendent is then conceptually available.

In several tractates of the *Enneads,* Plotinus describes the essential characteristics of the One-Only. Our ensuing statements will try to explicate them in detail, yet we must follow his example and take great care in discussing these characteristics; for it is misleading and incorrect to speak of such "attributes" except in a most precise way, albeit a somewhat strange and mysterious way. We hope to be able to unfold the peculiar way in which the transcendent One is *inclusive* of all and is the cause of all and everything that is, explaining as we proceed *how* and why it is properly described as a nondual fullness rather than as a nothingness (or monism). ✦

In tractate V.4.2, Plotinus refers to that which is above the Intellectual-Principle as "self-distinguishing throughout . . . that self-intellection which, stemming from an inner consciousness, takes place . . . in a mode other than that of the Intellectual-Principle." In VI.8.16 (Deck translation), he calls it an eternal "wakefulness" (or a consciousness) that cannot be separated from the totality of its indeterminate contents—contents that are not other than or different from the container.

On page 32 of *Nature, Contemplation, and the One,* John Deck asserts that "the One is called an eternal super-knowledge." He is referring to his translation of tractate VI.8.16, 31–36:

> If, now, its [the One's] act does not become but is always, and is a kind of wakefulness which is not other than the one who is awake, being a wakefulness and an eternal super-knowledge, it will *be* in the way it is awake. The wakefulness is beyond being and Nous and intelligent life; the wakefulness is itself. [Anthony emphasized the phrase "is a kind of wakefulness." —eds.]

Deck asks, is this "a wakefulness which is a *super*-knowledge, or a wakefulness which is *above* knowledge?" Bearing in mind that there are other interpretations, Mr. Deck, a sincere and discriminating devotee of the sage, tends to identify the One's wakefulness with its super-knowledge. He says:

> Plotinus goes on to say that the One is "not, so to speak, imperceptive, but everything of it is in it and with it [i.e., it is entirely self-contained]; it is entirely self-discerning; life is in it and all things are in it; and its self-knowledge is itself, a self-knowledge by a kind of synesthesis being in eternal stasis and in a knowledge otherwise than knowledge according to the Nous." (John Deck, p. 33, quoting V.4.2, 16–20)

✪ EDITORS: For Plotinus, the One or the ultimate principle has a positive reality even more profound than that of Being; it is in no way an abstraction. Being is a lower order of infinity, a contraction or determination of the absolute infinity of the One.

▲ PLOTINUS, V.4.1: . . . it must be authentically a unity, not merely something elaborated into unity and so in reality no more than unity's counterfeit; it will debar all telling and knowing except that it may be described as transcending Being . . .

✦ PLOTINUS, V.2.1: 'The One is all things and no one of them'; the source of all things is not all things; and yet it is all things in a transcendental sense—all things, so to speak, having run back to it . . .

"Thus knowledge," Deck concludes (p. 33), "which appears in a way in nature, more fully in soul, perfectly in the Nous, is not absent from the highest 'nature,' the One: the continuity of knowledge is not abruptly broken in the ascent from the Nous to the One."

Deck's concentration here is intensely and sharply focused on the true meaning of the One. His is a magnificent exposition in positive terms of the master's teaching on the One. He has seized the jugular vein of the doctrine when, a few lines later, he states further that "the self-identical synesthesis ▼ of the One can be a super-knowledge in a positive sense." (p. 34)

▼ Editors: In a note he adds at this point in his own text, John Deck suggests that "*synaisthesei* conveys the notion of self-perception, or better, self-contained perception." The alternate reading "*syn aisthesei* would mean merely 'with perception.' " (John Deck, p. 38 n. 27)

The statement that the One "will *be* in the way it is awake" is a particularly emphatic assertion of the identity of wakefulness (consciousness) with existence. Now we are not discussing the identity of being and knowledge such as is found in the Intellectual-Principle, but the true identity and intact perfection of the One. As our discussion develops, we will perceive how all the so-called contents of this super-intelligence are as it (the One). But first, we must fully encounter the uniqueness of this conception of the Absolute. Our minds should be brought to a halt by what we are being asked to consider. Pure unconditioned Being in all its universality is being conceived here as [in] identity with intrinsically appropriate super-intelligence: eternal stasis reigns. We are presented with a Supreme Intelligence that does not loosen, dwindle, or divide itself, nor is it modified or spread out among the lower orders of reality. This is the initial premise. Appreciating this intact perfection is a necessary and absolute prerequisite to arrive at a proper understanding of emanation.

A pause is required here so that we may intently examine this nugget of gold, for it is a capital point, and we must insistently push our minds in this direction. First, What is the nature of the utter simplicity and complete perfection of the One? Only after we have firmly grasped this nature can the second question arise: How can this unique and unknowable Absolute be brought into relationship with anything else? The question may become meaningless if there is not anything else, that is, if there are only illusory modifications of itself. But assuming that there are other principles—not necessarily outside the One—the question will naturally arise as to the nature of the relationship between the One and the others. And it is to this issue that we will direct our attention.

Plotinus clearly states that the One's simplicity is not that of a spatial point or a quantitative unit to be enumerated with others, "nor is its impartibility that of extreme minuteness; on the contrary, it is great beyond anything, great not in extension but in power . . ." (VI.9.6) In a preliminary fashion he dismisses the elementary associations that the word 'One' may have. (We will find, as we go along, more and more of what he does mean by the One, and then we can compose the meaning of the word according to his statements about it. We are not to bring *our* definition of the One, and the psychological associations it may have for us, to this term.) Now this very positive description of the Unity's power as beyond any determination or limitation means that unlimited power should be understood as an indivisible aspect that is identical with Unity. In this sense, it is beyond God, Divine Mind, or Being. Such a total affirmation is not only beyond any combination of the possibilities or determinations in the Intellectual-Principle (which includes the totality of all essences and Ideas) but points beyond to an ultimateless Ultimate "utterly a self-existent, with no concomitant whatever. This self-sufficing is the essence of its unity. . . . supremely adequate, autonomous, all-transcending, most utterly without need." (VI.9.6) It is the Alone. This One is not, therefore, an abstraction from other principles. On the contrary, all other principles are *literally* divine outflowings from the super-abundance of its Perfection. ⊞

⊞ PLOTINUS, V.2.1: Seeking nothing, possessing nothing, lacking nothing, the One is perfect and, in our metaphor, has overflowed, and its exuberance has produced the new . . .

The later principles are all dependent on It, but the One depends on none. For any principle or being to be at all, it must first be a unity: The first principle of all principles is Unity itself, the principle of all principles. A similar definition of this simplex is provided by René Guénon in *The Reign of Quantity.*

> If true unity is also to be described as "simple," that word must be understood in quite a different sense, so that it conveys only the essential indivisibility of true unity, and so as to exclude the idea that unity is in any way "composite," and this implies that it cannot rightly be conceived as made up of parts of any kind. . . .
>
> On the other hand, although the principial unity is absolutely indivisible, it can nevertheless be said to be of an extreme complexity, since it contains "eminently" all that constitutes the

> essence or qualitative side of manifested beings, when considered from the point of view of a "descent" into lower degrees. (pp. 94–95)

Similarly, the Plotinian affirmation and commentaries on the nature of the One do not preclude the possibility that the unity of every principle—every principle such as Number, Being, Life, Intellect—is paradigmatically included in the One's super-knowledge:

> What is present in Intellectual-Principle is present, though in a far transcendent mode, in the One . . . (Plotinus, VI.8.18)

> The First, then, must transcend this principle of life and intellect which directs thither both the life in itself, a copy of the life-giving principle in the First, and the intellect in itself which is again a copy, though of what original there we cannot know. (Plotinus, V.3.16)

The precise way in which these beings or unities are unoriginatedly and indivisibly actual within the One is the thread theme to develop as we proceed. However, let us first continue to look into the nature of this "wakefulness which is not other than the one who is awake": pure being-consciousness.

Obviously, no symbol is adequate to this Absolute. But we are preoccupied with an understanding of our understanding of it, and Plotinus assures us that "there is a certain rough fitness in designating it as unity." (VI.9.5)

We are articulating in words and symbols our conception of the Absolute, but must keep in mind that while trying to make sense of Plotinus' remarks we remain confined within the sensible and should not assume our understanding of them to be an intimate acquaintance with the Reality that is beyond it. There are many who assume that the correctness of conceptual schemata within their imagination is a certification and domination by the Reality, and there are even others who go further and insist that imagination is that Reality.

In confronting such obstacles, the sage may more easily dissolve the difficulties with pregnant silence than with words. Such eloquent silence should not be taken as agnostic or as indicating a lack on the sage's part, for the sage will speak or not speak according to the needs of the era and the status of the individuals asking the questions. Not all knowledge is appropriate for all persons, or at any time. In times and circumstances such as our own, however, more harm may be done by *refusing* to speak of the Unity or Absolute God: Such refusal may too easily invite intellectual skeptics to assert that even the sage cannot go beyond the imagination, cannot know Truth or directly "experience" it, that it does not exist, and so forth. ▽ [6, p. 303; 7, p. 305]

We cannot in fact approach analytically the perfection, the self-sufficing completion of the One. The nature of Unity precludes any such possibility.

> Thus, when we say that universal Possibility is infinite or unlimited, it must be understood that it is nothing other than the Infinite itself, envisaged under a certain aspect, in so far as one may say that there are aspects to the Infinite. For the Infinite is truly "without parts," and strictly speaking, there can be no further question of a multiplicity of aspects existing really and "distinctively" within it. It is we who in fact conceive of the Infinite under this aspect or that, because we cannot do otherwise . . . (René Guénon, *Multiple States,* p. 33)

In the strict sense, then, our approach is incorrect; but, based on Plotinus' reasoned revelations, we will conceive of the One as having indivisible aspects that we can examine one at a time in an effort to make them at least vaguely accessible to our understanding. We must learn to accommodate our mental processes to this expanding vision. So we will proceed in the unorthodox fashion of subjecting each of these indivisible "aspects" to itemization and description and will try to understand each and every one of them in a certain relative manner.

Then we may take either of two approaches to begin adjusting for our initial error. We may, on the one hand, speak of how these various aspects of reality fuse and interpenetrate one another. Or, through a second approach, we may try to think of each mode or aspect of the Unity as the entirety itself and then see how they *must* necessarily merge as undifferenced Unity. Once we have done all this, and *only* after we have done all this, we cancel out the whole process of reasoning that has brought us this far and seek to plunge beyond our own rationality: By a leap of the creative imagination, through inspiration from the

▽ EDITORS: While we investigate the nature of the One, we also question what understanding it is possible to have about the One itself. We may have doubts whether it is useful to say anything at all about the Absolute, or possible to understand anything in a positive way about it. However, Plotinus reassures us that we can say something meaningful and useful.

life of the Knowing Principle, we are to immerse ourselves intently and silently in this notion of an infinite life that is eternally and intrinsically complete and perfect.

Any description of this perfect One is admittedly "as if"—to accommodate that level of mind that is constantly in the process of instantiation and transformation. The net result, however, of this exercise in ingenuity, this attempt to see the complete Perfection from various angles, is an approach to one of the most mysterious, difficult, and yet absolutely pivotal ideas in the Plotinian teaching—that is, Power or the Dyad.

▼ EDITORS: Nonduality presents an inclusive view of reality: Manifestation and multiplicity are not extrinsic and illusory (as in a pure monism), nor are they principles other than and separate from the One (as in dualism).

Power

Plotinus characterizes the infinite and indeterminate Power of the One as both identical with the pure undifferenced knowledge of the One and radiating from it. On the one hand, the Power and the undifferenced knowledge are synonymous: The One remains a simplex. On the other hand, he distinguishes them: The One is the source of all distinctions and substance found in the All. Any approach to defining the One and explaining how it produces the All (how the many arise) must call upon its Power to simultaneously fulfill and sustain this double view, the ultimate paradox, which is the nondual Absolute. ▼ The way in which it is able to do so becomes available through a proper understanding of this Power.

> There is in everything the Act of the Essence and the Act going out from the Essence: the first Act is the thing itself in its realized identity, the second Act is an inevitably following outgo from the first, an emanation distinct from the thing itself. . . .
>
> So it is in the divine also: or rather we have there the earlier form of the double act: the divine remains in its own unchanging being, but from its perfection and from the Act included in its nature there emanates the secondary or issuing Act which—as the output of a mighty power, the mightiest there is—attains to Real Being as second to that which stands above all Being. (Plotinus, v. 4.2)

Power, as spoken of here, has this twofold meaning. As a "completing power" (VI.7.40) which produces no new thing within the One, it coincides with and is indistinguishable from the One; secondly, that same Power flows forth or emanates from the One (V.4.2). Excluded from the Godhead is any personal volition; the process is completely impersonal.

In tractate V.1.5 (MacKenna translation, first edition), Plotinus says that "the Dyad is, as regards that sphere, undetermined—representing as it were, the underly (or Matter) of The One." [8, p. 306] The Power which goes forth or emanates from the One, i.e., the Intellectual-Principle undetermined, is the matter ▪ which will be shaped by the indivisible aspects or the transcendent models which are found in the One. All our discussion of "aspects" of the One refers to an internal distribution of that Power, a Power which is conceived as not other than the One. All other powers and principles are derivatives from it and imitative of it.

Thus, we have the Power which is emanated from the One and which is the potentiality for Intellectual-Principle. This potentiality gets determined by the transcendent paradigms within the One, so that at one and the same time this Power becomes the subject and its determination, the object.

This emanating Power contains and reflectively includes the transcendent models or paradigms and the Ideas. Therefore, we may speak of this Power as an externalization or efflux of the complex interiority of the indivisible aspects of the Supreme Perfection. This outflow of Power thus includes self-determination—that is, it is the Intellectual-Principle, stamped from within, and in this manner may be said to be molded or be the "matter" of the One. Consequently, the Intellectual-Principle, in its entirety, will exhibit the transcendental paradigms or indivisible aspects of the super-knowledge from which all the unities within Being are suspended. ●

We are tentatively establishing this verbal distinction between the One and its issuing act for the sake of clarification, for these higher principles will be majestically duplicated within the realm of universal manifestation and then again within its parts. Consequently, this distinction will ultimately give rise to the practical truth alongside the Only-Truth, and if it is not made available to us here will be the cause of much confusion and semantic

▪ EDITORS: Here "matter" refers to the indeterminacy of Power, its receptivity, and its ability to include and convey or "exhibit" the complex nature of the One. In each hypostasis the word "matter" has a different meaning.

● EDITORS: In "The One: Symbolization" pp. 58–66, Anthony shows how these transcendental paradigms comprise what he calls the Triad.

obscurity in the relative sphere. [9, p. 307] For understanding the nature of these principles and their precise application to the practical questions of philosophy is necessary in order to answer those questions.

The implications of this view of the One-and-its-Power can be easily illustrated and later on will be compared with other possible views. If on the one hand, for example, the world exists, if it is the result of a real power, and if this power is other than the One, then there will be an ultimate metaphysical dualism between the One and the products of that Power. If, on the other hand, the Power is illusory then any products of this Power—from our universe to all the various levels of reality, including even the Divine Mind (Intellectual-Principle)—will be illusory. Neither of these two alternatives—metaphysical duality or illusory being—is acceptable for Plotinus.

For Plotinus, then, the One-and-*its*-Power stands first and foremost above all and simultaneously sustains and includes lesser levels of reality which are neither self-sufficient nor illusory. Through an understanding of this principle of Power, we can first make available to our minds what we refer to as the double truth—that is, the nondual reality does not exclude the dualistic but includes and supports this value which is not annihilated by a blind and omnivorous, affectively supported intellect.

In tractate v. 4.1, Plotinus tells us that

> If the First is perfect, utterly perfect above all, and is the beginning of all power, it must be the most powerful of all that is, and all other powers must act in some partial imitation of it. . . . the First Good, the Power towards all, how could it grudge or be powerless to give of itself, and how at that would it still be the Source?

In tractate v.5.11, he says that "this unseen First is the source and principle of Being and sovran over Reality." And in vi.9.1 he begins a discussion of the One with the statement, "It is in virtue of unity that beings are beings."

We can see from these and many other passages that, for Plotinus, perfection and self-sufficiency imply an inconceivably vast abundance of power. When its concentrated fullness is described as an utter stillness, that stillness is not meant to imply any sort of corpse-like passivity or immobility, nor is it to be thought of as a luminously quiescent matrix of mere potentialities that must be brought to actuality by another agent. The One is fully actual. Inwardly pregnant with all that is ultimately true, it "produces" by its very perfection and infinite Power, not through any need for a completion to be attained beyond or outside itself. It does not "need" creation; "creation" is the inevitable result of its super-abundance. Not only has it no need of its products, it has no "need" even of producing.

Throughout all the production that inevitably ensues, the One remains unmoved; all possible production is already transcendentally actual and accomplished within its own infinite Power. This is Truth. The truth of Intellectual-Principle is in the One; the truth of Soul is in the One.

In further considering this Power which we conceive as identical with the One's own nature, a twofold character of the One is observable: 1) the utter stillness or passive perfection of the Transcendent, and 2) its infinite Act, its sheer indeterminacy, the profundity of its infinitely unlimited activity as a self-tendance eternally achieved.

> That his being is constituted by this self-originating self-tendance—at once Act and repose—becomes clear if we imagine the contrary; inclining towards something outside of Himself, He would destroy the identity of his being. This self-directed Act is therefore his peculiar being, one with Himself. Thus, He created Himself because his Act was inseparable from Himself. (Plotinus, VI.8.16)

Here we find a presentation of these two ultimate principles, act and repose, within the self-nature of the One. Unavoidably polarized by our own thought to make them available to our understanding, these two ultimate complementaries must nonetheless be recognized as *coinciding* within the One's absolute completion. This paradoxical theme is an exquisite mystical vagary, one that is universally diffused in statements of the perennial philosophy. Our appreciation of it can be broadened or deepened by correlating it with similar views such as passive and active perfection, substance and function, yin and yang, and so on. The words of Lao Tzu have a similar flavor:

> How unfathomable is Tao! An infinite depth, the Source of all that is—the Ancient Progenitor, before all things. Yet, how pure and still is Tao! (*Shrine of Wisdom,* "The Simple Way of Lao Tzse," v, no. 18, p. 10)

When Plotinus attempts to describe to us something about the eternal status and completeness of the "interiority" of the Simplex, his speech accommodates our limited mode of understanding. That which is eternally achieved is expressed in terms of temporality and causality. But these terms must cancel themselves out if we are to be graced with some intuition into the timelessly fixed, perfect, and complete interiority of the One. ▼

▼ PLOTINUS, V.4.1: Standing before all things, there must exist a Simplex, differing from all its sequel, self-gathered not interblended with the forms that rise from it, and yet able in some mode of its own to be present to those others . . .

If we do not keep this in mind, we will arrive at the mistaken notion that the Eternal One is in the process of evolving and improving along with the development of the World-Idea. In seeking, by means of these passages, a glimpse of the incomprehensible nature of the One, the intellect must be like a bird which leaves no trace in flying through the air: The intellect's own operations must dissolve back into the stillness into which it is advancing.

It is axiomatic, then, that the One-Only and the totally complete (Nondual!) perfection must be taken as literally synonymous. Within the One itself there can be no changes or processes, no operations, arrangements, or combinations of any sort—not even the universal relativity of Forms found in the Intellectual-Principle. Bearing all this in mind, we must hold firmly that the One is in no way becoming anything. Therefore, when Plotinus speaks of the perfection of the One and the Act included within it, we are not to conceive that "first" stillness reigns and "then" it is rearranged, as though a fixed number of blocks in a container are shuffled about until the contents are perfectly ordered according to some plan that involves time. Such thinking is inadequate to the notion of the One's perfection and the Act that is indivisibly included within it.

Emanation and Illusory Modification

It may be possible to supply our demanding, egotistical intellect with a description or definition of the One that will swallow up the intellect. Let us consider an example. [10, p. 310]

If we could take the totality of all Ideas—as we can, for instance, with ten or a hundred of them, and permutate and combine them so that each and every possibility is actualized and fully exploited, we would arrive at a sum total that represents in essence their final product. Further, if we could then think of this final product as the undifferentiated, homogeneous unity of the substance of these Ideas, we would have produced in our thought an inconceivable, unthinkable, ungraspable, indescribable intelligence that would repudiate all its prior elements, including the power necessarily involved. In this coinciding of unlimited function and infinite substance, active and passive perfection are identical and we now have a view that permits us to conceive of Nirguna Brahman or the mystery of the One.

But let us reverse the position. Let us assume, or rather reaffirm with the philosophic mystics, that the First—pure intelligence—is, in other words, that the Idea of Ideas *is*. What consequences would follow? Would the other aspects of Brahman be transformations of the First, or would these lesser principles be part of the First, thus making of the One a thing of parts, or would the lesser principles be illusory manifestations of it? ⊞

Let us consider two examples as possible analogies for understanding how Absolute Mind⊕ can give rise to a many. The first is the much used example of pure gold being made into various trinkets. [11, p. 310] In this example, Mind itself has the power to modify or transform itself into these various forms, and Pure Mind, or gold, through its own functioning, has a continuous identity in the various appearances or aspects of itself [the trinkets]. Its being or substance (gold) and its function or production (the forms) are only nominally distinct. All the trinkets are actually only the gold—which stands here for the Absolute—and there has never been anything but the gold, the forms being illusory modifications of it.

As a second example, consider a very hot semi-transparent body and think of the glow or radiation of that body as its emanation. We must imagine that the body has its source of energy internal to and inseparable from it, heating it to the point of incandes-

⊞ Editors: None of these options exactly fits the relation Anthony will describe.

⊕ Editors: Absolute Mind here is a synonym for the One.

cence and causing it to emit a light. Imagine that the power source of this body is infinite and unfailing. Now the emitted light, an offspring of this body, will disclose a great deal of information about the structure of the glowing body. The spectrum or precise amount of light emitted at each frequency can be analyzed, and the spectral structure will depend only on the nature of the glowing body. The modern description of the emission in terms of quantum mechanics, with its very Pythagorean use of numbers, would characterize in precise terms the structure of the glowing body. We can quite accurately say then that the emitted light is a direct emanation of the structure of the glowing body which is revealed through spectral analysis of the light. At the same time, we can see that the emission is quite clearly distinct from the original.

In this example, the glowing body (for instance, a star) is the analog to the One, with the infinite energy representing its inherent, inseparable Power. The glowing body itself is unavailable for analysis, although it has a complex internal structure. The radiating glow is analogous to the emanation of the One—the Intellectual-Principle—which, as in the example, includes a structure that is coordinated by Number. In the example, the emanation is distinct from the original. Yet the spectrum of the emanated light does reveal to us something of the inner nature of the glowing body, which was otherwise unavailable. Just as the physicists study the emitted light to investigate the internal nature of the radiating body, that nature being homogeneous with the body, so the philosopher contemplates the Intellectual-Principle, the emanation of the One, to obtain some insight into the inaccessible interiority of the One.

In these examples we find models for two views of how the many arise from the One. The first example, in which the modifications of gold are only illusorily distinct from the gold substance, illustrates one of the primary Vedantic theories—that Brahman remains intact while the illusory forms present Brahman in apparently distinct guises.▼ This example, however, properly seems to apply to the One itself, as remaining in itself. The inner nature of the One is such that, in spite of its immense complexity, it is like the gold taking on an infinity of apparent completions or aspects (shapes), but remaining always itself. There are no other principles available in this view, for the One is self-constitutive. ⊞

▼ Editors: This theory, *vivarta*, is a key idea to be explored in subsequent pages.

⊞ Editors: In this example, an application of *vivarta* theory, the "aspects" of Brahman are not real "transformations" of Brahman, they *are* Brahman.

In trying to understand the inner intelligence of the One as remaining the One itself, we can resort to the example of the gold, for there is nothing that the One could become. Regardless of whatever forms the gold assumes, it is always the gold, and its forms are illusory [insofar as they appear as anything distinct from or other than the gold]. A consequence of this would be that any appearance—including the ego—is the illusory modification of the One.

The second example illustrates the view of emanation: The glowing light issuing forth is distinct from the original, and yet the glow exhibits the very formal characteristics that are united with but hidden in the nature of the original. In this theory we can consider other levels of reality as distinct emanations from the One, as traces of the One, and this does not in any way diminish or detract from the One's intact identity. ⊕

A delicate balance is required here. Thought is walking a razor's edge, for our understanding may miss the mark and fall on either side. On the one hand we can introduce a hiatus △ that leads to metaphysical dualism—which is proper and appropriate for the intellectual understanding, for the double standpoint demands recognition. On the other hand, thought must confront a monolithic monism, a unique and unknowable all-encompassing mind from which everything arises, the causeless cause; this view is appropriate to the absolute stillness of the advanced contemplator—the initiate—that state of mind without thought. The unique contribution of the philosopher-sage is the simultaneous grasp of these two views that exemplify the paradoxical nature of reality—the One and universal relativity. Adopting the poise of his understanding will supply our own with completion and balance. To understand that perfection engenders, that the eternally perfect emanates and engenders universal principles eternally which in their totality constitute manifestations at all levels from Being onwards, will satisfy the most encompassing and demanding intellects. ◈ [12, p. 311]

From and within the unlimited view of the Ultimate Reality itself, there is no other. However, from the point of view of the others, which may or may not be in identity with that ultimate viewpoint, there is the absolutely relative viewpoint that must be considered

⊕ Editors: "Emanation" declares that the "lesser" principles are not "parts" of the One, are not illusory, and are not separate realities from the One.

△ Editors: In an earlier version of this paragraph, Anthony had added that the hiatus is "between the One and the All."

◈ Plotinus, v.1.6: All existences . . . produce . . . from their essence, in virtue of the power which must be in them—some necessary outward-facing hypostasis continuously attached to them and representing in image the engendering archetypes. . . . Again, all that is fully achieved engenders: therefore the eternally achieved engenders eternally an eternal being.

alongside the nondual. The truth of Mind, the nondual, *and* the truth of its expressions is the paradoxical view of Reality, either universal or relative.

The monistic school of Advaita Vedanta skillfully and persistently hammers into us the one-sided point of view of passive perfection. Our purpose is not so much to criticize Advaita as to indicate the partiality exhibited in that view and its lack of ultimate discrimination. The following quotations from two well-known Vedanta interpreters present an explanation and application of the *vivarta* theory, or the theory of illusory manifestation, to which we take exception. ▼

▼ Editors: Anthony here critiques *vivarta* theory as incomplete, and as misleading when misapplied. He was aware that not all interpreters of Vedanta make the applications he is critiquing.

> ... *Brahma* modifies Itself in diverse ways (in the indefinite multiplicity of universal manifestation), without the aid of instruments or external means of any kind whatever (and without Its unity and identity being affected thereby, without it being possible to say, therefore, that It is modified in reality, although all things only exist in effect as Its modifications). Thus the spider spins its web out of its own substance, subtle beings take diverse (incorporeal) forms, and the lotus grows from marsh to marsh without organs of locomotion. That *Brahma* is indivisible and without parts (as It is), is no objection (to this conception of universal multiplicity in Its unity, or rather in Its "non-duality"); it is not Its totality (eternally immutable) which is modified in the appearances of the World (nor any of Its parts, since it has none), but it is Itself viewed under the special aspect of distinction or of differentiation, that is, as *saguna* or *savishesha:* and, if It can be viewed thus, that is because It comprises all possibilities within Itself, without their being in any sense parts of Itself. (René Guénon, *Man and His Becoming,* p. 83)

> The Samkhyas and those who are in sympathy with their view of transformation characterize the world as a transformation of Primal Nature, as curd is of milk. When a thing attains a state which is different from its present one, it is called transformation; when a thing, while not abandoning its prior state, appears to be of a different state, it is known as illusory manifestation. "That change, which is of the same grade of reality as the thing, is transformation; what is not of the same grade of reality is illusory manifestation; or, change, which is of the same nature as the cause, is transformation . . ." Brahman can be neither the originating cause nor the transformed cause. What is partless by nature cannot originate something *de novo;* nor can it get itself trans-

formed into something else. If Brahman were to abandon the earlier form and get transformed into another form, then, subsequent to creation, there would be destruction of Brahman of the form of knowledge and bliss. If, again, at the stage of dissolution, that Brahman, which is devoid of the form of knowledge and bliss, be transformed into Brahman that is knowledge and bliss, even thus, since at any moment that Brahman might transform its nature, there is the contingence of non-release. Nor is there any evidence in respect of transformation. Scripture in the text, "The unborn self, the great, the firm" predicates immutability of the self as opposed to transformation. And Brahman is immutable because it is partless. (T.M.P. Mahadevan, pp. 193–194, quoting *Siddhantalesasangraha,* v.i, v.ii and Brh. Up., iv, v. 20)

Let us put our understanding of this philosophy in words along with the admonition of Plato that above all we do not become misologues [*Phaedo* 89d and *Republic* 411d], for although reason cannot take us to the Absolute, neither will it mislead us. We cannot in all honesty offend the principle of non-contradiction and yet assume that we have properly interpreted these "sacred texts." To go beyond reason does not mean that we have to deny reason.

When we base our reasoning on Plotinus' understanding of the nature of the One, it becomes clear that the *vivarta* theory is properly applied to the One itself. In his description of the Intellectual-Principle and its relation to the One, the Intellectual-Principle is not an illusory manifestation but rather a transformation of the power of the One, a distinct emanation—that affirms the miracle or gift of Real Being.

... the secondary or issuing Act which—as the output of a mighty power, the mightiest there is—attains to Real Being as second to that which stands above all Being. (Plotinus, v. 4.2)

In the higher philosophic wisdom revealed to the individual intellectual soul of Plotinus, there is knowledge of the realm of the three primals, and something of—a revelation of—the holy stillness of the One. This ultimateless no-thingness, the only instance of perfection, is beyond the authenticity of the principle of Being. This One-without-a-second, which in its simplicity is synonymous with an ocean of possibilities timelessly, integrally perfect, is spoken of by Plotinus from many different points of view; but he always emphasizes that those indivisible aspects of the Absolute are in reality illusory only "in" the One itself. It seems that something about the interiority of this Simplex is revealed to Plotinus, for he speaks of the One's inner perfection and defines and separates it *in thought* from the Intellectual-Principle and Soul.

From the first standpoint then, illustrated earlier by the example of gold and its forms, the One is perfect and is the simplicity of infinite "knowledge," which is identical with the transcendent paradigms of Nous, Being, Life, and so on. These paradigms may also be considered as indivisible and unconditioned aspects of unity—or the viewing of unity in a unique way, in a certain individual aspect, such as its infinite being, or the infinity of its intellect or its life. (Each modality or aspect of this unity includes all the others, but secondarily, because the angle from which we view it has already predetermined in what way we shall perceive the One. For instance, when we think of the One as infinite Life, then infinite Being, infinite Intellect, is subsumed under the infinite Life.) Regardless of how these aspects of the One are combined and arranged and rearranged, in the end these views amount to an illusory modification or division of the One's innate perfection.

▼ Editors: This important paragraph addresses two errors to avoid while distinguishing between Mind (or the One) and its expressions. On one hand, we must not think that the distinction between them is only nominal: It is a real distinction. On the other hand, we must not think of the expressions as utterly other than the Absolute, which would make them illusory.

Here, *and only here*—in the utter simplicity of the innately perfect—is the application of the *vivarta* theory or illusory manifestation fully justified. Only the One bears the full weight of the implications in this theory. We cannot in all sincerity apply the theory of *vivarta* to the relation that exists between universal manifestation and the One: Mind-and-its-expressions requires that we maintain the distinction or lose our balance; it does not assume that there is anything else besides Mind.

The Intellectual-Principle is not merely nominally different from the One. Still less can we declare the Intellectual-Principle and its Divine Ideas to be an illusory manifestation of the One, unless—through semantic obscurity and existential legerdemain—we insist on denying the experienced world, which is there for the sage to see along with Brahman. ▼

The earlier quotation from Guénon presents the point of view that reduces everything to Absolute Mind or its passive perfection; even its ability or power to modify itself is there subsumed under its passive perfection. In the quotation from Mahadevan, the distinction between the theory of *vivarta* (illusory modification) and the transformation (*parinama*) theory is made more explicit. This difference has been expanded into two conflicting theories—that is, the Vedanta and Samkhya respectively.

If we attempt to resolve this conflict from the viewpoint of Plotinus' double act, we would be able to say that Absolute Mind or the One itself always remains Self-intact, indivisible, while its issuing act or Power is the emanation that gets transformed into the Intellectual-Principle, etc. Both these points of view, as previously pointed out, must be held in balance.

The intact perfection of the One always remains what it is. This is what is referred to as its *passive* perfection. Within this intact perfection is that complexity of the interiority of the One, and regardless of how we conceive of it—any sort of arrangement or rearrangement of the transcendental paradigms—it will always remain indivisible in its unity. It is only here that the *vivarta* theory is applicable. ⊞ Then Plotinus speaks of the issuing Act (i.e., its *active* perfection) which goes forth; and reflected within this Power are the One's indivisible aspects which now become Intellectual-Principle. This emanation can be conceived as the One transforming and diversifying itself. These two viewpoints, that of the absolutely Absolute and the absolutely relative, are an expression of its double act.

⊞ Editors: Through these last five paragraphs, Anthony entirely supports applying *vivarta* theory to what Plotinus calls the "passive perfection" of the One. But he takes exception to interpreters who use it to suggest that what Plotinus calls the One's "active perfection" is somehow unreal or illusory. For Anthony as for Plotinus, both perspectives—the One's passive perfection and its active perfection—are equally necessary to approach an understanding of how from One many arise.

Further, Anthony considers *vivarta* misleading (and basically irrelevant) when applied to the level of universal manifestation. Applied in that context, it would deprive the distinctions within manifestation of the authenticity they derive through participating in the One's eternal offspring, Intellectual-Principle and Soul. He will develop this theme more in the chapters on Intellectual-Principle and Soul.

Primal Hypostases

If we now take the standpoint of the three primal hypostases—the One, the Intellectual-Principle, and Soul—in terms of the descriptions that have been given, we have to consider them as distinct realities of different degrees. Plotinus' principle of emanation must be applied and must be understood to mean that the eternally achieved gives rise eternally to an eternal emanation. The Divine Mind is a reflex; but in striking contrast to the *vivarta* theory, it is an image of an infinite original of such power and perfection that the very copy, so-called, stands a reality.

The Intellectual-Principle and its emanation, Soul, are not illusory in any sense; they are also eternal, and they are not annulled or canceled out. ▼ ◘

Even if we view the three primals as forming an integral whole, which is often referred to as the Absolute when considered from the side of manifestation, the distinctions cannot be dissolved so that the Nous and Soul, so carefully defined, become illusory principles. In our understanding of the metaphysical Infinite, we retain the view of the One as the pure and only perfect reality, as well as the view of real and distinct emanations from it. We do not violate the One's sovereignty by granting to each of the other levels of reality its proper status; they do not become nil and void in the face of the One which they eternally contemplate.

It is true that these distinctions of the three hypostases are given to the novice as a help in understanding. It is true that later on, when the intuitive life is practiced, they may be disregarded and all three may be considered as God or the Absolute. But this is not the issue here. What is important to remember is that these distinctions are going to provide us with the necessary foundational principles that must be applied first to universal manifestation and then again applied to that portion of universal manifestation referred to as Body—thus providing us with the necessary means to understand that the cosmic circuit can announce or reflect back to us the meanings that are in the intelligible world. Furthermore, the logical deduction that follows from the original premises will permit us to point out the non-difference between the unity of a god and the body that represents that god. There is no hiatus between the immediate efficacy of its will and its execution, even though we have established a hierarchy of causation and cognition.

The Bhagavad Gita propounds a view of Brahman as having four aspects, the first three of which are transcendent and correlate to the One, Intellectual-Principle, and Soul. ● [13, p. 312] This teaching corroborates the Plotinian doctrine in that these three are defined and distinguished one from the other prior to their immanence in *Vishvasrit,* or the fourth aspect of Brahman, which Plotinus calls Zeus, the Demiurge. Neither in the *Enneads* nor in the Gita are these three confounded with each other or reduced to one homogeneous simplicity to which the *vivarta* theory could be applied.

▼ Plotinus, v.2.2: There is, accordingly, from the first principle to ultimate an outgoing in which unfailingly each principle retains its own seat while its offshoot takes another rank, a lower, though on the other hand every being is in identity with its prior as long as it holds that contact.

◘ Plotinus, iii.2.2: The Intellectual-Principle, then, in its unperturbed serenity has brought the universe into being. . . . For the Emanation of the Intellectual-Principle is Reason, an emanation unfailing as long as the Intellectual-Principle continues to have place among beings.

● Editors: Anthony referenced this point of view to the work of T.S. Row (*Notes on the Bhagavad Gita*) and I.C. Sharma. Row holds that this view, while not explicit in the Gita, is "therein indicated with sufficient clearness." (p. 12) Sharma asserts (p. 92 of his essay) that the Brahmanic analysis of the Bhagavad Gita is that same as that of the Vedas. Figure 31, p. 71, compares the Plotinian fourfold with the view Sharma provides.

In brief, Plotinus' position is that it is because the One remains itself, perfect, that its emanants are guaranteed to stand as eternal realities. We cannot say that because the emanants are real, the unity of the One is shattered, nor that because the One must remain intact the emanants must be illusory. Vedantins may interpret "the One without a second" to mean that there is only Brahman and that all else is illusory manifestation; they may even say that there is nothing else. [Cf. Chandogya Upanishad, VI.ii.1.] But we will say instead that there is nothing else *like* Brahman, for it is the only perfection and all the others are its expression.

To go further, we cannot apply the principle of *vivarta,* as we have understood it, within any of the levels below the One. The eternal ordering and interrelation of the aggregate of Beings in the Intellectual-Principle does not permit us to say that each Idea of the total of Ideas is an illusory modification or manifestation of the Intellectual-Principle, for the Intellectual-Principle *is* the totality of these eternal, distinct, real Ideas. Similarly, in the level of Soul where there is a multiplicity of unit souls, characteristically defined powers and reason-principles distinguish each soul from the others, although souls also each include one another. Nor can the principle of manifestness—(the Overself) *Asmita*—be submerged and dissolved, for this would leave the individuality unexplainable.

We come next to universal manifestation. Perhaps here, more than anywhere else, to insist that the universe and the things within it are illusory manifestations or modifications of Brahman—rather than expressions of Brahman—is to reduce to rubble and chaos the Divine and spiritually ordered relativity of existence. In other words, we are not being offered (through *vivarta* theory) an explanation and some understanding of the Divine theophany—but a partial and incomplete expression of its truth. Emanation properly understood will provide us with an understanding of this ultimate Truth in its fullness. The erasure of all frameworks would result in the meaningless assertion that the ego does not exist, or, that it is absolute. As expressing its prior principles in image or imitation being, the universe is not an illusion, but is the perpetually becoming body of the Divine, without beginning or end. Within any of the levels of manifestation the presence of the living forms or reason-principles, the outward functioning of wisdom intelligence, prevents us from speaking of these as illusion. And these principles provide us with emanated levels of reality that are frameworks within which truthful statements can be made. Consequently, it is meaningless to say that "nirvana is samsara."

Perfection [the One] has no tenses, nor is it even the timeless, ordered simultaneity of the eternals as found in the Intellectual-Principle. It is pure, unconditioned positivity, uncomposite fullness, beyond the isness or the characterization of any particular form or Idea. Yet it contains in a transcendental mode—that is, in a mode other than that of the Intellectual-Principle—all that is ultimately true. How?

> In this way: the intellectual object is self-gathered (self-compact) and is not deficient as the seeing and knowing principle must be—deficient, I mean, as needing an object—it is therefore no unconscious thing: all its content and accompaniment are its possession; it is self-distinguishing throughout; it is the seat of life as of all things; it is, itself, that self-intellection which, stemming from an inner consciousness, takes place in eternal repose, that is to say, in a mode other than that of the Intellectual-Principle. (Plotinus, v. 4.2)

This self-contained Mind-in-Itself is the source from which arises what Plotinus refers to as the second Act. The "earlier form" of the double act is the realized identity of the One itself. That is, active and passive perfection, as identical, is the nature of the One; and this is the first Act. "The second Act is . . . an emanation distinct from the thing [the One] itself," (v. 4.2) Plotinus says; and he refers to this as "the output of a mighty power, the mightiest there is." This second Act is the Intellectual-Principle, and it may be conceived of as the outgoing efflux or emanant, which is distinct but not separated from its all-mighty prior. In v. 1.6, Plotinus refers to this as its outward-facing hypostasis. The path from metalogical infinity to Being is established in this peculiar way, and there is no hiatus between them.

The ultimate truth of the lesser levels of reality is to be found in the One itself; but through the process of emanation there is a transmission, so that traces of this truth are revealed in the lesser principles. The similarity between this Absolute Mind or superknowledge and the *Parabrahman* of the Vedantins can hardly be overlooked. Yet it cannot be said [as some Vedantins say] that this production of eternal Being is an illusory modification of the Principle of all principles, suggesting impotence of this mighty power to engender eternal realities that in their turn produce the whole system of Nature in which they are immanent. The perpetual validity of the two views of truth is based on this production, which includes within itself the ground for a variety of ontologies and epistemologies. Nor is it possible to conceive the unoriginated as "the only transmigrant," ▽ buttressing its perfection by creating illusory manifestations that it then must reabsorb, as if it were incapable of producing eternal offspring. The One is immune to such solicitations of official orthodoxy.

If the Absolute knowledge and the absolutely relative knowledge persist side by side, paradoxically and perpetually, then the application of this truth to the realm of contingency provides answers to certain "unaskable questions"—answers that are in direct confrontation with the official religious dogmas. For example, if the Absolute can grant the eternal gift of Being to the Soul, Soul in turn will manifest eternally. As authentic essence the Soul includes a principle of manifestation, and to claim that Soul is reabsorbed when it achieves recognition of its true Being is to deny its status as an authentic essence capable of engendering perpetually a reflex of itself. Consequently, self-realization does not necessarily entail the cessation of its manifestation. ⊞ The Buddha or a sage will continue to reappear periodically, for it is in the very nature of Soul to be represented by an ego. It is the very nature of Soul as an authentic essence to be a metaphysical wanderer in the infinitude of God's Being.

▽ Editors: This reference is to Ananda Coomaraswamy's description of Ishvara as the "one and only transmigrant," meaning that Ishvara, appearing as the individual *jivas,* appears to become individuated and enjoy experience.

⊞ Editors: Anthony is addressing this remark, about Soul as including a principle of manifestation, to those who take enlightenment to mean the cessation of rebirth.

The One: Symbolization

The symbolism proffered to us in the cosmic mandala is the most comprehensive and integral approach to philosophy that at present is available. It is within the compass of our discursive understanding, yet transcends its limitations and reveals the possibility of a more universal language. It includes the subject-object dichotomy yet suggests a more intimate resolution. Its amplitude may resolve some of the most perplexing and profound problems of existence, being, and knowledge. Its very nature demands that human preconceptions be put aside. The attempt to perceive through this manner produces thought that forces us to exclude boundaries—truly a trying to understand rather than a claim to have done so. Philosophy as wonder—and the insight into that wonder—will procure for us the inspiration for our daily and ordinary lives.

Overview: Metaphysical Chart

Many who try to penetrate the thought of Plotinus find his ideas too vast and intricate to grasp. With the verbal distinctions of the previous section now available, we have in mind a visual schema that will allow us to: 1) itemize and isolate the principles from one another for individual study, 2) follow these same principles through the various interrelations that exist in and among them, 3) show how they are involved in universal manifestation, and 4) spread out the whole system to provide a comprehensive view of the totality.

For instance, when Plotinus speaks of a principle which transcends Being, we can readily translate these remarks into a pictorial form. For example, in v.1.10 he writes:

> There exists a Principle which transcends Being; this is The One, whose nature we have sought to establish in so far as such matters lend themselves to proof. Upon The One follows immediately the Principle which is at once Being and the Intellectual-Principle. Third comes the Principle, Soul.
>
> Now just as these three exist for the system of Nature, so, we must hold, they exist for ourselves. I am not speaking of the material order—all that is separable—but of what lies beyond the sense realm in the same way as the Primals are beyond all the heavens; I mean the corresponding aspect of man, what Plato calls the Interior Man.

These four principles can be diagrammed as a circle divided into four quadrants as a simple static representation of his meaning. [*fig. 6*]

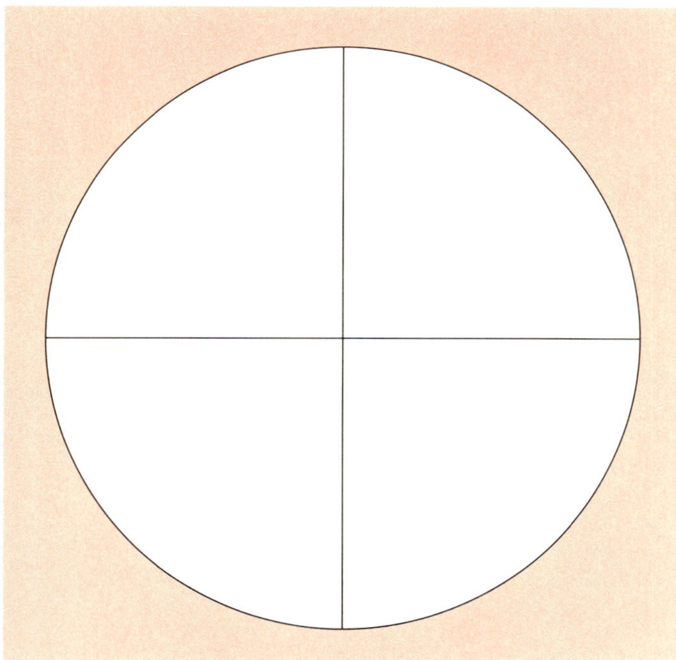

Fig. 6

The first quadrant represents the One; the second Intellectual-Principle; the third symbolizes Soul; and the fourth the system of Nature, or Universal Manifestation in its totality—that is, the Body of God with a Unity, Being, Soul, and the Visible Cosmos. [*fig. 7*]▼

This schema will be more than a useful device to objectify our thinking only. Later on, as we shall see, it will become a mandala in the true sense of the word.

The diagram will aptly illustrate our twofold viewpoint. We may look upon the whole circle as a symbol of the One itself—all and everything is included. Again we can look at the divisions within the circle as those principles that emanate from the One. These two points of view are simultaneous in the chart.

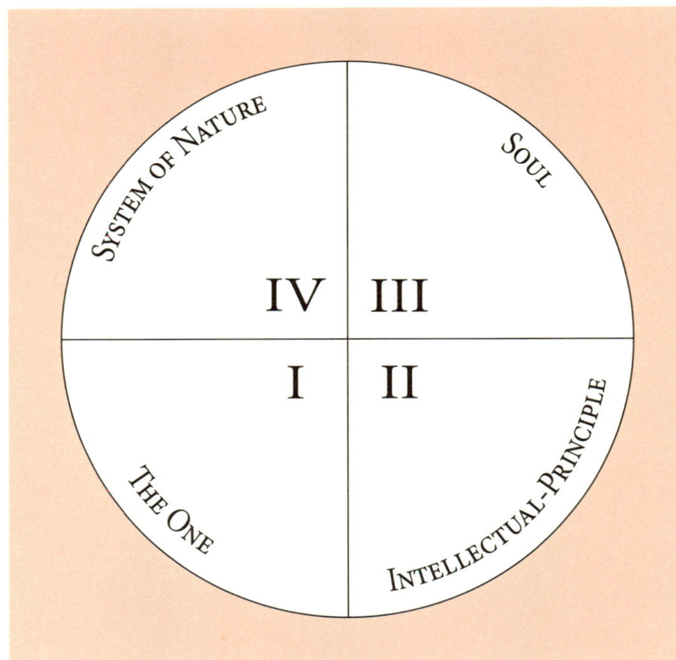

Fig. 7

▼ Editors: The Plotinus quote (v.1.10) and Anthony's remarks suggest that the three unmanifest principles of Unity, Intellectual-Principle, and Soul, which exist in their own right, can also be applied to the fourth quadrant, resulting in an image of the cosmos which itself contains a unity, intellect, soul, and body. Anthony elaborates this idea of the fourfold later on pp. 67–71.

⊞ Editors: Anthony uses the dynamism of the circle as both emanating from a central point (as the One) and as contracting from the infinite space. The first image seems to suggest the point of view of emanation of principles from the One. [*fig. 8*] Reversing the order, the entire background suggests the omnipresence of the One and the other principles are seen as successive circles nested within it. [*fig. 9*]

As Plotinus observes in v.5.9:

> . . . Soul is not in the universe, on the contrary the universe is in the Soul. . . . Soul is contained in Intellectual-Principle and is the container of body.

continues on p. 49

The illustration might also be presented in a much more dynamic fashion: the unknown center radiating four concentric circles, the first two innermost circles representing [Being and] the Intellectual-Principle, then Soul, and lastly, Body. [*fig. 8*]

> There is, we may put it, something that is centre; about it, a circle of light shed from it; round centre and first circle alike, another circle, light from light; outside that again, not another circle of light but one which, lacking light of its own, must borrow. (Plotinus, IV. 3.17)

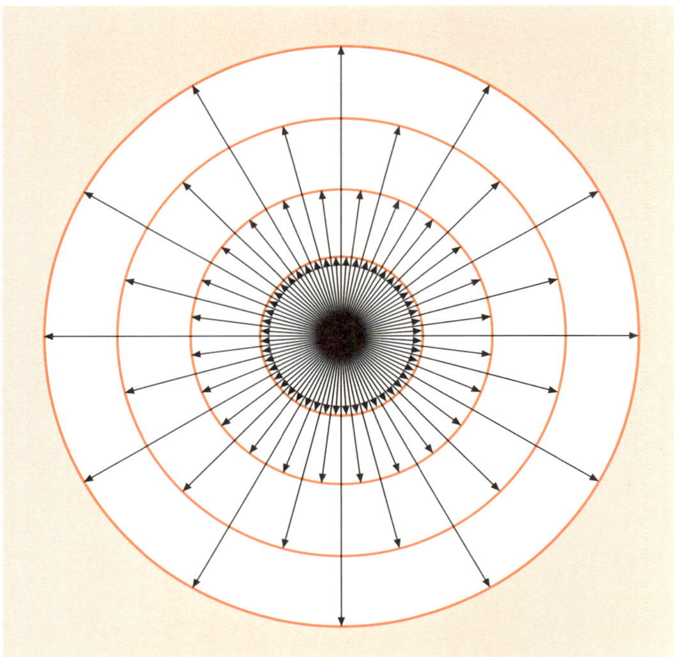

FIG. 8

continued from p. 48

The Intellectual-Principle in turn is contained in something else; but that prior principle has nothing in which to be: the First is therefore in nothing, and, therefore, nowhere. But all the rest must be somewhere; and where but in the First?

. . . the First is neither remote from things nor directly within them; there is nothing containing it; it contains all. It is The Good to the universe if only in this way, that towards it all things have their being, all dependent upon it . . .

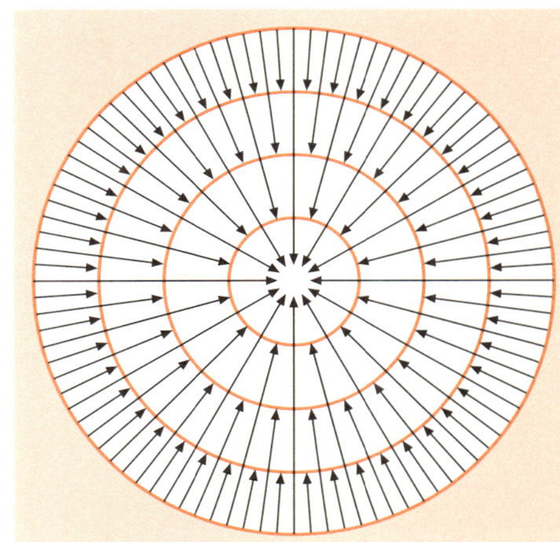

FIG. 9

Superimposing the perspectives of rings and quadrants visually suggests the paradoxical nature of reality: intelligence and its activity, or substance and function simultaneously—which is a prerequisite for metaphysical understanding. [*fig. 10*]

The ring view in its dynamism connects or relates the substance—it offers an approximation to the dynamism which connects together the quadrants or intelligences.

The static integrity of each of the quadrants permits us to speak of it as substance (i.e., intelligence), whereas the four circular forms represent the dynamism or the functioning of this intelligence. Taken in themselves, either of these points of view is an extreme; our method is to superimpose or to fuse the two images so as to produce a symbol that contains both points of view simultaneously, expressing the paradoxical nature of reality.

▼ EDITORS: Ptolemaic astrology assigns the planets special functions in relation to specific signs of the zodiac. Those positions are called "dignities." Ptolemy explicitly speaks of "rulerships" and "exaltations." Anthony extends the notion of the dignity beyond these two, to include their opposite positions: "detriment" opposite rulership, and "fall" opposite exaltation.

Anthony uses the planetary glyphs (☉ ☽ ☿ ♀ ♂ ♃ ♄ ☊ ☋) to symbolize the active perfection of reality. Each of the planetary beings embodies a divine function which ultimately is rooted in the One. These deities are placed according to their dignities in the appropriate divisions of the circle (which represent reality as passive perfection). The meanings of these dignities will be detailed in a few pages.

⊞ EDITORS: In figure 11, the simultaneity is represented by the relation of the planetary symbols and the divisions of the circle.

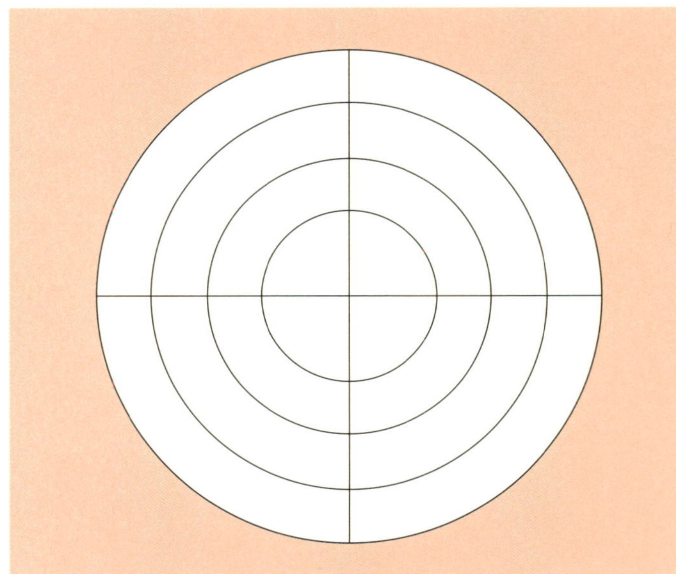

FIG. 10

Again in v.1.8, Plotinus speaks of the "Primal" principles which surround or adorn each hypostasis:

> This is the explanation of Plato's Triplicity, in the passage where he names as the Primals the Beings gathered about the King of All and establishes a Secondary containing the Secondaries and a Third containing the Tertiaries . . .
>
> Thus Plato knows the order of generation—from the Good, the Intellectual-Principle; from the Intellectual-Principle, the Soul. These teachings are, therefore, no novelties, no inventions of today, but long since stated . . .

We can symbolize the Primals as and by the dignities ▽ as we find them in the astrological mandala, with the following results. [*fig. 11*] Plotinus' fundamental premise about the simultaneity of the double act is available in this bare sketch. ⊞

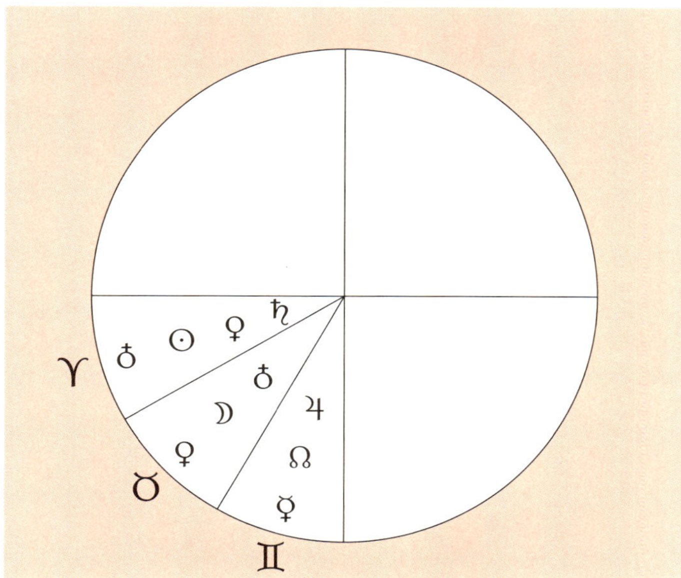

Fig. 11

An emphasis is made that the first three principles are pure hypostases. What we should note in tractate v.1.8 is the grouping of these principles with their derivatives, the establishment of priorities or an order among them. Plotinus is laying down, in a general way, a solid foundation for the teachings of Neoplatonism. These abstractions need more thought-coverings in order for them to appear to the dimness of our mental eye. Therefore we use the astrological symbolism of mundane astrology and the significations it ascribes to the planetary dignities, and from these borrowed meanings infuse more substance into these conceptions.

Thus we will associate the Primals with the first quadrant, situating them in the same locus as the "King of All," the One, about which they are gathered. Similarly, the secondaries are associated with the second quadrant, and the tertiaries with the third. The fourth quadrant will be used to represent these spiritual agencies at work within the system of Nature.

If we allow ourselves the freedom to expand the correlation with the dignities that we find in the astrological mandala, the outermost circle would be filled in by all the rulerships, the second circle by the exaltations, the next by the detriments, and the innermost with the falls. We now have at our disposal a chart that diagrams the All and its Source. [*fig. 12*] ▽

▽ Editors: This placement of dignities in distinct rings helps to further differentiate and pictorially organize the principles within each level of reality. All the rulerships are placed in the outermost ring to represent Being and Life; the exaltations in the second ring represent Intellection; the detriments in the third ring represent Soul and Nature; and the falls in the fourth represent Body. Thomas Taylor discusses details of Being, Life, Intellect, Nature, Soul, and Body in his Introduction to *The Six Books of Proclus* and *On the Theology of Plato* (pp. x–xii).

Although Anthony used the inner-to-outer order of the rings previously in this section, he uses the order of outer-to-inner in almost all his other diagrams. From here on we will use the outer-to-inner order.

With this mandala at our disposal, we have available a device by which we can investigate and illustrate all the subtle intricacies of the Neoplatonic tradition. And like a general with his maps planning for battle, we can pinpoint with little flags our understanding of the chief ideas, their interrelationship with one another, and the tremendous complexity that ensues—especially if we try to grasp it all at once. As we pick up one quotation and attempt to define our understanding pinpointed in the diagram, we can then put it down again within the matrix of this web of relationships without fear of it slipping away.

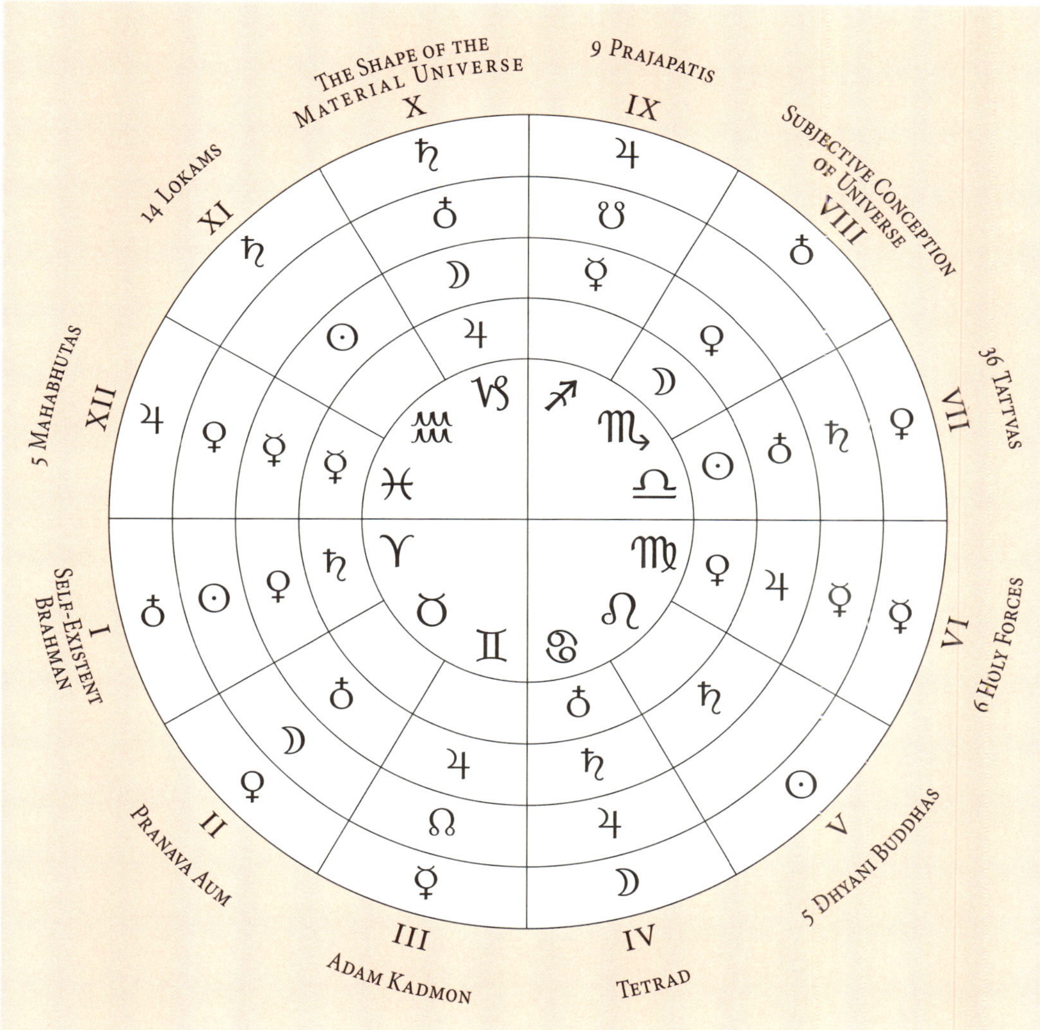

FIG. 12: THE METAPHYSICAL CHART

*B*RING THIS VISION *actually before your sight, so that there shall be in your mind the gleaming representation of a sphere, a picture holding all the things of the universe moving or in repose or (as in reality) some at rest, some in motion. Keep this sphere before you, and from it imagine another, a sphere stripped of magnitude and of spatial differences; cast out your inborn sense of Matter, taking care not merely to attenuate it: call on God, maker of the sphere whose image you now hold, and pray Him to enter. And may He come bringing His own Universe with all the gods that dwell in it—He who is the one God and all the gods, where each is all, blending into a unity, distinct in powers but all one god in virtue of that one divine power of many facets.*

PLOTINUS, V.8.9

THE ONE or the absolutely transcendent can be conceptually more available when the cosmic mandala is fully expanded and amplified. Where the totality of significations is spread before us in pictorial form with the assigned meanings of each, then and only then will a unified presentation of humanity's rich philosophic inheritance become usefully depicted. The implications are profound and descend even into the subterranean recesses of our functioning—its value observable until human intuitiveness has sufficiently evolved to dispense with it.*

The One is to be thought of as a continuum within which diaireses [division] may be conceived without sundering the unity.

ANTHONY DAMIANI,
NOTES AND PAPERS, P. B-2

ONE, DOUBLE ACT, TRIAD

First house

Starting with our opening quotation, v.4.1, we will now identify the Absolutely Transcendent with the first mansion, the Idea of the Ideas, or Absolute Mind:

> Standing before all things, there must exist a Simplex, differing from all its sequel, self-gathered. . . . Untouched by multiplicity, it will be wholly self-sufficing, an absolute First . . .

All the references made above regarding the One's utter simplicity and inviolable transcendency will apply to the 1st house. [*fig. 13*] ▼ ◘ [14, p. 313] This is the intact superknowledge, which cannot even be spoken of, as suggestive of the identity of passive and active perfection.

In VI.9.6, Plotinus goes on to remind us that absolute self-sufficiency characterizes it.

> This is utterly a self-existent, with no concomitant whatever. This self-sufficing is the essence of its unity. Something there must be supremely adequate, autonomous, all-transcending, most utterly without need. . . . The sovranly self-sufficing principle will be Unity-Absolute . . .

This most positive affirmation about That which is essentially beyond statement brings language to a breaking point. In summary fashion we are simply reviewing what we have previously said regarding the simplicity of the One, but it is essential that this point be emphasized and attached to the proffered symbolism.

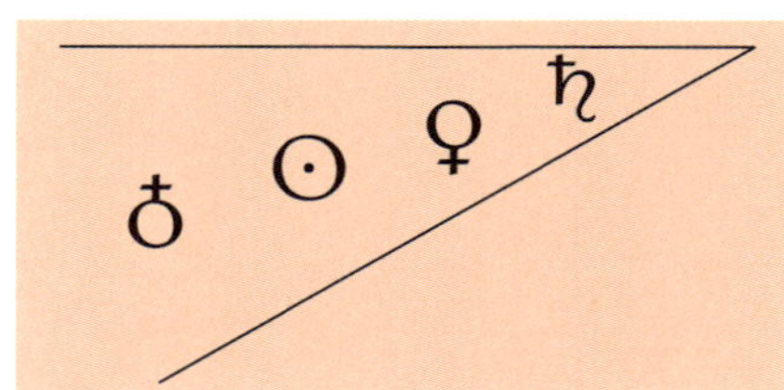

FIG. 13: THE ONE-ONLY

Dyad, double act

Figure 13 suggests that the 1st house—the passive perfection of the One and its unfathomable depths of power (symbolized by the four principles) as identical with it—is purely One. This precisely illustrates Plotinus' meaning in VI.8.16:

> That his being is constituted by this self-originating self-tendance—at once Act and repose—becomes clear if we imagine the contrary; inclining towards something outside of Himself, He would destroy the identity of his being. This self-directed Act is therefore his peculiar being, one with Himself. [*fig. 14*]

▼ EDITORS: Symbolizing the One as the first of twelve sections of the chart does not supersede or invalidate representing the One either as the first quadrant or as the whole chart. It is simply another way to view it.

In VI.7.40, Plotinus refers to this illusory division within the One's interiority. He points out that the operation of its power produces no new thing within the One, and he distinguishes that power, which is identical with the One, from the emanating or issuing power which comes to be the Intellectual-Principle.

> ... it is an activity of that source perfecting the potentiality there, not by producing anything but as being a completing power to the principle in which it inheres.

⊞ EDITORS: In this metaphysical part of the work, Anthony is not using the terms "house" and "mansion" as they are used in mundane astrology. He uses them here as generic terms for sections of the twelvefold division of the circle. He also uses the term "sign" in a principial manner, to designate these same divisions insofar as the twelve sections may be correlated with the signs of the zodiac.

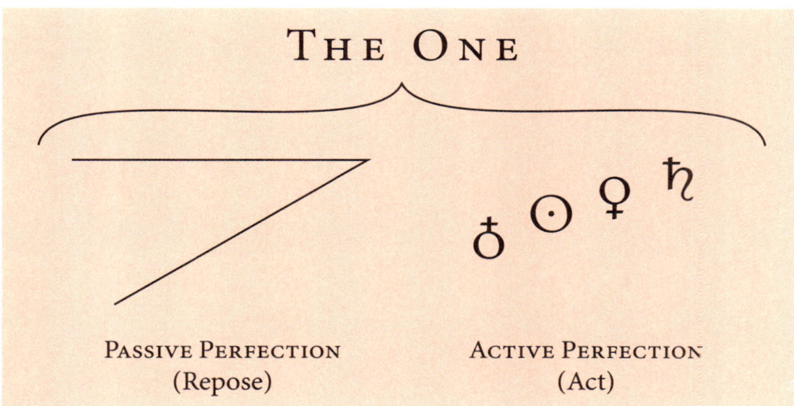

FIG. 14

Now it is with the power which is identical with its source, which is a completing completed power within that source, that our difficulties will arise. For we are being asked to incorrectly conceive this power (which is indistinguishable from the One) as spread out among or within its infinite complexity and totally present within each of its indivisible aspects.

It is the same power which simultaneously exhibits Infinity and Bound. ▽ In other words, the passive and active aspects of the One are being distinguished by the 2nd and 3rd houses with their respective powers. In so "factoring out" the illusory components of this indivisible One, we get an expanded version of its interiority. ⊞ [*fig. 15*]

If we think of the 2nd house as unlimited substance, and the 3rd house as infinite function, these two are not different from the 1st house, for in dividing the One by itself it remains unalterably itself. ⊕ Now let us apply this understanding directly to the diagram, so that we can properly achieve an appreciation of the first quadrant as the Unity.

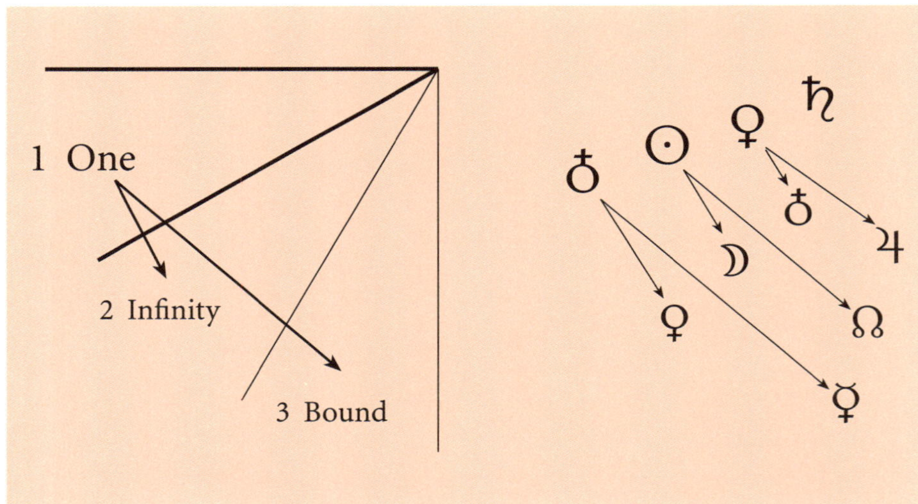

FIG. 15

▽ EDITORS: In the *Philebus* (23c and 27b, T. Taylor translation, v. 4, pp. 494–495 and his footnote) Plato uses the terms "infinite" and "bound" (Hackforth, in *Plato: The Collected Dialogues,* uses "the unlimited and the limit") to define the two principles that unfold the One. The Infinite is the One's actuality or power; Bound measures out the Absolute. In his footnote to 23c, Thomas Taylor says: "If beings, therefore, subsist from *bound* and *the infinite,* it is evident that these two must be prior to beings, or, in other words must be superessential. Hence, as *bound* and *the infinite* are superessential, Socrates with great propriety says that 'God has *exhibited* them.' For their procession from the highest God is ineffable . . ."

⊞ EDITORS: The passive and active aspects are not illusory, but the ultimate distinction between them is.

⊕ EDITORS: To get the perspective of the whole first quadrant, the passive (house) and active (dignity) aspects of the One are each visually "expanded." (Cf. p. 54: The One is "conceptually more available when the cosmic mandala is fully expanded and amplified.") Anthony expands or "factors out" the components by applying both aspects of the One to each aspect. First he says (two paragraphs earlier) that "we are being asked to incorrectly conceive this power . . . as spread out among or within its infinite complexity and totally present

The 2nd and 3rd houses show the double act distinguished within the One. The One divided by itself is still itself.

The One or the coincidence of the double act in its realized self-identity distinguished within the distinctionless whole—or the Power as identical with the indivisible infinite complexity of the One—is here shown by figure 16.

The juxtaposition of the views of the One as house 1 and as the entire first quadrant is an example of the double act as remaining in the One. ▲

Through a simultaneous combination of both these points of view we may glimpse the Triad as the meta-ontic, a-temporal "completion" or the "interaction" of the One's Active and Passive Perfection. The Triad so constituted is an "expanded" version of the

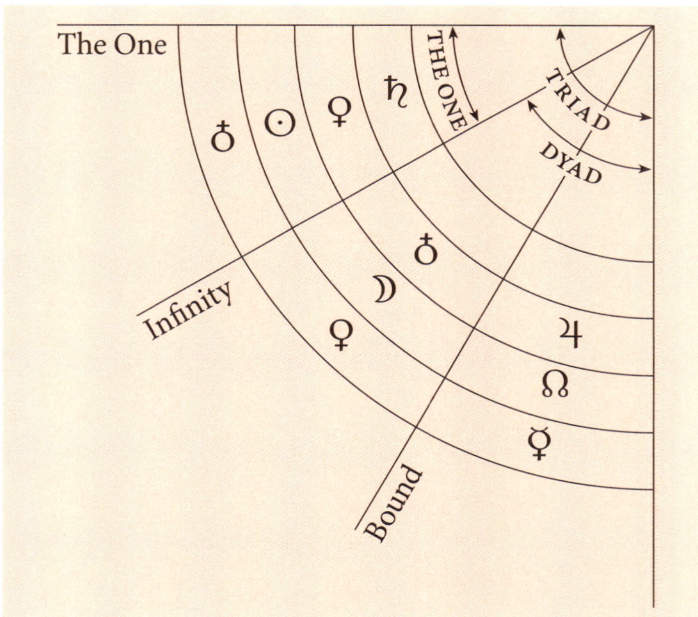

Fig. 16: One, Dyad, Triad

within each of its indivisible aspects." Then he conceives this "power" as the "doubling" exhibited as a) the 1st house and b) the dignities, and literally spreads this doubling out as all the components of the first quadrant. As a result, the 1st house, "repose," is expanded into "incorrectly conceived" components: the 2nd and 3rd houses, respectively passive and active aspects of repose. Anthony calls the 2nd house "unlimited substance" or "passive," and the 3rd house "infinite function" or "active." Similarly, the dignities, originally the "active" aspect of the One, are expanded into "incorrectly conceived" components: passive and active aspects represented by the dignities in houses 2 and 3, respectively. Subsequent pages give more meanings for these dignities.

▲ Editors: This fragment of an *Astronoesis* working draft (Notes and Papers, p. O127) elaborates the relationship of house 1 and the entire first quadrant:

> In our symbology, the One as first house and the One as first quadrant are complementary positions which are not two, but are the One intelligence. Each aspect of unity must itself be described in a nondual way as intact identity and infinite internal complexity. Repose and Act are the two modes under which we are compelled to see the One. For a monistic Vedantist, *Maya* is illusory, and so is its product. For Plotinus, the Power is assimilated to the Absolute, and thus will explain how from One the multiplicity arose.

One, revealing the indivisibility of its aspects and the supposed interaction that could take place among them. [*fig. 17*] ▼

There has been no change; no "creation" has taken place: The One's own essence stands ineffably "completed," its unfailing self-tendance was and will be abidingly self-fulfilled. This essence, which is the very being of the One's absolute positivity, is both passive *and* active.

▼ Editors: Triad is a term Anthony uses for the "expanded" view of the One as first quadrant, revealing the infinite complexity of its aspects (called unities).

The Triad

From the sheer intensity of this perfection, there emanates, as a distinct and lower existent, the Intellectual-Principle, the aggregate of eternally-existing Ideas and Beings. In the Intellectual-Principle, we find the eternal ordering of the essential forms and their gnostic

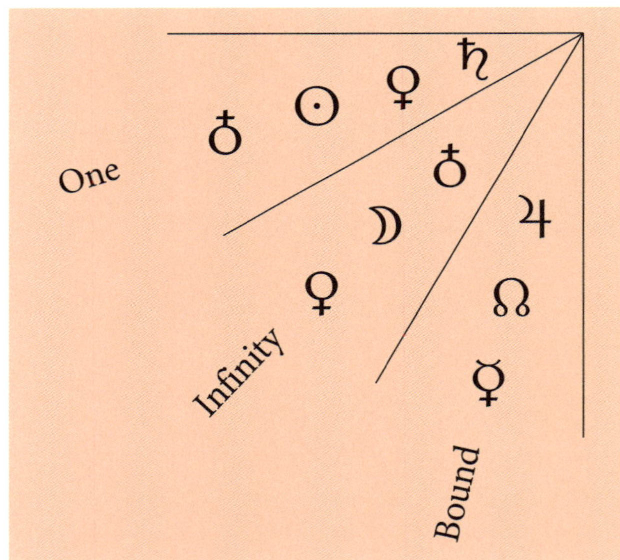

Fig. 17: The One as Triad

powers. But beyond and above even the eternal self-ordering of the Knowing Principle itself [the Intellectual-Principle], we find the utterly undifferenced prefiguration of all that lesser principle's activity in the supremely encompassing self-fulfilled super-intelligence of the One. In contrast, Intellectual-Principle contains all the compossibilities of Being. The One's own essence, however, must carefully be distinguished from all these real distinctions, and may strictly be called universal Being, pure positivity, beyond the authentic forms of any and all determined Being. The paradigms of all facets of Being infinitely coincide in the Perfection of the One.

The Triad is conceived as a completed completing power within the fullness of the One's infinite Being. We must make sure that the words do not deceive us. They should not be allowed to suggest temporality, causality, or change, for we are speaking about the One as a totally accomplished and complete perfection. When we distinguish its active and its passive perfection, we are still speaking of Absolute Mind itself: a no-thingness, paradoxically a plenum, self-contained. There never could have been a beginning, nor will there ever be a lapse from this fullness that is the immaculate, inviolable virginity of the Idea of the Ideas. Universal possibility ever remains meta-ontically achieved.

Within this infinite and undefiled continuum, we conceived of an ever-accomplished Act. Its outward facing hypostasis or radiation is Power, the Void-mind's utterance, within which the Good imparts to itself (that is, its Power) this shoreless ocean of wisdom—the "Idea of the Good"—which reflects something of the interior thrice-unknown darkness. It is to these unfathomable glimmerings that we address ourselves so that we may speak of the paradigms or transcendent models [of the Intellectual-Principle that are to be] found in the One. We will call this [transcendent model] the Triad. Each partless part of the One is conceived of in turn as the totality of the One, and the Triad is in a sense the illusory divisions of this Unity. This amplification permits us to conceive of the facets of the One as a prefiguration of the Intellectual-Principle. Or, conversely, the aggregate of Ideas and powers, which in their totality constitute the Intellectual-Principle, may be seen as a reflex of the excellencies found in the One's effulgence.

We can now arrange the variety of individual unities within the One to suit our purposes:

We can arrange the unities in a form similar to a Kabbalistic tree. [*fig. 18*]

Or we can rearrange the Triad in this fashion. [*fig. 19*]

These diagrams will allow us to more easily perceive the transcendental paradigms of Nous, Idea, etc., which Plotinus refers to in VI.8.18:

> What is present in Intellectual-Principle is present, though in a far transcendent mode, in the One . . .
>
> The Supreme is cause of the cause: it is cause pre-eminently . . . for in it lie the Intellective causes which are to be unfolded from it, author as it is not of the chance-made but of what the divine willed . . .

We are now faced with trying to understand the Triad: how it more than any other of the aspects of meta-ontological infinity provides us with the modeless distinctions, an eternal simultaneity, and timeless stasis. The One-continuum is a totally complete Perfection—Absolute Mind. There could never have been nor will there ever be a lapse from this Fullness, the immaculate Virginity of Isness.

So perhaps it will be wise for us to examine the actualized possibilities which are in the Triad, remembering always that these remain in the realm of the unmanifest or, more precisely, are pre-ontological.

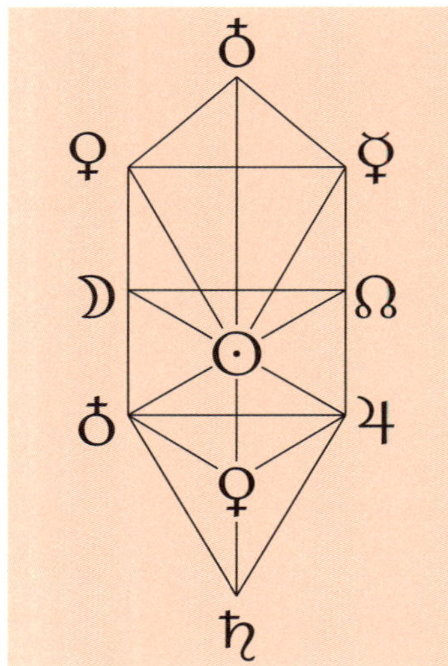

FIG. 18

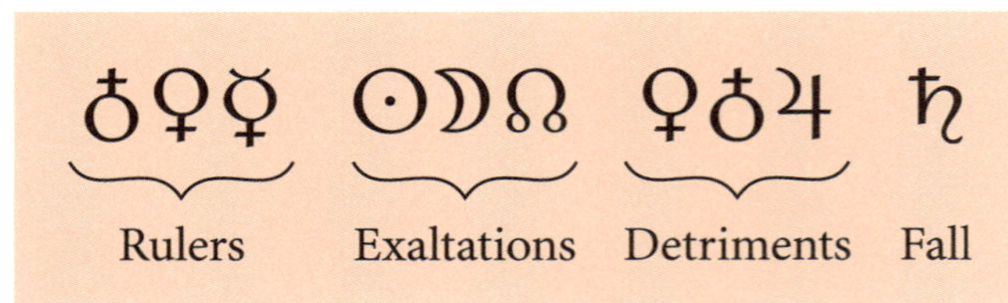

FIG. 19

As illustrated in figure 20A:

♂ Mars in the 1st house (Aries) represents the power and principle of Individuality. Manifesting, as it must, its infinite fullness, this necessarily reveals the double act.

♀ Venus in the 2nd house (Taurus) represents the totality of essences—the Being of infinite possibilities—as the objective aspect of this infinite fullness.

☿ Mercury in the 3rd house (Gemini) is the pure Being of Subjective Wisdom Consciousness.

Thus ♂ Mars, in manifesting the double act within the Transcendent, reveals the paradoxical coincidence of the opposites which constitute it, or which may be conceived as an illusory division of itself. [*fig. 20B*]

These three are described in Orphic theology as Universal Being or Goodness, Universal Soul, ▽ and Universal Mind.

In this way we can see how ♂ ♀ ☿ can symbolize the One. ⊞

▽ Editors: The term Universal Soul is used in this one instance as by G.R.S. Mead (see the diagram on p. 58 in *Orpheus*). Plotinus uses the term differently, and in the rest of the *Astronoesis* text, Anthony follows Plotinus' usage. (cf. IV.4.10)

⊞ Editors: In natal astrology the planets are interpreted as psychological functions, and reflect at a lower level the principles in the One. Used at the level of the One, these symbols represent the One's active nature.

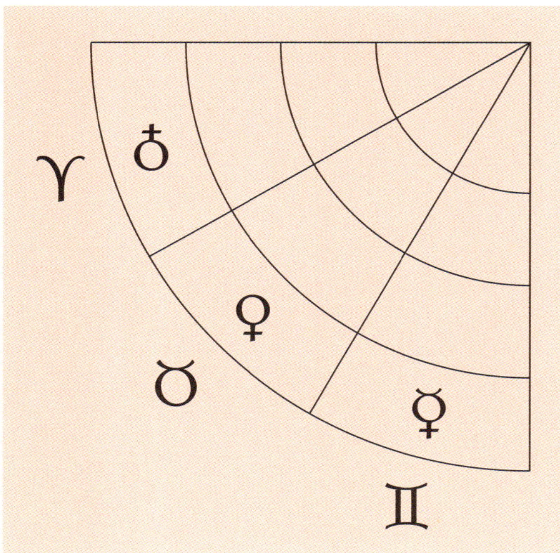

Fig. 20A

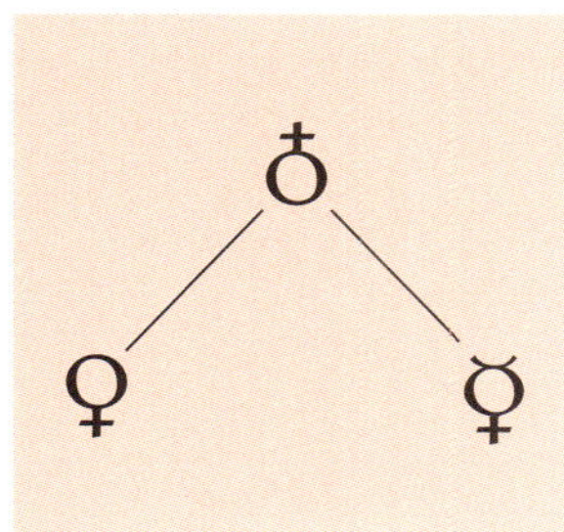

Fig. 20B

The ☉ Sun in the 1st house (Aries) represents the Intellect of the Transcendent One, the author and paradigm of all determined Being which we will find in the Intellectual-Principle. It also exposes a version of the double act within the Transcendent. Two complementary aspects of the Unity's undifferenced super-intellection are revealed.

On the side of Repose, we see the ☽ Moon in Taurus, which represents the *act* of the infinite possibilities.

On the side of Active Perfection, we see ☊ the Head of the Dragon ▼ in Gemini, representing the eternal order which reigns among the two foregoing possibilities [i.e., ☉♈ and ☽♉: This latter is equivalent to the principle of intellectual "numbering" which represents Order within the One. [*fig. 21A*]

▼ EDITORS: The symbols ☊ (North Node, Dragon's Head) and ☋ (South Node, Dragon's Tail) represent directions in space determined by the intersection of the Moon's orbit and the plane of the ecliptic. Throughout the text, Anthony uses the abbreviated terms "Head" (☊) and "Tail" (☋).

Thus, the ☉ here manifests or includes in its act the totality of the intelligence of these essences ☽, and the perfection inherent in their mutual reciprocity ☊. This ☉ is equivalent to the Intellect of the One and the author of all Being. [*fig. 21B*]

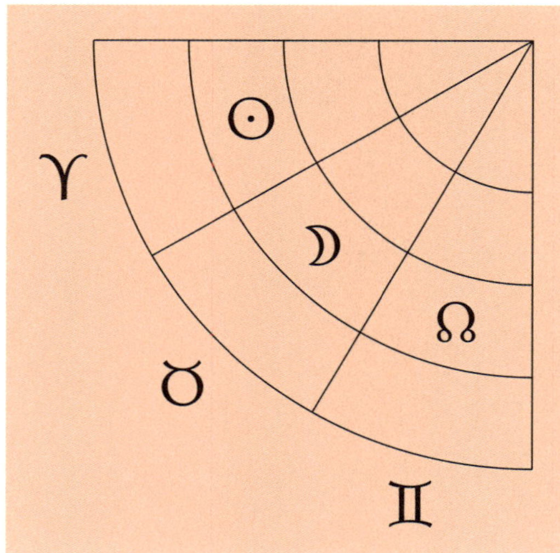

FIG. 21A

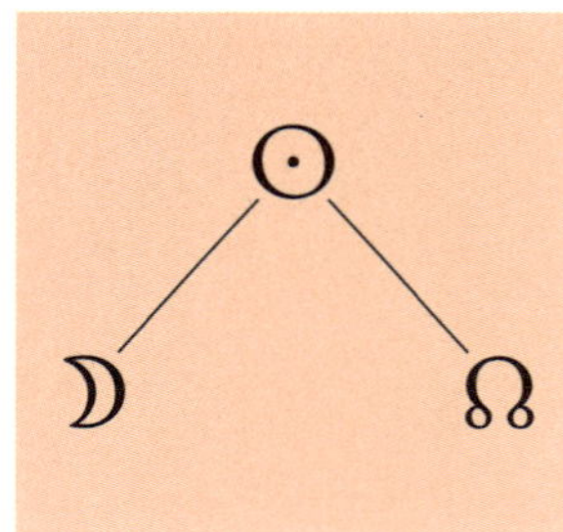

FIG. 21B

♀ Venus in Aries represents the Infinite Nature, the uncircumscribed operational activity of the One, the unlived and unpolluted fountain of Life.

♀ Venus is projected into "view" by the totality of power inherent in the nature of these infinite possibilities—♂ (Mars in Taurus) conjoined with the undivided Demiurgic power of the Subjective Wisdom Consciousness ♃ (Jupiter in Gemini). Or we may say ♀ is the unity of the ♂ power of the distinct essences combined with the ♃ Demiurgic Activity, the One's infinite formative power. [*fig. 22A*]

These three, ♀ ♂ ♃, in their totality are what may be referred to as the infinite operation of the One. [*fig. 22B*]

EDITORS: Note that Venus in Aries ♀♈ represents a different principle than Venus in Taurus ♀♉; Mars in Taurus ♂♉ represents a different principle than Mars in Aries ♂♈. The differences in these principles becomes clearer in the text as dignities are shown to represent various powers of Oneness.

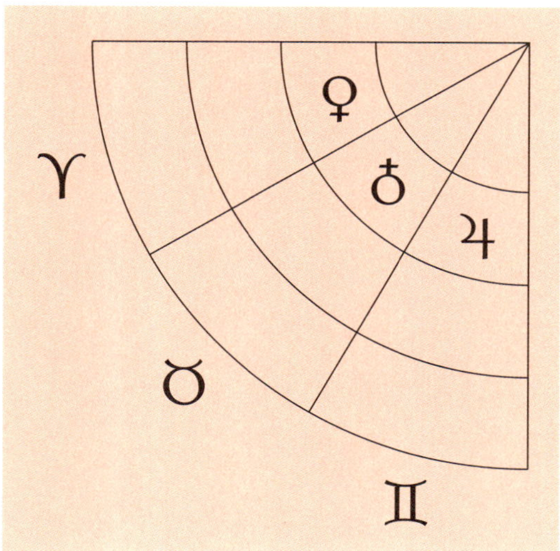

FIG. 22A

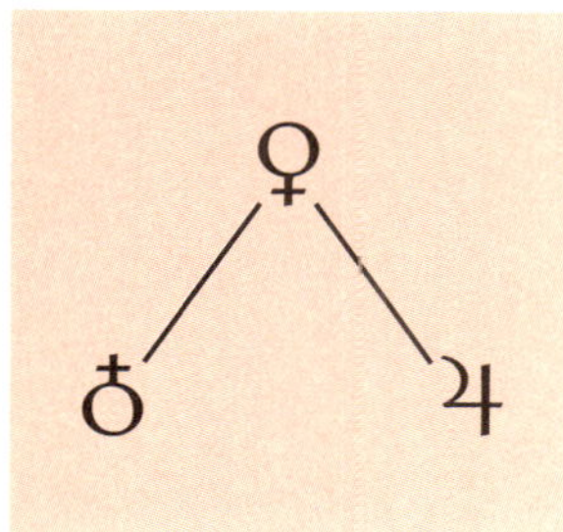

FIG. 22B

Throughout this internal self-distribution of its own power, the One remains intact, stable in its unchanging identity. The reciprocal and mutually interpenetrating aspects of this undifferenced continuum we are incorrectly expressing as the identity and indivisibility of the One. One multiplied by One infinitely equals One. This inner completion is the Act of the One remaining in itself. Stillness reigns.

Finally ♄ Saturn in Aries, the fourth of the dignities in the 1st house, represents the principle that is perhaps the most mysterious of all: the *Mysterium Magnum* (of Paracelsus), the Principle of Difference of the Platonists, and also the *Mulaprakriti* of the Vedantins—which "throws a veil" over the Transcendent. And is this principle of matter also the *Mulaprakriti* of the Samkhyans? [*fig. 23*] ▼ [15, p. 314]

▼ EDITORS: Some of these ideas may refer to principles which *unfold from* Saturn in Aries rather than applying directly to Saturn in Aries itself. More on Saturn in Aries appears later in this chapter (pp. 72–75).

Is this the matter that is co-eternal with God? The clues offered on the nature of this principle are scanty and meager and we will follow this up later.

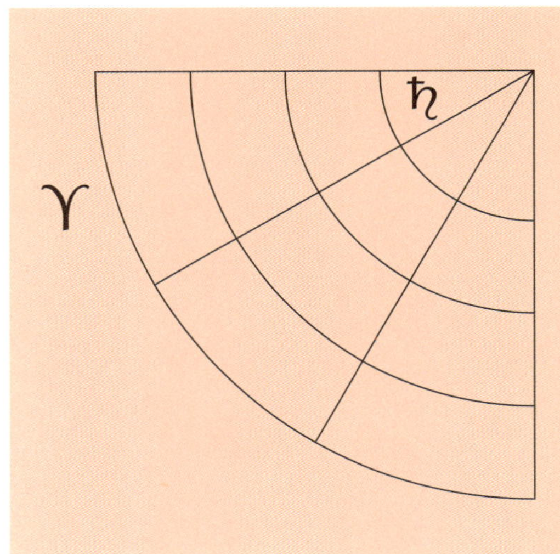

FIG. 23

In summarizing the analysis of the One as constituted of the double act, we have tried to tentatively understand each of these principles within this distinctionless whole, allowing us to assign certain notions to each principle. ☐

As figure 24 symbolizes:

♂ ♀ ☿ Mars in Aries, Venus in Taurus, Mercury in Gemini together represent the One.

☉ ☽ ☊ Sun in Aries, Moon in Taurus, Head in Gemini represent the Infinite Being of the One. Further analysis allows us to distinguish within the Being of the One, the ☉ Sun as the paradigm of Being, the ☽ Moon its actionless activity or Power, and the ☊ Head as the numbering principle within it which assigns specific Ideas their powers.

♀ ♂ ♃ Venus in Aries, Mars in Taurus, Jupiter in Gemini would be the Life or Absolute Soul of the One, which is also the hyparxis of individual souls and Universal Soul. That is, an individual soul is constituted of a portion or spark of the One's infinite

☐ EDITORS: Saturn in Aries is not analyzed here because its double act isn't differentiated in the One. In terms of the diagram, Saturn in Aries (♄) is the only one of the four dignities in the 1st house which does not have two other dignities distinguishing its passive and active perfection.

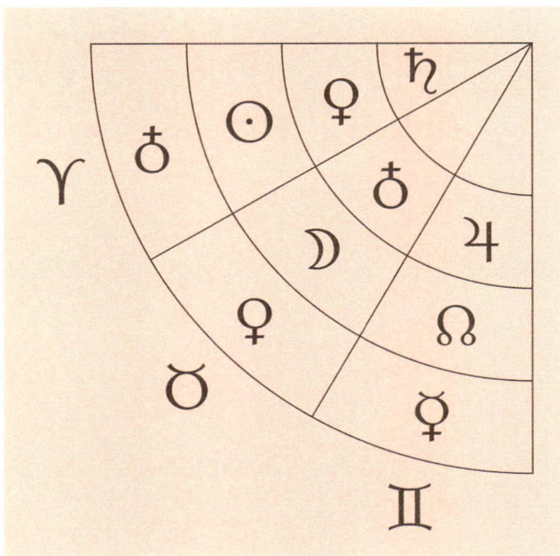

FIG. 24

nature combined with certain super-essential possibilities, just as Universal Soul has for its a priori the Infinite Nature of the One and certain super-essential possibilities.

With figure 25 we can say of this universal Being that it is the self-sufficing pure intelligence, eternally self-causing itself, intrinsic perfection without any intermediary. The One is its own final (♁♈) efficient (☉♈) instrumental (♀♈) and material (♄♈) cause. [Cf. Aristotle's *Metaphysics* 983a, L.24–34.]

When we apply the Idea of the fourfold to the One as represented by the Triad, we are, of course, trying to understand it in a unique perspective, which does not prevent us from applying the other numbers to it.

So finally it comes to this: that the Pure Intelligence as a unity paradoxically reveals itself to our understanding as the measureless Power that supplies actualization to the infinity of possibilities; or the reverse view—which is just as inconceivable—that the possibilities in their totality, through their inherent self-knowing, are that same indescribable Power. This is basically the concept of the Triad. ▽

These primitive and fugitive sketches of the unities may allow us to grasp something of what Plotinus refers to as the transcendental paradigms of Being, Life, Nous, Number, etc., having their Source in the One-Only. Only here is it appropriate to speak of each in all and all in each. [16, p. 314; 17, p. 317] The fertility of this idea should unfold as our analysis of the One proceeds. Plotinus was obviously communicating a philosophic wisdom within which the reconciliation of the "Who Am I" and the "What Am I" could take place. ⊞

▽ EDITORS: To summarize this analysis: Plotinus offers a view of the Absolute One as One, a view of the One as power, and a view of the One as a super-essential wisdom integrating indivisible unities. From the perspective of One as absolutely one, there is no other description. From the perspective of the One as power, we get a paradoxical and indeterminable view of infinity. From the perspective of the Triad, we find infinite indivisible unities. That there are different perspectives from which to view the One, or different ascents to the Absolute, rather than just one, is in fact the perspective which we are calling infinite wisdom or the Triad. Each perspective of the One is valid and necessary to keep alongside the others.

⊞ EDITORS: The terms "who" and "what" indicate that the I has its source in both knowing and being. "*Who* am I?" refers to the practice of tracing the I as subject to its source in the higher consciousness. "*What* am I?" refers to the practice of tracing the I as existence to the Ideas or World-Idea. The terms are used by Paul Brunton throughout his *Notebooks*. Cf. especially volume 6, part 1 (The Ego), p. 7, no. 36.

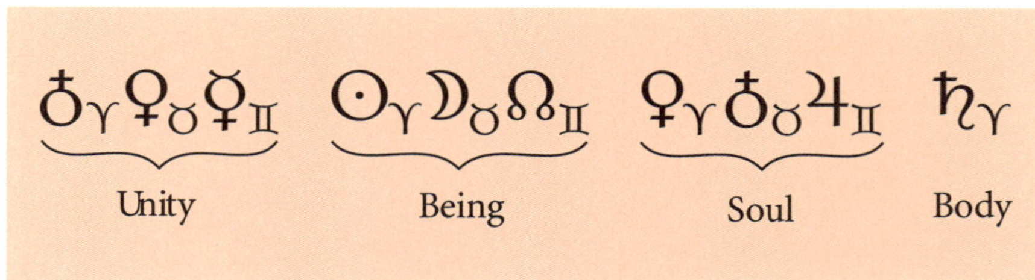

FIG. 25

The Fourfold

Examining our diagram of the 1st house, we notice the four dignities. These indicate or suggest that the Absolute Perfection has, in a mode peculiar to itself, a fourfold nature. [*fig. 26*]

Riding on the formalism of the chart we will see that 4 is the symbol of manifestation *par excellence,* and that its applicability to every level may be noticed by the following considerations.

The fourfold nature of the One has been somewhat anticipated in our previous discussion of the three primal hypostases and their embodiment in the system of Nature. This was symbolized as a circle divided into four quadrants. The first quadrant refers to the Unity of the One, the second quadrant to its Being, the third to its Soul, and the fourth quadrant to its Body. [*fig. 27*]

EDITORS: The two diagrams of figure 27 illustrate the correspondence between the two views of the One as transcendent and immanent and how each of the dignities in the 1st house becomes the leader of a quadrant.

Continuing the application of the principle of the quaternity, it must also be applied to universal

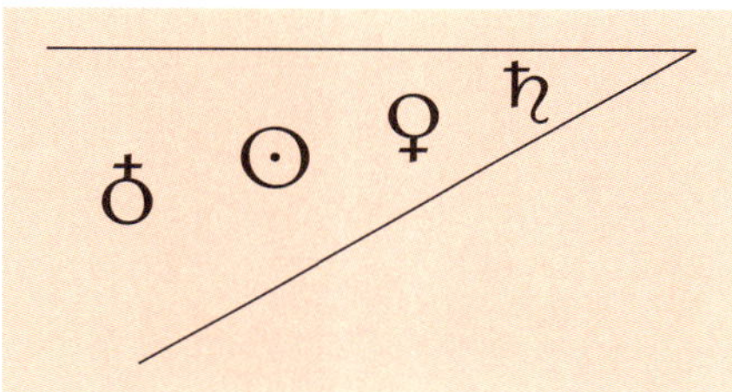

FIG. 26

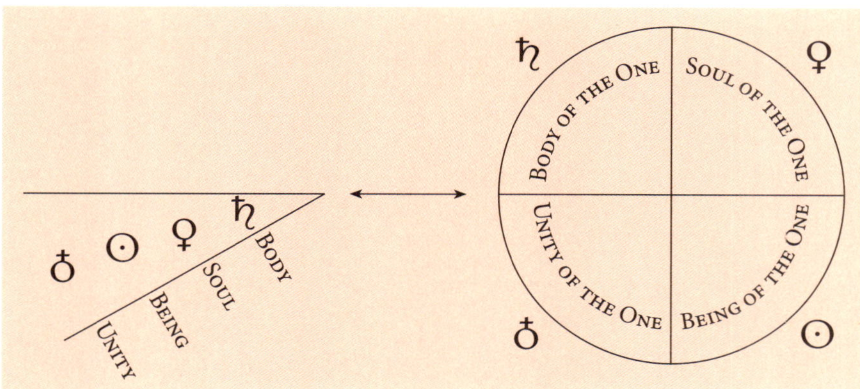

FIG. 27

manifestation. Here again we can say that the latter has a Unity, Being, Soul, and Body. [*fig. 28*] Within the totality of universal manifestation, the infinite variety and complexity of the three primal hypostases will be reduplicated. [13, p. 314] There is the implication that the entirety is the Sacred Body of God and His abode, which is fraught with Divine significance: a theophany of lustre and light—Intelligence—radiating His presence throughout and invisibly describing Himself with various living qualities in patterns made by the clusters and galaxies of starlight. And the breath of vital wisdom speaks to the archetypal imagination and its powers resulting in the form of the Cosmos.

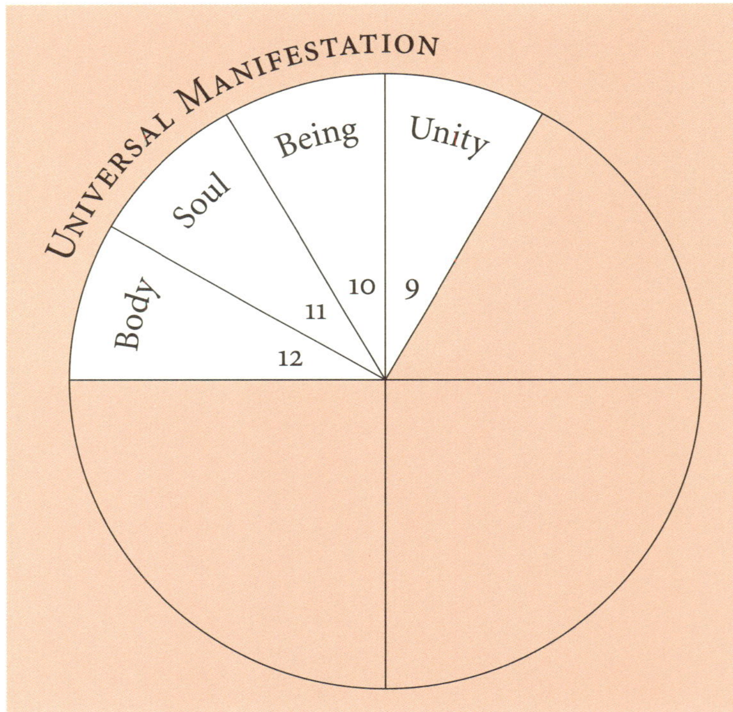

Fig. 28

Further, the fourfold Idea is again applied to the sensible portion of universal manifestation [represented by the 12th house]. Thus we may speak of this realm having a Unity, provided by the Life of Universal Soul which manifests itself or is veiled behind the five Great Elements, an Intellect which we can correlate with the inerratic spheres, within which we can situate an individual cosmic soul which manifests its vision of that Intellect, and finally, the corporeal world. [*fig. 29*] ▼

Again, this fourfold Idea can be applied to Mind's Absoluteness or its integrity of perfection (*turiya*), from which emanate the other three states of consciousness, each one of

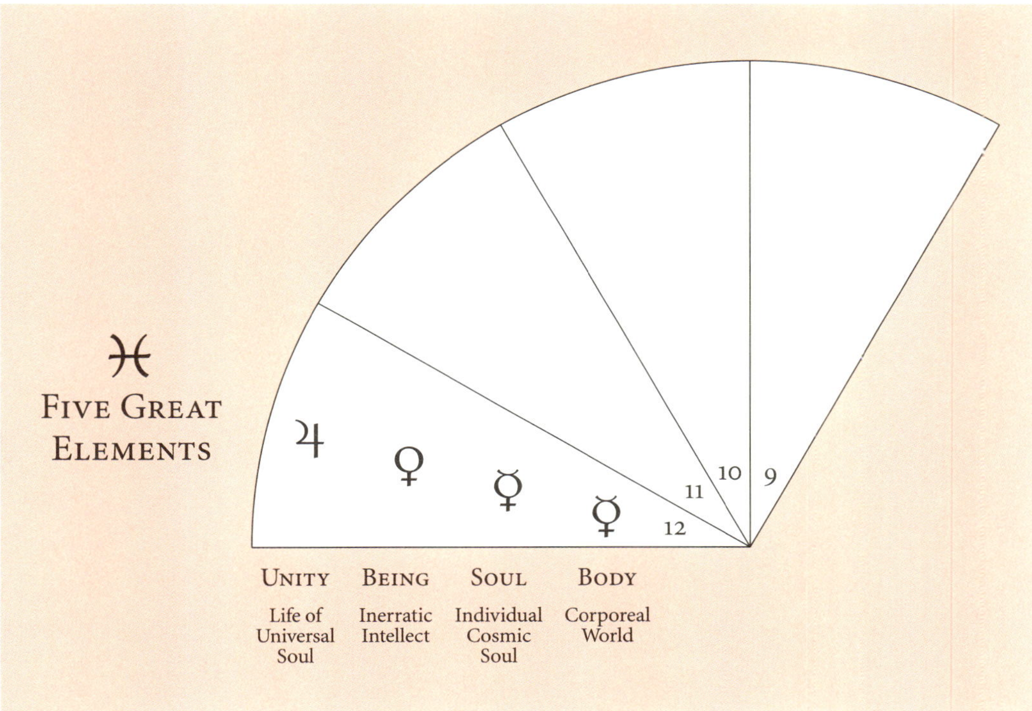

Fig. 29

▼ Editors: The Great Elements are the substance of any and every sensible universe. They are akin to what modern physicists call force or energy fields, and should not be confused with the elements in the periodic table of chemistry. The substance of the world is not fixed, but—as Plato recognized in the *Timaeus*—the world is always becoming to be but never truly is. These vast energy fields contain the traces of all life activity.

Each of the four dignities in the 12th house represents a deific function that unifies, organizes, and manifests the sensible forms on the basis of these great force fields.

♃ as the life of Universal Soul unifies the vast chaos of the elements into a suitable ground for manifestation.

♀ determines how the Ideas are instantiated in or bound to the elements. "Inerratic" means without error, and Platonists use the term to indicate the realm of the stars, the intelligence of the cosmos.

☿ in the Soul ring represents individual cosmic soul. Each star, including our Sun, is a cosmic soul, unfolding its vision through its own planetary system.

☿ in the Body ring represents the corporeal world.

Each individual human participates in these four levels of the cosmos. Chapter IV, The System of Nature, explores these points in detail.

which has a fourfold nature. ▼ For example, *vaishvanara* or wakeful consciousness has a fourfold nature; that is, we can refer to wakeful-wakeful consciousness, wakeful-dream, wakeful-sleep, and awakened-wakefulness or that fourth state of wakefulness which is identical with the Demiurgic Life and *turiya* consciousness. [*fig. 30*]

▼ Editors: In the terminology of the Mandukya Upanishad, the four states are called *turiya* or pure consciousness, *prajna* (sleep), *taijasa* (dream), and *vaishvanara* (wakeful). As "pure consciousness" is not an adequate English equivalent, we will retain the Sanskrit term *turiya.* Its meaning is close to the idea Plotinus expresses in vi.8.16, that the One *is* in the way it is Awake. The presence of the four rings in the first quadrant indicates that the One prefigures the other levels within itself in a transcendent and mysterious mode. Moreover, the terms "sleep" and "dream" are used only analogically. *Prajna,* for example, indicates a state of knowledge, not one of blank unconsciousness as in our usual sleep.

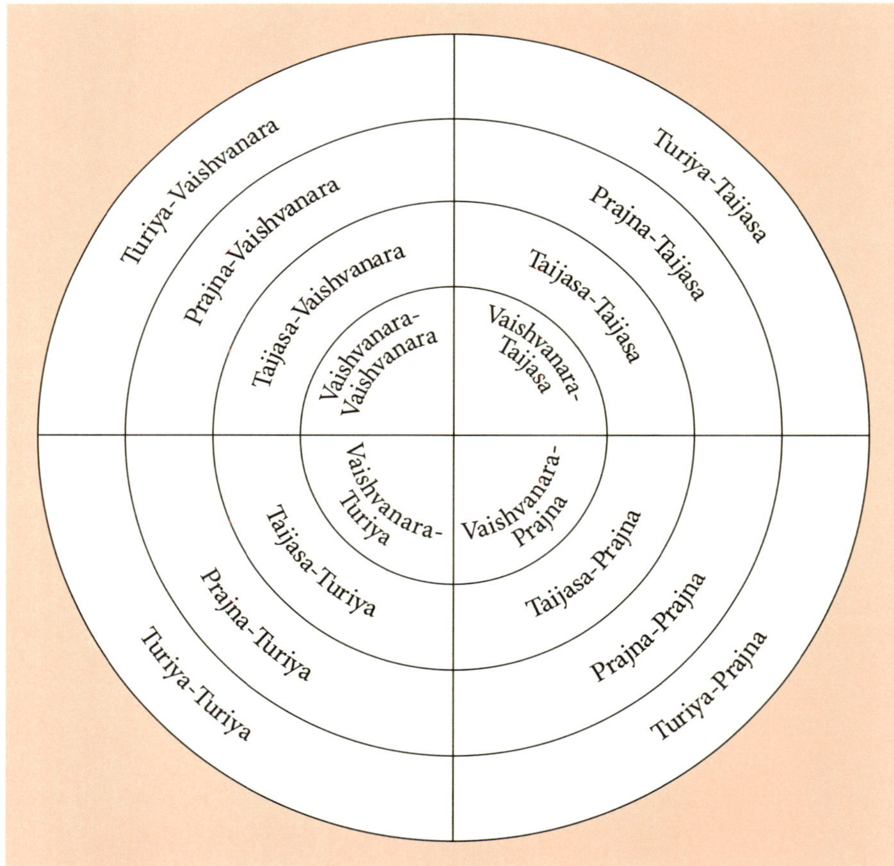

Fig. 30

The following table compares the Plotinian and Vedantic classification of principles.

PLOTINUS	
The three pure hypostases	**These three principles as they are applied and exist in Nature:**
1. The One	4. The Unity of Universal Manifestation.
2. Intellectual-Principle	Its ordered Intelligence or Being
3. Soul	Its Life or Activity
These three principles are enumerated by Plotinus, and in their totality can be referred to as the Divine Hypostases. From this side of manifestation all three together can be conceived of as the Absolute.	The Body of Universal Manifestation, within which will be included: a. Unity of the sensible / corporeal b. Intellect of the sensible / corporeal c. Soul of the sensible / corporeal d. Body of the sensible / corporeal
THE VEDAS	
(which include Brahmanic and Upanishadic literature) propound the fourfold aspects of Brahman:	
1. *Avyaya Brahman*—Infinite Beyond, Beyond Being, without any qualifications . . . 2. *Akshara Brahman*—Indestructible One, the Ultimate Cause, not yet differentiated from Pure Existence 3. *Atmakshara Brahman*—The Cosmic Self with the potentialities of creation, preservation, and destruction, not yet manifested in Cosmic Creation . . .	4. *Vishvasrit Brahman*—the projection of the *Atmakshara* into the entire creation of time, space, causality. As such, *Brahman* is the Cosmic Form—the World Soul. This fourth is also called *Virat Purusha.* (Points 1–4 above are cited from I. C. Sharma, p. 90) Consequently Universal Manifestation will have immanent in it a) Unity, b) Being, c) Life or Soul, d) Body.

FIG. 31: THE DOUBLE STANDPOINT REQUIRED BY PHILOSOPHY IN PLOTINIAN AND VEDIC TERMS

Something of the One is revealed through its emanation: the "Body" of the Lord, pure Space, represented by the fourth of the intelligibles in the 1st house—Saturn in Aries ♄♈. Perhaps we can follow the lead offered by Leo Schaya, in *The Universal Meaning of the Kabbalah.*

> Therefore it is the existence of a 'second' in the midst of the 'One without a second' which makes the only reality mysterious; in the end, all spiritual search comes to the question: by virtue of 'what' (*mah*) does cosmic existence become concrete? The Kabbalah answers: this 'what', *mah,* is divine immanence, the last *Sefirah, malkhuth,* the 'kingdom' of God, which produces, encircles and penetrates the whole of creation. *Malkhuth,* situated at the lower extremity of the 'middle pillar', is not only the 'recipient' of all the emanations of the *Sefiroth* on the luminous or intelligible 'right side', and on the dark or unintelligible 'left side', but also of the central *Sefiroth,* the highest of which is *kether,* the superintelligible principle. In other words, *malkhuth* is the 'descent', the *shekhinah* or omnipresence of the supreme, which is manifested on the one side by intelligible emanations united in the universal spirit and on the other by unintelligible emanations, the darkness of which becomes concrete in creative substance. (p. 57)

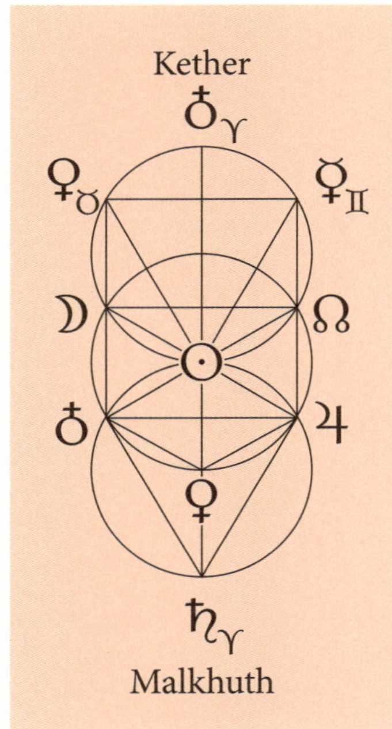

Fig. 32

We can easily transpose the Kabbalistic tree into a more suitable framework for ourselves by diagramming the ten intelligibles (dignities) of the first quadrant into the form of the Kabbalistic tree, as illustrated in figure 32.

This second which is in the midst of the "One without a second," this "what," *malkhuth,* which produces, encircles, and penetrates the whole of creation, can easily accommodate that mysterious section from v.5.8, wherein Plotinus says:

> No doubt it is wonderful that The First should thus be present without any coming, and that, while it is nowhere, nowhere is it not: but wonderful though this be in itself, the contrary would be more wonderful to those who know. Of course neither this contrary nor the wonder at it can be entertained.

In v.5.9, he continues in the same theme:

> That Source, having no prior, cannot be contained: uncontained by any of those other forms of being, each held within the series of priors, it is orbed round all, but so as not to be pointed off to hold them part for part; it possesses but is not possessed. Holding all—though itself nowhere held—it is omnipresent, for where its presence failed, something would elude its hold. At the same time, in the sense that it is nowhere held it is not present: thus it is both present and not present; not present as not being circumscribed by anything; yet, as being utterly unattached, not inhibited from presence at any point. That inhibition would mean that the First was determined by some other being; the later series, then, would be without part in the Supreme; God has His limit and is no longer self-governed but mastered by inferiors.

Although the One in itself is not in space, and in this sense it is nowhere—for we described it as being supremely self-adequate; thus there is for it neither coming nor going—it is also present everywhere because that everywhere is the "place" wherein actuality (or the emanations) will occur. This principle—the "ground" for the emanation from the Unity—Schaya refers to as "not a distinct substance, but rather the undifferentiated, uncreated principle of all substance which in no way emerges from the infinite and indivisible unity of the created causes: this principle envelopes [*sic*] them and is yet hidden within them . . ." (p. 59) This is the paradigm of that principle Plotinus refers to as "Authentic Non-Existence" (II.5.5), the Principle of Difference (II.4.16). He refers to it in II.4 as the base of the intellectual realm (and all the other levels as well). In II.4.5,

Plotinus also speaks of this principle as a "cleavage," distinguishing itself from the First by a motion that results in sheer indeterminacy or matter, and requiring the Unity to give it determination.

This determination, which we refer to as the "Body" of the One, has space as its closest analogue. To conceive of this space as the Body of the Lord is certainly to dematerialize whatever associations the word "body" may suggest. A closer equivalent would be "the formlessness of emptiness," its ungraspability, the dimensionlessness. As Plato pointed out [in *Timaeus* 49a], the receptacle is available only to a kind of spurious reasoning. At any rate, confrontation with it is not possible; always receding and elusive, it may be referred to as an authentic principle which has the "Being of non-Being." ▼

▼ Editors: Because Saturn in Aries (♄♈) is authentic emptiness, it serves as the "mirror" of non-Being (see Plotinus III.6), in which the emanation of all the prior formless principles is reflected. When Plotinus says that the One is "both present and not present" (V.5.9), he is referring to the mystery which allows for both the omnipresence of the One and its "otherness."

It is this principle, which is an aspect of the One itself, in and through which the emanation will appear. The *Mulaprakriti* of the Samkhyans, the root through which and in which emanation occurs, has already implicit all the intelligibles: It is not the nothingness of matter.

The uttermost outward-facing aspect of metaphysical infinity, Saturn in Aries represents the veil of unintelligibility thrown by the Transcendent over itself. ⊞ It is the inner and the outer, without any in-between; that is, if one conceives of inner and outer as opposites, one misses the mark. We can say the same of infinite space that we say of the Absolute, and apply to it the same epithets—unseizable, ungraspable, etc. Perhaps more than any other aspect of metaphysical infinity, this principle illustrates the distinctionlessness of the Transcendent's double act.

⊞ Editors: The Absolute is the unknowable darkness, without any second. In order for anything to be, this Absoluteness has to be veiled, the darkness covered over, so that the many Ideas can be revealed. The veiling seems to us like a manifesting, because metaphorically the One must be veiled so the many can appear. Saturn in Aries (♄♈) is this kind of difference, allowing something to appear without being separate from the original.

Nonetheless, however incomprehensible our survey has been, there have been schools of thought which insist upon the possibility of this principle's differentiation into various grades of matter, each grade being a distinct stadia of this principle of authentic non-Being. [*fig. 33*] ⊕

It seems clear now why *malkhuth,* or the principle of non-existence, is referred to as the "foundation" of the (intellectual) worlds. Space which cannot be located is the "place" which permits the One to be everywhere, without itself being anywhere. From here, the outflow or emanations of the One can proceed. ▲

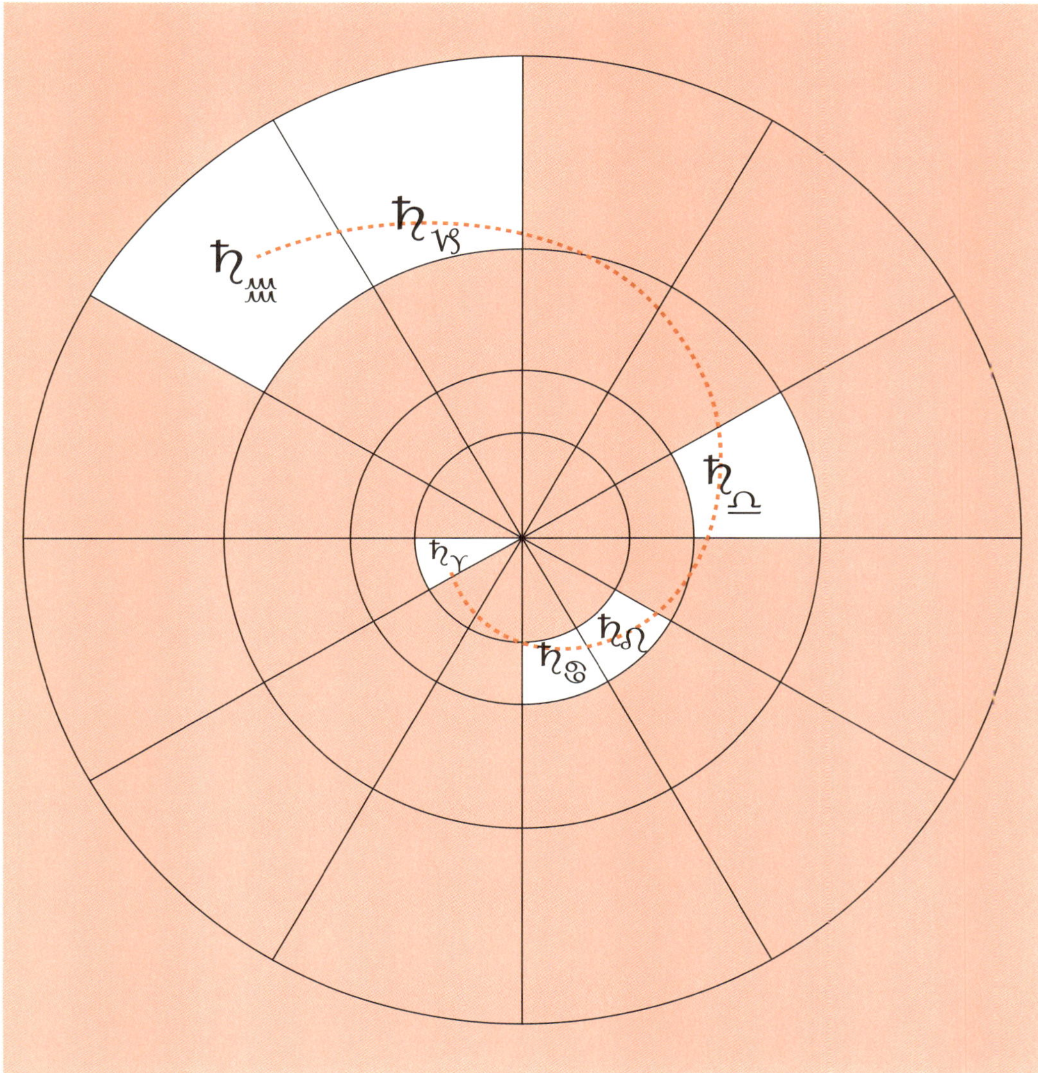

Fig. 33

⊕ Editors: Saturn (♄) spiraling through the astrological rings illustrates "grades" of matter. If we follow the sequence of Saturn (♄) dignities from Aries to Aquarius: Saturn in Aries (♄♈) is the principle of Difference, Saturn in Cancer (♄♋) is (p. 73) "the base of the intellectual realm," Saturn in Leo (♄♌) is the tendency for differentiation that characterizes Soul in the intelligible realm, Saturn in Libra (♄♎) is the distinguishing of the entire spectrum of outward-facing reason-principles or *tattvas,* Saturn in Capricorn and Aquarius represent *prakriti* as the potentiality for manifestation (♄♑) as well as the actual function (♄♒) of manifestation. In his essay "Studies in Samkhya Philosophy," K.C. Bhattacharyya writes: "Prakriti then presents, both potential being and actual function." (para. 129)

△ Editors: This means that now the Intellectual-Principle as a multiple organ of perception can arise. This "outflow" gives rise to a new way of seeing the mandala where the One appears as the unity of Ideas and its unknowability is hidden. Saturn in Aries (♄♈) allows the Intellectual-Principle to be distinct from the One, which is necessary for it to be a self-determining distinct existence in its own right. At the same time, the One is present to the Ideas. The Saturn (♄) series, as illustrated in figure 33, provides a basis for similar self-defining activity of all levels of emanation.

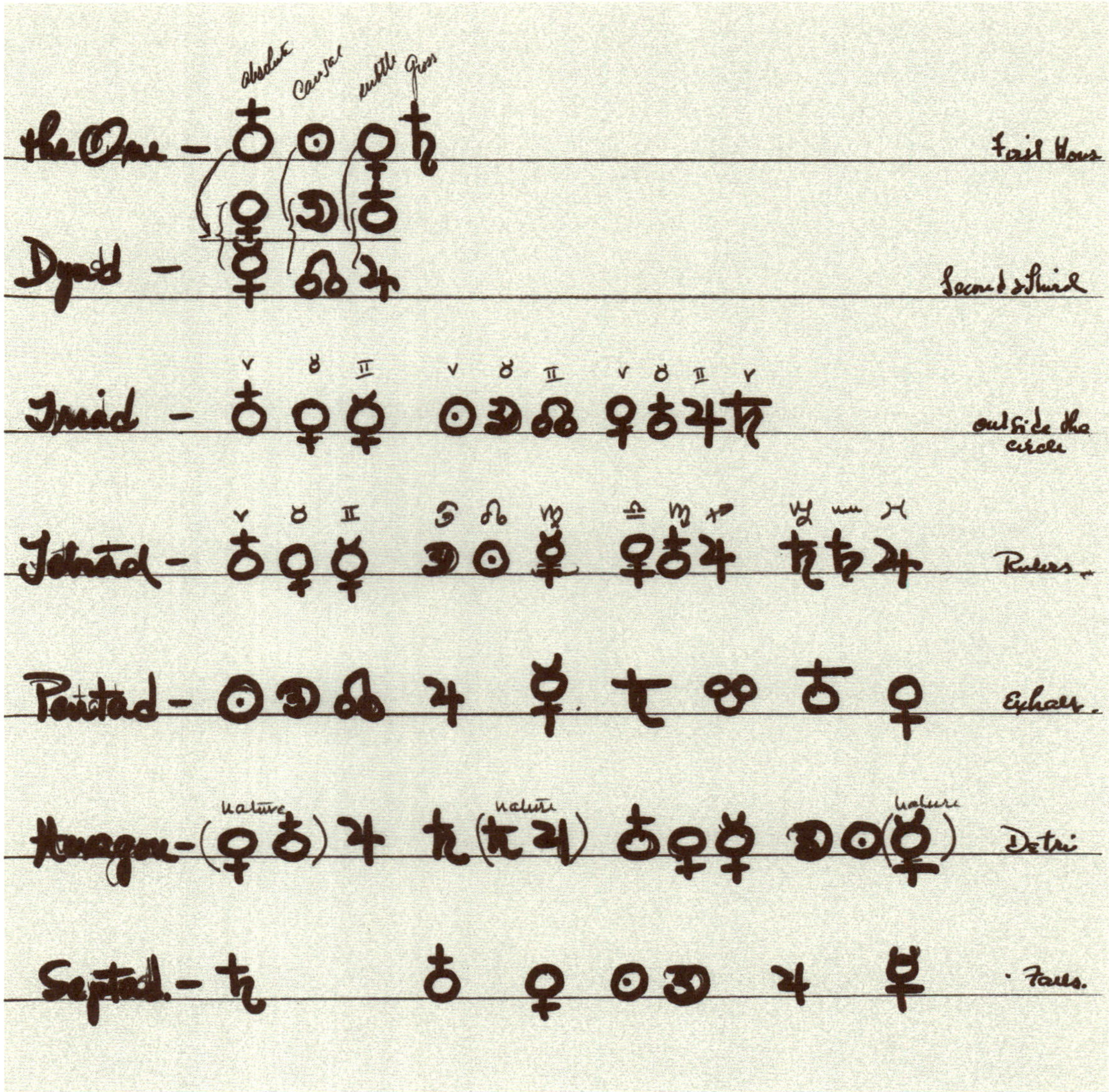
absolute
causal
subtle
gross
the One -
Dyad -
Triad -
outside the circle
Tetrad -
Rulers
Pentad -
Hexagon -
nature
Detri
Septad -
Falls

Interlude

NUMBER

Editors' Introduction to Number

A Qualitative View of Number

Context: Neoplatonic Number Symbolism

Throughout *Astronoesis,* Anthony uses a qualitative aspect of Number whose meaning relates directly to the symbolism of the mandala. Motivated by Platonic ideas about numbers as ordering principles in the cosmos, he uses the ten number principles as paradigms of order for his own investigation of reality. Plato is said to have written over his academy gate, "Let none ignorant of geometry enter here" but he was not referring to the quantitative mathematics with which we are generally familiar.

The Neoplatonists, and the Egyptians before them, thought of number in a *qualitative* way. Each number had a certain quality and Number as a whole represented the harmonic principle of both inner reality and outer order. Today, most of us who are not mathematicians have no "view" of mathematics at all, much less a qualitative one. Even mathematicians generally do not consider number in a qualitative way.

How does anything of what is now called mathematics relate to the mathematics of the Neoplatonists? Certainly, many mathematicians and physicists would agree that Number is the underlying structuring principle of the universe. If we view the world as within intelligence or mind, the ordering principles of mathematics appear as fundamental properties of mind. Mathematics can then be seen as one way humans view the world: literally as an organ of perception. As we then deepen our investigation of reality, we use an immaterial reasoning that can contact the qualitatively operative ordering principles in Soul and Nous. No longer are we dealing with quantifiable units.

Exploring mathematics as a description of the universal system of Nature wherein all beings are in harmonic interrelation, we meet such notions as "the harmony of the spheres."

> The being we are considering is a living unity, and therefore necessarily self-sympathetic: it is under a law of reason and therefore the unfolding process of its life must be self-accordant: that life has no haphazard, but knows only harmony and ordinance: all the groupings follow reason: all single beings within it, all the members of this living whole in their choral dance are under a rule of Number. (Plotinus, IV.4.35)

Mathematicizing in this way is a natural part of attuning to the wisdom inherent in the universe itself. In ascending to the speculation of the Ideas that govern the universe, it is also useful to cultivate a view of sound and words as related to vibrations at subtler and subtler levels, such as found in Kashmir Shaivism's doctrine of vibration. [18, p. 319]

In his work *Plato: The Written and Unwritten Doctrines,* the noted modern scholar J.N. Findlay refers to evidence that Plato was seeking, late in his life, for a method of "arithmetizing" the Ideas that comprise the wisdom of what he called the Divine Mind. Anthony mathematizes these Ideas in direct relation to the mandala symbol, and particularly the zodiacal mandala. The unique combination of number principles within the astrological symbolism of this mandala provides a foundation for a new and far-reaching expression of the Neoplatonic form of the perennial wisdom.

To more fully appreciate the number meanings that the mandala symbolism and astrology provide, a perspective on mathematics that corresponds to the qualitative language of the Soul is required. [19, p. 319]

Number as a Way of Seeing

Understanding qualitative meanings of individual numbers at a psychological level is similar to interpreting symbolic meanings of picture glyphs, such as hieroglyphics. As Robert Lawlor points out in his introduction to *Symbol and the Symbolic* by Schwaller de Lubicz, a picture or hieroglyph can be described literally or symbolically. A jackal, for example, is an animal with four legs, a certain kind of fur, etc. But to interpret the meaning of the jackal symbolically is to see through the physical image to the function which the jackal, as an eater of carrion, symbolizes—that of digesting leftovers. Psychologically the jackal mind performs the function that helps us break down and digest experience, and the Jackal God indicates the universal principle of dissolution and regeneration. The jackal represents this faculty of digesting experience, which we are to recognize and develop in order to transform ourselves. Thus, Lawlor writes: "We must worship the jackal function in us and find out from it . . . the precise timing and laws applicable in . . . transmutation . . ." (p. 15)

In a similar way, each number has its precise function when thought of in a qualitative or symbolic sense, and we can identify these number functions in us. Each offers a unique "way of seeing" experience.

To view the world through the number 1, for example, is to see unities. Nothing can be at all unless it is a One. A tree, a plant, a person, an idea, must first be a one, be a whole, be itself. We can say nothing of anything unless we first notice *that* it is, a *single* thing. The oneness of a being is also its core, and includes everything of the essence of that being. Because the essence is never used up in its instances, no matter how many, the oneness may be said to be infinite.

> . . . a seminal notion like Absolute Unity both contracts into a nucleus in which naming and saying cease, and expands into a fully developed structure in which all the possibilities of Being are distinguished and yet held together indissolubly. (J.N. Findlay, p. 247)

Further, for this essential core to manifest, it must express itself. It must in some way become other than itself, much as a ray directed outward from a point. As soon as we try to grasp or express something about our experience, we make distinctions between this and that, I and world. As we do this, we are viewing experience through the number 2. Two is a lens through which we see things in their complementary attributes such as whole and part, substance and function, motion and rest, yin and yang, discrete and continuous, and so on. This view of things through the perspective of the number 2 also reveals the nature of reality as Power.

We get a different view through the 3: a mediating, relating, or synthesizing factor which effects an indissoluble reconciliation of these opposites. We comprehend and reconcile the opposites into a new whole. In Hegel's dialectic, for example, the triad of position-negation-sublimation indicates the dynamics of knowledge. We view the world through the lens of the number-meaning 3 when we understand in this relational or synthesizing way. Whereas two complementary items always indicate a kind of dynamic interaction (i.e., 2), 3 is stable.

Each number-symbol provides a way of analyzing our experience. Each is a function we can cultivate to see the world (through complementarities, dialectical synthesis, etc.). Thus, we first view reality as unity, then as power (the meaning of the 2), then as intelligence (3), etc.

The One and the Ten Divine Number Principles

For Anthony, the levels of Number—monadic, psychical, essential, and divine—correspond to four levels of reality. Furthermore, Number is not subsumed under either of the two great categories of substance or function. Nor is it a hypostasis. Number provides us with a way of viewing reality based in part on reality's mysterious power to explicate and articulate itself. In the context of the One, the Number principle is prefigured as the self-"synesthesis" of the One, prior to Intellection.

Number is a pivotal element in Anthony's complex, positive conception of the One. His application of the ten number principles to the One constitutes a dialectic of unity which parallels, yet is more complex than, the discussion of the One found at the end of Plato's *Parmenides.* He uses the number principles to order his discussion of the interiority of the One.

Though the One is absolutely infinite, the number principles direct our attention to specific glimpses of It. Each number structures our understanding of the One and its power in a unique way. Each of the number qualities provides a way of viewing the perfection of the One itself, revealing something of its inner nature and its emanation. Each of the numbers is applied to the One and then merged or dissolved, but only after it reveals to us something true of the One's nature.

In this "interlude," Anthony's spartan descriptions of the ten number principles apply to the One itself. In the next chapter on the Intellectual-Principle, his elaboration of the twelve signs will help articulate the distinct number principles.

Number

Four Levels of Number

The importance of Number as the structuring principle of ontology should not be underestimated. It figures into the internal structure of every essence as well as every thing in so far as material objects are capable of reflecting the order inherent in Ideas. Understanding these divine "qualities" is a prerequisite to comprehending the rationale behind the particular ordering of Ideas in the Intellectual-Principle, Soul, and the system of Nature. The three pure hypostases and the system of Nature suggest the four corresponding levels of Number, respectively designated as Divine, Essential, Psychical, and Monadic.

The Pythagoreans considered *Monadic numbers* to be quantitative and nothing more than the image of essential number. Monadic numbers are composed from a certain number of units irrespective of their nature. For example, five is a name assigned to a collection of any five things. There is nothing in the five sensible objects which demands that we consider their "fiveness" as essential to their existence. Taking one away only affects their quantity and not their quality. Monadic number is employed in enumeration and in calculation.

Psychical number structures and harmonizes the soul's multitude of reason-principles. ▼ Number in soul is purely qualitative, and is integral to the definition of reason-principles. Soul is the entire outpouring of Life from the Intellectual-Principle and Number in Soul orders this Life and its contents, an ordering which becomes evident in the unfolding of universal manifestation. ⊞ Plato in the *Timaeus* and Plotinus in III.6.4 (end) and III.6.5 spoke of the soul as a melodic principle in the following manner: The entirety of universal manifestation is eternally held in the Divine Mind's intellection, while

▼ Editors: Reason-principles are Ideas in the mode of Soul.

⊞ Editors: Some contemporary psychologists who have developed the work of Carl Jung recognize the presence of these numbers in the soul as archetypal patterns of relations. Marie-Louise von Franz explains this perspective in her book, *Number and Time.*

⊕ Editors: Psychical number operates clearly and explicitly in astrology. The ratios that relate the planets demonstrate Number's applicability to an unfolding individual life. For example, a "trine" (3) is formed by two planets having an angle of 120 degrees, and an "opposition" (2) is formed by two planets having an angle of 180 degrees. These ratios are understood as applications of the qualitative meaning of the numbers 3 and 2.

As ratio in mathematics relates two quantities, qualitative numbers in the rational soul relate the finite and the infinite. The term "rational" soul fits its nature as the intermediary and ratio of the finite and the infinite. In the *Timaeus* the Demiurge is said to insert the ratios and reasons into the soul. (35b–36d) Plotinus adds that the soul perpetually circles the Nous; its receptivity to the power of intellection gives rise to an image of the Divine Ideas and guarantees that the universe is unending:

continues on p. 83

soul introduces time, and through Number unfolds the unity of the world composition in orderly terms of melodies, periods, rhythms, etc., which represent a precise arrangement of the activities of reason-principles.

For an individual soul, Number is that power which orders the lifetime's sequence of events for the development of its vehicles. This ordering, revealed by the structure of the chart of the heavens cast at birth and by the angular motions thereafter, can be likened to the melodic principle as applied to the individual soul's expression of Life and each soul's unique experience of meaning. ⊕

At a third level of application, *Essential numbers* are the structural determinants in the Intellectual-Principle. "According to the Neoplatonist Syrianus, Pythagoras in the 'Sacred Discourse' states that 'Number is the ruler of types and ideas and the cause of the Divine and Angelic Progressions.'" (*Shrine of Wisdom,* "Symbols: Numbers," iv, no. 6, p. 119) The specific quality of an Essential number, its "just-so-ness," can be correlated with an Idea, and this correlation is what determines the arrangement found in the circle of Ideas. ▲ Essential number also orders the eternal activity (double act) of Intellectual-Principle. With reference to Soul, Essential number prefigures the arrangement of reason-principles that constitutes one soul as distinct from another. ◈

At the fourth level, *Transcendental* or *Divine numbers* are super-essential unities. Each is an infinite aspect of the infinite. Each is a unique way of viewing the absolute totality of possibilities.

The following quote expresses the idea that the individual properties of Number, the divine structuring principles, are derived from the One itself.

> Divine Numbers. The Monad of Monads is God as the ONE. In Him all Ideal Numbers subsist superessentially. They are Infinite and Divine because they symbolize the Powers and Attributes of the Infinite One. ["Ideal" here corresponds to "Essential" as used above. – eds.]
>
> Although each superessential Number is infinite this does not imply duplication or triplication of infinity, for the Infinite One is transcendentally beyond all ideas of increase or multiplicity.

continued from p. 82

The Soul exists in revolution around God to whom it clings in love, holding itself to the utmost of its power near to Him as the Being on which all depends; and since it cannot coincide with God it circles about Him. – II.2.2

▲ Editors: If we refer to the Mind which holds all the Ideas as a unity, then Essential numbers are inherent in the nature of the many Ideas in the Mind. The chapter on Intellectual-Principle develops the essential aspect of Number as applied to the Ideas. Each number correlates with an Idea-principle, creating the twelvefold circle as a mandala of the Ideas in the Divine Mind. Within the cosmos the Ideas are "mathematized" as reason-principles, which are represented in astrology by the 360 degrees of the circle.

◈ Plotinus, VI.6.9: Clearly Being is to be thought of as Number collective, while the Beings are Number unfolded: the Intellectual-Principle is Number moving within itself, while the Living-Form is Number container of the universe. . . .

Hence it is that the Forms (Ideas) have been described as Henads and Numbers. This is the authentic (essential) Number; the other, the 'monadic', is its image. The Authentic is that made manifest in the Forms and helping to bring them to be . . .

But even as Archetypes are Divine Thoughts of Ideal Perfections, so Divine Numbers are expressive of the Divine Infinite Powers according to which all things abide in, proceed from, and return to the Superessential Absolute ONE. (*Shrine of Wisdom*, "Symbols: Numbers," iv, no. 6, p. 119)▼

▼ Editors: An article by S.V. Ramamurthi similarly suggests that number applies to the progression from and return to the One: "Can the One be the subject of mathematics? Yes. This is the mathematics conceived in the . . . Upanishad[s]. . . . Mathematics deals with both the unfolding and refolding of self, with the passage of Brahman to Atman and back to Brahman. Mathematics yields not merely a static vision but also a dynamic vision. Advaita is thus not only the culmination of mathematics. It is, too, the foundation of mathematics." (p. 61)

The Kabbalists also regard the ten primal numbers as a fundamental approach to understanding the One. M.L. von Franz refers to the authority of Dr. Hurwitz:

It is common knowledge that in the cabala the ten primal numbers, the *sephiroth,* are considered to be emanations of divine immaterial origin in the sense that the "original one" remains far removed from all things and only manifests in each single *sephira.* The cabalistic doctrine of numbers is in itself a revival of ancient Gnostic ideas about numbers. In the latter doctrine numbers as *archōns* or divine powers, continually stream forth from the unknowable *archē* of the one. All are manifestations of single aspects of the primal one. For this reason the first ten numbers of the *sephiroth* system are designated by the word *sephira,* instead of the ordinary word for numbers (*mispar*). They are "spheres" and "totalities" of the unknowable primal one. (p. 83 n. 55)

In Plotinus' *Ennead* v.5.5 we find reference to the same notion:

And just as there is, primarily or secondarily, some form or idea from the monad in each of the successive numbers—the later still participating, though unequally, in the unit—so the series of Beings following upon The First bear, each, some form or idea derived from that source.

There is a clear parallel here to our discussion of the way that the power of the One imparts to each unity its characteristic super-essence. Each number has a peculiar just-so-ness which we have isolated as representative of its unity, [*fig. 34*] and this aseity of

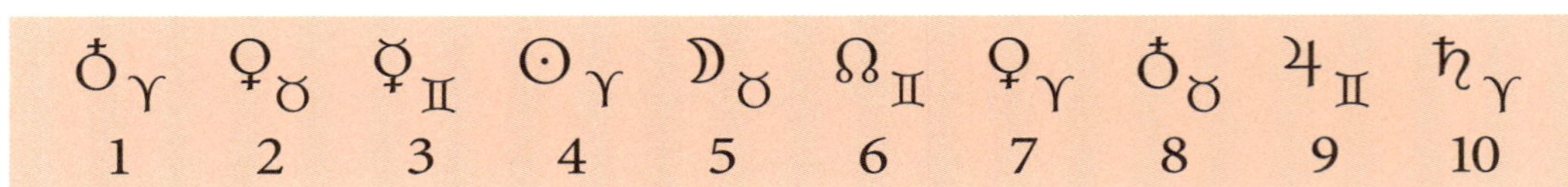

Fig. 34

the number is a unique quality derived from the One that reveals the One in a special way. Later on, the same analogy will be applied to show that each Idea in the Intellectual-Principle is a special and unique version of the whole. But before we apply these principles to the Intellectual-Principle or to any "one," we will concentrate on the meta-ontic properties of Number. Since the properties of Divine number derive from or mirror certain facets of the One, understanding them is a means of opening our gnostic eye to the immense complexity of the Intellectual-Principle.

The sequence of the Divine numbers is a succession of ways of seeing the complexity of the One. Each number is an impression of Unity and orders our understanding of the One and its power in a unique way. ⊞ Following the hint of the Kabbalists, we can correlate each of the Divine numbers with one of the spheres of totality (*sephiroth* or dignities) in our version of the Triad. [*fig. 35*]

⊞ Editors: Divine number transcends Being and multiplicity, yet structures the multiplicity of Ideas from Being throughout all the levels of multiplicity.

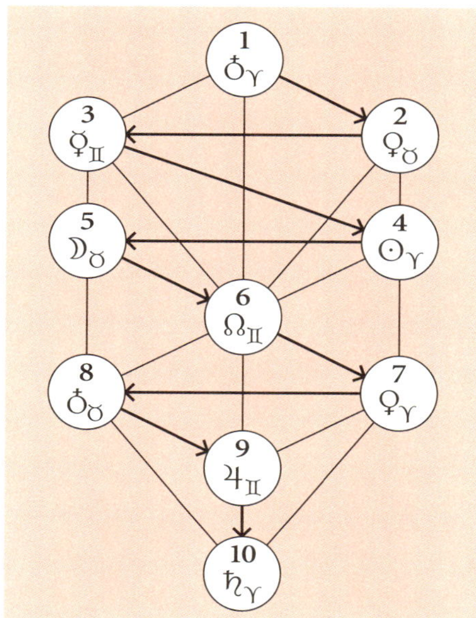

Fig. 35

The Ten Divine Numbers

In the same way that we treat Unity as having aspects and say that each of these aspects represents the indivisible Unity in a certain way, so must we see that Number as a principle is a unity with diverse aspects. What we are saying is that each of the numbers up to 10 represents a way of viewing the One itself. It would not be correct to conceive the operation as a factoring out of distinct possibilities contained within the One. We are simply seeing the One in a variety of ways, and it should not surprise us that we cannot perceive the totality of all the aspects of it at first glance. By definition, the contents of perception are determined by the selectivity of attention. The meaning of Number at the highest level would be that it contains an impression of unity.

One

The principle of the number 1 will bear the same connotations, or at least reflect the same remarks, made about the One-Only. The uniqueness of the self-sufficient transcendent, the infinite unity, and simplex are all synonyms for the meaning of One. Symbolically, this correlates with the view of the One as the Transcendent Idea of Ideas (the 1st house) in comparison to the One as immanent throughout [all twelve houses]. ▼

▼ Editors: Anthony presents the view of the One as one in his discussion of the self-sufficiency and transcendence of the One as "absolute positivity," "the only Perfection" (p. 24), and as the meaning of the 1st house (p. 54, fig. 13).

The Absolute Mind or meta-ontological infinity, the One, is complete perfection itself. When Plotinus presents us with the self-sufficient Unity, the transcendent Simplex and One-Only, we have the paradigm for the meaning of the number 1. One cannot be thought of in the quantitative way as the first in a series, or in relation to others, for it is from the One that all the others will arise.

> The One does not bear to be numbered in with anything else, with a one or a two or any such quantity; it refuses to take number because it is measure and not the measured; it is no peer of other entities to be found among them . . . (Plotinus, v.5.4)

The Oneness is not quantitative, or a point, for it is infinite perfection, and

> remains intact even when other entities spring from it. In the case of numbers the unit remains intact while something else [Anthony adds: (the Dyad)] produces, and thus number arises in dependence on the unit: much more then does the unit, The One, remain intact in the principle which is before all beings . . . (Plotinus, v.5.5)

In mathematically illustrating the identity of the Absolute, the choice has been to use the symbol "One" (1), and not "Zero" (0), for the latter was not available [in the West] at the time the Greeks were discussing these ideas. The Absolute, the One, is conceived as a fullness, and each number will derive from the One a unique intact aseity, will be some one thing (aspect of the infinite). The meaning of "One" as used by the Neoplatonists is equivalent to that which the Eastern seers assign to "Zero." The Eastern seers would say everything arises from the Void (Zero) and is Void. In the case of the Platonists we have an affirmative theology, and the other provides us with a negative theology. Both would insist that all things arise from the One (in Platonism) or the Zero, and both would insist that if you take fullness from the full, the full remains. Plotinus' remarks are very much to this point when he says that the many arise from the One but the One remains intact. Plotinus would say that which arises from the Void or universal relativity should not be considered monistically but rather from the aspect of simultaneous emanation.

So we have the One, which, strictly speaking, is the principle of oneness in that it is a necessary prerequisite. For anything to be at all it must first be a one.

Two

In its widest application, the 2 represents the double act of the One: infinite Power which exhibits two aspects, passive and active perfection, or substance and function. ⊞ The One reveals something of itself through this paradoxical nature, but it itself remains inscrutable. The "interrelationship" of power with the One, a completing completed power, is included in its total perfection. Two is the infinite indeterminate Power, inseparable from the One, and also exhibiting its inner nature through the dual modes of repose and act which are inherent in it.

⊞ Editors: Anthony's discussion of the infinite depths of power of the Dyad (pp. 30–34), and of the 2nd and 3rd houses as the One's active and passive perfection (p. 56–57 and fig. 15) illustrates the view of the One through the number 2. Figure 14 (p. 55) shows it in terms of the 1st house and its dignities.

This Power of the One is the symmetrical relation of the One as remaining intact and the One exhibited in its completeness as the source of all, or the Triad. Each of the indivisible unities and ultimately all Being will participate in this Power or in some imitation of it. The Dyad is the paradigm for the complementarity of the powers of function and substance or yin and yang, which are revealed in the levels of reality from the Triad down to the sensible world.

> If The First is perfect, utterly perfect above all, and is the beginning of all power, then it must be the most powerful of all that is, and all other powers must act in some partial imitation of it. (Plotinus, v. 4.1)

Power is identical with the One, and Power also issues forth or exhibits the One through the dual modes of repose and act which are nondualistically inherent in it. The Dyad is the paradigm for all the complementary modes of the expression of reality—active and passive, function and substance, sevenfold and fourfold, and so on.

We can list many ways of seeing the power of the Dyad, all of which present the double act—so simply put by the mathematicians—as 2 = 1 + 1.

> The 2 is the One plus its infinite Power.
>
> The 2 is the Repose and Act remaining in the One as identical with it.
>
> The 2 is the Repose and Act or Infinity and Bound exhibited as houses 2 and 3.
>
> The 2 is the indeterminacy and symmetry of the One as transcendent and as immanent. [8, p. 306; 10, p. 310]

Three

The number 3 reflects the description of the One as the Triad. ▼ Here we see the One not as the intact simplex, but as the source of all, the total of indeterminate possibilities and power, prefiguring all principles in itself in an unmanifest and undifferenced way. The internal completion in the One is symbolized as the interaction of the One and the Dyad. Neither the One nor the Dyad is strictly a Number principle, for the One is measureless and

▼ Editors: Anthony's discussions of the "super-knowledge" and inner intelligence of the One (pp. 25–26), and the Triad as the "expanded" version of the One and as the first three houses with dignities (pp. 57–58 and figs. 16 and 17) are applications of the meaning of the number 3.

without relation, and the Two is infinite power and substance. The Three is the principle of boundless indetermination, and suggests the super-essential principle of the Void, the 3rd house. Three is the stable and unchanging product of the One and Two, and is inclusive of the two priors and nondifferent from them.

This perfection can be conceived as having indivisible aspects or as illusorily divided. Each of these indivisible aspects or quotients may be conceived of as the One in its entirety in a certain mode, and each and every mode will be a transcendent paradigm for the unities that exist in Being, Soul, etc. The above considerations result in the notion or concept of the Triad.

> . . . it is therefore no unconscious thing: all its content and accompaniment are its possession; it is self-distinguishing throughout; it is the seat of life as of all things; it is, itself, that self-intellection which, stemming from an inner consciousness, takes place in eternal repose, that is to say, in a mode other than that of the Intellectual-Principle. (Plotinus, v. 4.2)

Four

The number 4 prefigured in the Absolute suggests that in its own mysterious manner the One includes the superior version of the quaternity in its utter perfection. Quaternity here refers to the three hypostases and their immanence in the system of Nature. ■ Also, it is the total synesthesis of what we understand as the separative states of consciousness—here referred to as the fourth state or *turiya*. ● [13, p. 312]

The fourfold nature of the One's intelligence serves as a model for any "one," which will extend through all the levels of reality, so that we are allowed to speak of a "one" as having Unity, Being, Soul, and Body. Within the Triad, 4 can be seen as the Substance which is the ground, or the passive perfection, of the double act.

> . . . The Good, cause of Being and Intellectual-Principle, is a light appropriate to what is to be seen There and to their seer; neither the Beings nor the Intellectual-Principle, it is their source and by the light it sheds upon both makes them objects of Intellection. . . .

■ EDITORS: Anthony's discussion of the meaning of 4 in relation to the principle of manifestation (pp. 67–69, including figs. 28 and 29) is an example of applying the principle of the number 4.

● EDITORS: *Turiya* literally means "fourth" in Sanskrit. Here the number 4 as applied to the One describes how the other three states of consciousness are prefigured in the One.

> Giving need not comport possessing; in this order we are to think of a giver as a greater and of a gift as a lower; this is the meaning of origin among real Beings. First there must be an actualized thing; its laters must be potentially their own priors; a first must transcend its derivatives . . . (Plotinus, VI.7.16–17)

The number 4 applies to our discussion of:

the fourfold nature of the One

the three hypostases as applied to the system of Nature

how these principles are duplicated in the sensible world

the quaternity as symbolizing the Absolute with its fourfold division of Consciousness

an excursis into the Hindu conception of the fourfold compared with Plotinus. ▼

▼ EDITORS: This comparison appears in figure 31 on p. 71.

Five

According to the editors of the *Shrine of Wisdom,* "The Pentad is the Number of Divine Incarnation, of the infinite monad [or one] manifesting in the Finitude of the Tetrad." ("Symbols: Number," iv, no. 16, p. 122) We endorse and fully accept this at the level of Being, for the statement refers to the presence of the One in the Intellectual-Principle or Being—even further, in any Idea within the Intellectual-Principle. In other words, the One is immutably present in each Idea. However, our interest is in pointing out that this principle is already prefigured in the One itself, as what allows us to say "indivisible aspects" of the One: The One is present in any "aspect" of itself; it is its presence in all aspects of itself that gives rise to the notion of the Divine Pentad—and thus we can refer to each aspect of the One as a pure indistinguishable one.

> Every particular thing has a One of its own to which it may be traced; the All has its One, its Prior but not yet the Absolute One; through this we reach that Absolute One, where all such reference comes to an end. (Plotinus, III.8.10)

The Perfection of any "one" is due to this transcendent principle: The One has the power to reproduce itself entire in any unity, each unity inheriting the total interior complexity of the One. Since each of the ten unities is an indivisible aspect, and the One can be totally present in each, 5 represents the indivisibility of the One's "aspects" and each aspect is a unique formless form. ⊞

Six

When we speak of the intact perfection of the One, we are referring to the completed and lawful order of the One's power evenly distributed throughout the totality of actualized possibilities it contains, and which we must consider as non-final or endless. ⊕ This is the transcendent principle of Number within the One which is of course reflected in the eternal ordering of the Supernals in Being.

Within the first quadrant itself there is already exhibited the unmultiplied and evenly distributed power of the One expressed by the view that the addition of One, Infinity, and Bound speaks to us of its unlimited function.

The Divine Hexad refers to the actualized totality of all the permutations and combinations of possibilities by the power inherent in the One: non-finality, infinitude. This permutability (symbolized by the number 6) is the principle or the very essence of order in the One. Therefore, Number within Pure Being is the completed ordered Act and the very substance of activity found within it—the completed completion.

> When it takes lot with multiplicity, Being becomes Number by the fact of awakening to manifoldness; before, it was a preparation, so to speak, of the Beings, their fore-promise, a total of henads offering a stay for what was to be based upon them. (Plotinus, VI.6.10)

⊞ Editors: Anthony's discussion of the meanings of the individual unities in the Triad, each conceived of as the One in its entirety in a certain mode, is an application of the principle of the number 5 (see pp. 61–66, figs. 20–25).

⊕ Editors: Anthony's discussion of the internal order of the One—its "timelessly fixed, perfect, and complete interiority" (p. 34)—applies the principle of the number 6.

Seven

Seven (7) is that aspect of the indivisible unity which we can refer to as its functioning or outward-facing power but resident within itself. ▼ ⊞ Seven is an exemplification of the principle of essential production or the outward-facing hypostasis by and through which the One exhibits everything which is in itself to itself without introducing duality.

> Given this immobility in the Supreme, it can neither have yielded assent nor uttered decree nor stirred in any way towards the existence of a secondary.
>
> What happened, then? What are we to conceive as rising in the neighbourhood of that immobility?
>
> It must be a circumradiation—produced from the Supreme but from the Supreme unaltering—and may be compared to the brilliant light encircling the sun and ceaselessly generated from that unchanging substance.
>
> All existences, as long as they retain their character, produce—about themselves, from their essence, in virtue of the power which must be in them—some necessary, outward-facing hypostasis continuously attached to them and representing in image the engendering archetypes . . . (Plotinus, v.1.6)

The 7 reveals the functional aspect of the One, just as the 4 is its substantial expression. This functional aspect is unbegotten and unbegetting, virginal in its indivisibility. [16, p. 314; 17, p. 317] The colors of the spectrum reveal the properties of pure light and can be distinguished, yet you cannot take away any one of them and still have pure light. And again, the seven notes of the musical scale objectify and unfold the content of the key note—for they are indissolubly linked with one another and are manifesting it. Perhaps the best example is the seven Chaldean planets [of astrology] as the outermost and utmost expression of the rational soul.

▼ Editors: Anthony's use of the analogy of a glowing body emanating light (pp. 35–37), to show that the outward-facing emanation of the One does reveal its inner nature, is an application of the number 7.

⊞ Editors: While 5 establishes a view of the One as unique unities and 6 views the internal order or structure of the unities, with 7 we get a view of the One through the fullness of seven functions symbolized by the planetary dignities—its motionless motion, or unlived life.

Eight

Eight (8) is the principle of infinite divisibility which makes the One totally available to any and every part of itself. ⊕ The entirety of the required number of formless principles is signified by the number 8 or Divine Ogdoad. It indicates the activity of the super-knowledge which perfects its source not by producing anything but by being the completing power within itself. The fictional divisibility or illusory modifications within the One permits us to speak of the indivisible aspects or partless parts of the One. [20, p. 319]

The number 8 represents the infinity of possibilities actualized—contained in the indeterminacy of the One. Therefore, 8 is said to be the most divisible, and signifies the infinity of attributes of the One, the variation of which is utterly indeterminate. A binary tree diagram [*fig. 36*] is sometimes used to indicate an infinity of possible paths which seem to involve infinite choice; but as all paths are simultaneously actualized in the One, any "choice" is already in conformity to the One. All is accomplished, yet its infinite indeterminacy implies an absolute freedom. △

> . . . we appeal to the character of this power, not subject to quantity but such that though you divide it mentally for ever you still have the same power, infinite to the core; in it there is no Matter to make it grow less and less according to the measured mass.
>
> Conceive it as a power of an ever-fresh infinity, a principle unfailing, inexhaustible, at no point giving out, brimming over with its own vitality. (Plotinus, VI.5.12)

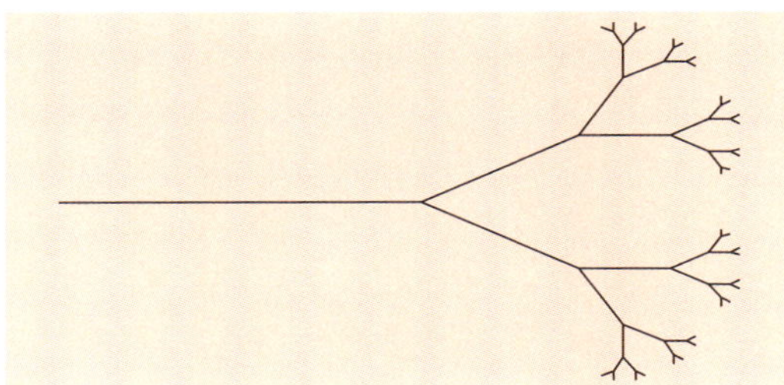

FIG. 36

⊕ EDITORS: Anthony's "gold and trinkets" analogy (pp. 35-37) is an example of seeing the unities in the One through the number 8.

△ EDITORS: Infinitely divisible, the binary tree pictorially represents the number 8, which Anthony described, in an earlier draft, as "the infinity of modes or aspects within the One."

Nine

Nine (9) is a principle of completion which retains all within the source, the pre-ontological or pre-ideational completed actualization of all the possibilities. ▼

> The Supreme is cause of the cause: it is cause pre-eminently, cause as containing cause in the deepest and truest mode; for in it lie the Intellective causes which are to be unfolded from it, author as it is not of the chance-made but of what the divine willed: and this willing was not apart from reason, was not in the realm of hazard and of what happened to present itself but of what must needs be, since hazard is excluded from that realm. (Plotinus, VI.8.18)

The 9th principle is said to return all things to itself. "All things" refers to the infinity of attributes which a Transcendent One must include—the unfailing divisibility of its infinite principles according to the view of 8. The 9 indicates the completed-completion, the One's double act within itself. The 9 is said to circulate all things in itself, which is an artificial attempt to conceive the boundlessness and infinity of the One as having boundary—it bounds without limiting. So, the view of infinite divisibility and infinity is not left unbounded or unordered. We may conceive of the 8 as a line or infinite continuum, and the 9 as a circle which bends upon itself—which is a different notion of infinity. The description of God as "a circle whose circumference is nowhere and whose center is everywhere" is appropriate to the meaning of 9.

In linguistic terms, we may say 9 is a definition without a negation, or a definition defined, in which the infinity of attributes are returned each to themselves—to the One. Nothing can be added to, or go outside of the One, or Authentic Existence. The three hypostases and their immanence in the lower phase of Soul—that is, the circulation of all the principles of reality within the Universal Constructor—are prefigured in the Triad.

▼ EDITORS: "The peculiar way in which the transcendent One is *inclusive* of all and is the cause of all" (p. 25), as including all within itself prior to emanation, is a view of the One through the principle of 9.

If we use the analogy of the mind thinking its thoughts, 9 indicates the principle of thought—and all thought, which means all manifestation, is actually of the nature of Mind. In this way we can understand the meaning of circulating all things back to the Good, "through which all things abide in, proceed from and return to the . . . One." (*Shrine of Wisdom*, "Symbols: Number," iv, no. 16, p. 119) Plotinus' statement that the One "is the measure and not the measured" (V.5.4) is appropriate for describing the meaning of the number 9.

Ten

Ten (10) presents to us the presence of a second or "otherness" within the One without a second. ■ The emanations of all the prior formless principles are reflected in the mirror of non-Being, and appear as other, yet while allowing the omnipresence of the One and its identity or sameness with the appearance.

> No doubt it is wonderful that the First should thus be present without any coming, and that while it is nowhere, nowhere is it not . . .
>
> Holding all—though itself nowhere held—it is omnipresent, for where its presence failed something would elude its hold. At the same time, in the sense that it is nowhere held, it is not present . . . (Plotinus, v.5.8–9)

This is the transcendent principle from which is derived the principle of Difference, by which the existence of universal manifestation is possible, and [which] is the basis for the entire fourth quadrant or body of the One. ●

■ Editors: Anthony's discussion of the principle of difference and otherness as Saturn in Aries ♄♈ in the Triad (p. 64, fig. 23 and pp. 72–75) illustrates the meaning of the number 10.

● Editors: The Pythagoreans considered 10 to be the most complete number, not only because of its arithmetical property (10 = 1 + 2 + 3 + 4), but because the principial 10 is what allows non-Being or non-one, or this world of becoming, to occur.

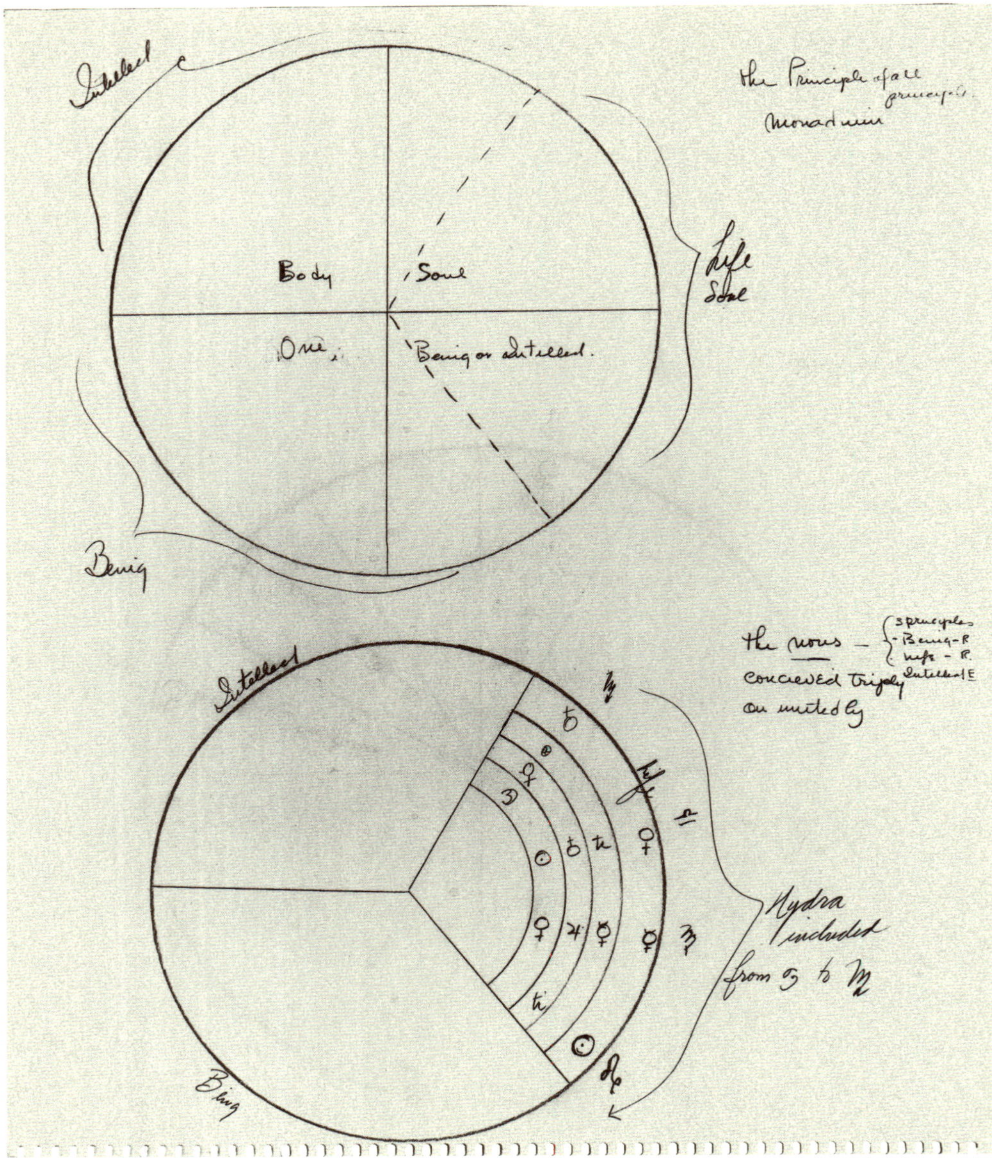
Intellect
the Principle of all principles
monadum
Body
Soul
One
Being or Intellect
Life
Soul
Being
the nous
conceived triply or unitedly
Intellect
Being
Hydra included

Chapter Two

THE INTELLECTUAL-PRINCIPLE

An Outline

▼ Editors: All references in parentheses within this outline are to Plotinus' *Enneads.*

The One as source of all contains transcendentally, in its inner nature, all the principles which are to be distinguished from it. (v.2.1) ▼

In the undifferenced knowledge of the One are prefigured the Intellectual-Principle and Being, the Ideas and Numbers, so in essence the unity of the Intellectual-Principle is to be found in the One. (v.4.2, vi.8.18, vi.9.9)

The Power of the One, the indeterminate Dyad, is so great that the Intellectual-Principle, although an image One, stands as authentic reality (in eternal actuality); this "outward" expression of the One paradoxically stands as a distinct hypostasis midst the One without a second. (v.1.6, v.4.2, vi.9.9)

The radiant emanation or the divine outflowing is now unfolded. Within Being arise the first distinctions, knowledge and its objects, the One-Many. (v.1.10, v.6.6, vi.7.15)

The Intellectual-Principle or Self-thinking Thought knows the One in essence. (v.3.11, v.5.2)

But the inability to grasp the Infinite One in its totality even by the Intellectual-Principle and its repeated attempts to do so, accounts for the innumerable successive impressions of that One as universe after universe is unfolded and the planned evolution or Idea of manifestation progresses. (iii.8.8, v.3.10, v.3.11, vi.7.16–17)

The principle of Number, which is also transcendentally in the One, will form the basis for the ordering of the Ideas in the Divine Mind. (v.1.5)

Number presides over the unfolding of the Intellectual-Principle, arranging and coordinating the Ideas and their Intelligible powers, or substance and function. This principle of Number will coordinate the specific relations of the Ideas with the appropriate powers or dignities (rulerships and exaltations) which essentialize and singularize the Ideas. (VI.6.10, VI.6.15)

We symbolize the unity of Divine number as the "Head of the Dragon" (☊) in the Triad. (VI.6.9)

The Intellectual-Principle is a unity-in-duality of Knowing and Being. In the Intellectual-Principle the relation of the Ideas and their power is made explicit. They are its own divine content which is itself self-gnostic. It is that which it knows. (V.1.4, V.3.5, V.6.1, V.6.6, V.9.5, VI.7.16)

Each Idea is an Intellectual-Principle entire, a special and unique form of it, a duality and unity with a specific internal structure of its own determined by Number. The totality of Ideas as self-gnostic and self-distinguishing gives rise to the notion of the separative intellect rooted in the one unified Divine Intellect. (V.9.6, V.9.8)

Plotinus attempts to reveal something of the essence of the Divine Mind by a discussion of the categories or the constitutive principles of Being: isness, motion, permanence, difference, identity. Sometimes he speaks of five categories and sometimes of six. The nature of these principles is the cause of some ambiguity. Nonetheless, these are not parts of Being but Being itself. Mind thinks, which is an eternal act; this is divine ideation. It is a permanent reality and has Power or motion and is stable essence. ("Mind" equals isness, "thinks" equals Power or motion; its thoughts are self-identical and different from one another, their permanence is their stable essence. In their totality they are identical with Being or Intellectual-Principle.) (V.1.4, VI.2.7–8, VI.2.15)

There is no notion of becoming or potential in the Intellectual-Principle; all is actualized, and all is one eternal life. (III.7.4–5, V.1.4, V.2.1, V.8.4–5)

Every entity in the empirical world partakes of the Idea to the extent it is able. (V.8.2, VI.2.20, VI.2.22, VI.7.2–4)

The Intellectual-Principle

The Good comes to it [Intellectual-Principle] not as it was in its primal state but in accord with the condition of the Intellectual-Principle. The Good is the source from which the objects of contemplation come to be seen in the Intellectual-Principle; Intellectual-Principle has produced them by its vision of the Good. . . . the Good bestows what itself does not possess. From that Unity came multiplicity to Intellectual-Principle; it could not sustain the power poured upon it and therefore broke it up; it turned that one power into variety so as to carry it piecemeal.

All its production, effected in the power of The Good, contains goodness; it is good, itself, since it is constituted by these things of good; it is Good made diverse. It might be likened to a living sphere teeming with variety, to a globe of faces radiant with faces all living, to a unity of souls, all the pure souls, not the faulty but the perfect, with Intellect enthroned over all so that the place entire glows with Intellectual splendour. . . .

At first it was not Intellect looking upon the Good; it was a looking void of Intellection. We must think of it not as looking but as living; dependent upon That, it kept itself turned Thither; all the tendance taking place There and upon That must be a movement teeming with life and must so fill the looking Principle; there is no longer bare Act, there is a filling to saturation. Forthwith it becomes all things, knows that fact in virtue of its self-knowing and at once becomes Intellectual-Principle, filled so as to hold within itself that object of its vision, seeing all by the light from the Giver and bearing that light with it.

Plotinus, vi.7.15–16

OUT OF THE OCEAN OF INFINITE POSSIBILITIES present in the ineffable undifferenced intelligence of the One, there pours forth a radiant display of glorious revelations—the Divine Mind and its Ideas. These luminous existences, emanating on the surface of its voidness, exhibit the inner nature, the inherent perfection of the unique and unknowable God. The Power of the One is so great that this Divine Mind, though an "image One," is itself authentic reality. This "outward" expression of the One paradoxically stands as a distinct hypostasis midst the One without a second.

> Seeking nothing, possessing nothing, lacking nothing, the One is perfect and in our metaphor, has overflowed, and its exuberance has produced the new: this product has turned again to its begetter and been filled and has become its contemplator and so an Intellectual-Principle.

> That station towards the One . . . establishes Being; that vision directed upon the One establishes the Intellectual-Principle . . . (Plotinus, v.2.1)

The outpouring of power at once takes on determinations and expression as the Intellectual-Principle. [*fig. 37*]

> . . . the Intellectual . . . announces in its own being the entire content of the Good, that prior of all, locked in unity, of which this is the expression already touched by multiplicity. (Plotinus, v.9.2)

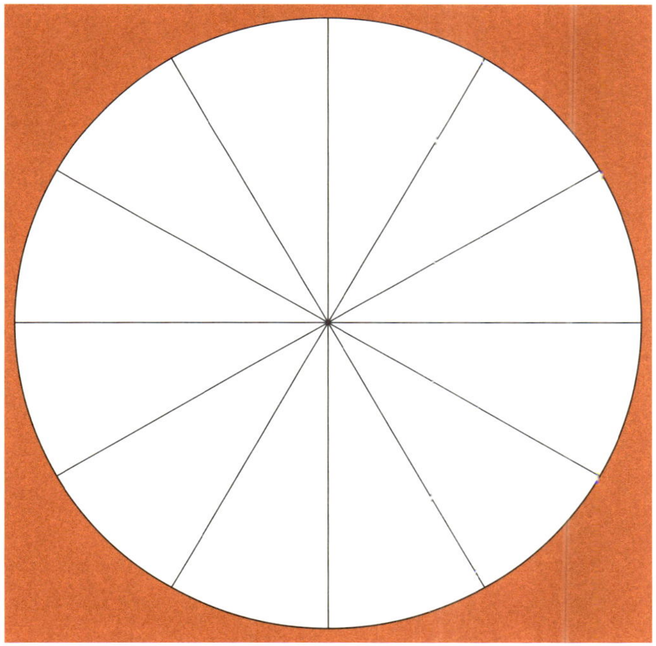

FIG. 37

The Intellectual-Principle is Intellection pure and simple, the ecstatic activity of Ideation; the thoughts that it thinks are the many membered parts of itself. This activity of Ideation does not cancel out that uniqueness, its indivisible self-identity; it does not introduce any element of duality or predication into itself, but it does have many thoughts (parts) and these together with its self-identity constitute a whole. This oneness is both the same as and different from the sum of its parts. So the consequence is that a "one" is in fact and in actuality a whole composed of unity and multiplicity.

> Contemplation and its object constitute a living thing, a Life, two inextricably one.
>
> The duality, thus, is a unity; but how is this unity also a plurality?
>
> The explanation is that the object of Contemplation is not a unity: even when that object is the One itself, it is not seen as a unity; if it were, the Intellectual-Principle cannot exist. (Plotinus, III.8.8)

The concentrated unity of the One's intelligence is reflected in the Intellectual-Principle, which Plotinus characterizes as no longer a unity but rather a "One-Many" (V.3.15) or a "unity in duality" (V.6.1), for within it arise the first distinctions of knowledge and its objects (Being). ▽

In the Divine Mind there is a vision of Ideas: Subjective and objective, as intelligible powers and ideal substances, are only logically distinguished as the knowing and the known.

> When a thing is a Being, it is also an Intellectual-Principle; when it is an Intellectual-Principle, it is a Being; intellection and Being are co-existents....
>
> Thus all is dual [Anthony adds: within the Intellectual-Principle]: the unit is a duality and yet again the dual reverts to unity. (Plotinus, V.6.6)

Plotinus also states (V.1.4, V.3.5, and VI.7.16) that intellection and Being are complementary realities in the Intellectual-Principle, each constitutive of the other, both rising from their common source, which we can suppose to be none other than the One. Plotinus is following the line of thought in *Republic* 508e where Plato says that the Good imparts

▽ Editors: Plotinus characterizes the three primals in V.1.8: "The Platonic Parmenides is more exact; the distinction is made between the Primal One, a strictly pure Unity, and a secondary One which is a One-Many and a third which is a One-and-Many; thus he too is in accordance with our thesis of the Three Kinds."

⊞ Editors: For Plotinus and Plato, Being is not the ultimate nature of reality. The ultimate or absolute reality, the One or the Good, is beyond Being, although sometimes a facet of the One is referred to analogically as Universal Being or Authentic Being. Being refers to the state in which authentic existences are themselves, distinctly exhibiting their own characteristic nature.

Plotinus goes further than Plato, in making the Intellectual-Principle self-determining as well as dependent on the Good or One. The Intellectual-Principle establishes itself as distinct from the One by exercising its Intelligence. Simultaneously, it knows itself through being established as distinct Being. Exercising its vision upon itself, this undetermined life brings forth the trace of the Good as the multiplicity of beings. Knowledge and Being thus form the inseparable unity in duality of a vast intelligence reflecting upon its prior and bringing itself to birth.

"truth to the objects of knowledge and the power of knowing to the knower." (*Plato: The Collected Dialogues,* Paul Shorey trans.)

What the Divine Mind knows, it also is, and it knows all that it is entirely, nothing remaining outside its vision. "If this identity does not exist, neither does truth . . ." for the principle which *knows* the truth of a being must *be* that being, otherwise what it holds is only an image of the Truth, "something different from the realities." (Plotinus, v.3.5)

The Ideas which are the Intellectual-Principle's substance, and the intelligibles which are the Ideas' very activity, can be symbolized as the twelve signs around the wheel with their rulers. Both are grounded in the unitary consciousness, *turiya.* We then see that the totality of the signs and rulers is collectively the Intellectual-Principle, and that each sign is a particular facet of Divine Wisdom, a particular Intellectual-Principle or Idea. Each Idea has a certain power, for this is the mark of Real-Being, a power primarily to *be,* which forms an inseparable unity in duality—the rulerships in their totality being the collective power of the Intellectual-Principle as a One-Many of essentially self-gnostic Ideas. [*fig. 38*]

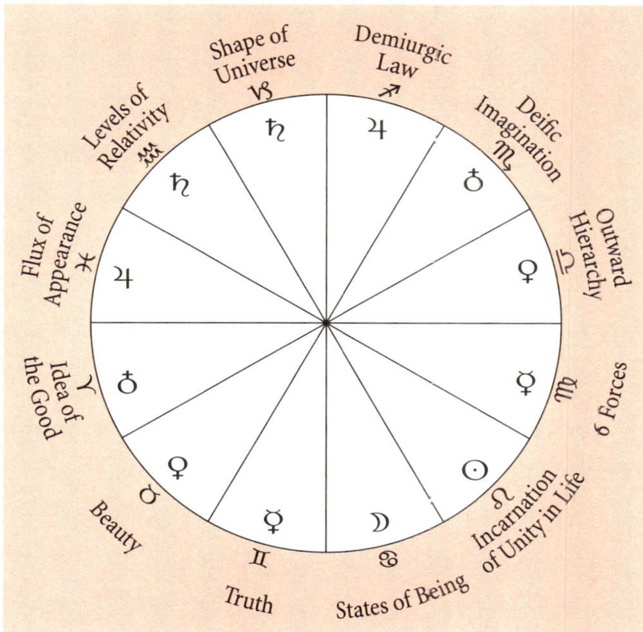

Fig. 38

> . . . this is pure being in eternal actuality . . . everything in that entire content, is Intellectual-

Editors: This formulation of the nature of Divine Mind precludes giving primacy to either knowing or being. When Being is considered as foremost, as in v.2.1, Plotinus says that the Intellectual-Principle must first Be as a distinct principle (hypostasis) in order to know itself, and by knowing itself it also knows the One. When he takes Knowing as primary, as in v.3.10, Intellectual-Principle is Vision reflecting upon itself, giving rise to Being. On the one hand, Intellectual-Principle is an emanant of the One, and all the principles of Being are prefigured within the One. On the other hand, Intellectual-Principle is self-determining, and its content is a trace of the unities in the One—organized by the principle of Number.

Editors: When we view the chart as the fourfold reality, *turiya* may be represented either by the first quadrant, or by the outermost ring, or by both. When we view the chart as the Intellectual-Principle, the whole circle represents *turiya,* the pervasive consciousness or Mind itself in which the twelve Ideas are "grounded."

Principle and Authentic-Existence. . . . Intellectual-Principle by its intellective act establishes Being, which in turn, as the object of intellection, becomes the cause of intellection and of existence to the Intellectual-Principle . . . (v.1.4)

Considering the whole chart, we could say that the Wisdom-Insight which has assigned to each sign of traditional astrology its corresponding dignities is the same as that which Plotinus expresses when he says that when a thing is an intellect, it is also a being, the two are never apart.

The unitary foundational consciousness of the Intellectual-Principle as the circle is unrecognizable unless the gnostic powers of each Idea, which distinguish one Idea from another, make available the aggregate of Ideas (Forms).

First we conceive of the encircling Consciousness as Intellect as a whole. Only after this conception can we see that each Idea is included within and supported by that whole. Therefore the prior provides the nutriment for each Idea, and consequently is the commonality of substance for all of them. But insofar as each individual existent Idea is different from another and has its own being, we can see that the Intellectual-Principle is a One-Many.

Unfolding of the Intellectual-Principle through Number

The principle of Number will guide us in further understanding how the supreme form, the Idea of the Good, arises from its formless source. [*fig. 39*]

When it takes lot with multiplicity, Being becomes Number by the fact of awakening to manifoldness; before, it was a preparation, so to speak, of the Beings, their fore-promise, a total of henads offering a stay for what was to be based upon them. (Plotinus, vi.6.10)

Fig. 39: Triad—partless parts of the One

The indefinite multiplicity of Intellectual-Principle's contents [the Ideas] and powers is coordinated and ordered specifically through the principle of Number, which has its highest subsistence transcendentally in the One.

In speaking of Number, we are telling of an inconceivable intrinsic order of the One's complex internality, a partless aspect of it and consequently the source of all the subsequent ordered multiplicity in Being.

Speaking incorrectly, the total of unities has been referred to as the undifferenced infinity. As previously explained, transcendental unities inbound with Unity, each within all and all within each, without any distinctions, do not give a participated Unity. That is, the adding together (sum) of transcendental unities does not give rise to Unity. It is simply a way of presenting the One in a variety of ways. As "within" the One, these unities exist in a relationless and modeless homogeneity—for this is the Solitary. ▼

▼ Editors: Here Anthony equates the *henads* mentioned by Plotinus in VI.6.10 (as quoted on page 104) with the partless parts or transcendent unities in the Triad.

In the symbolism of the Triad, the Head of the Dragon (☊) represents the principle of Number included in but not yet applied to Being. Here, as Plotinus says, Number cannot be applied, for in so doing it would give rise to the essential distinctions inherent only in the Intellectual-Principle.

Intellectual-Principle in knowing its very essence cannot grasp this Unity in its entirety and thus goes forth "amassing successive impressions, so that to it the object (One) becomes multiple," i.e., Ideas.

> Thus the Intellectual-Principle, in the act of knowing the Transcendent, is a manifold. It knows the Transcendent in very essence but, with all its effort to grasp that prior as a pure unity, it goes forth amassing successive impressions, so that, to it, the object becomes multiple: thus in its outgoing to its object it is not (fully realized) Intellectual-Principle; it is an eye that has not yet seen; in its return it is an eye possessed of the multiplicity which it has itself conferred: it sought something of which it found the vague presentment within itself; it returned with something else, the manifold quality with which it has of its own act invested the simplex.
>
> If it had not possessed a previous impression of the Transcendent it could never have grasped it, but this impression, originally of unity, becomes an impression of multiplicity; and

> the Intellectual-Principle in taking cognizance of that multiplicity knows the Transcendent and so is realized as an eye possessed of its vision.
>
> It is now Intellectual-Principle since it actually holds its object, and holds it by the act of intellection: before, it was no more than a tendance, an eye blank of impression: it was in motion towards the transcendental; now that it has attained, it has become Intellectual-Principle: always self-constitutive, it now, in virtue of this intellection, holds the character of Intellectual-Principle, of Essential Existence, and of Intellectual Act where, previously, not possessing the Intellectual Object, it was not Intellectual Perception, and, not yet having exercised the Intellectual Act, it was not Intellectual-Principle. (Plotinus, v.3.11)

Prior to this visioning, Intellect is a foreseeing, an indeterminate Power. ▼ Through the activity of Number the indeterminate Power is awakened and organized into a specific manifoldness by being variously characterized.

▼ Editors: Being emanates, but Intellectual-Principle is self-determining. See p. 101 for Plotinus on Being and Intellect in v.2.1.

That the Intellectual-Principle "knows the Transcendent [One] in very essence" and that this knowing constitutes the essence of the Intellectual-Principle can be perceived and understood in terms of the Dragon's Head ☊ in the Triad, which stands for the transcendent model of the order found in the One. The Dragon's Head or Number is in fact the "impression . . . of unity" which the Intellectual-Principle or the second Hypostasis has as its source or essence. [*fig. 40*]

The Dragon's Head ☊, therefore, symbolizes the unity of Number and the principle of Intellection in the One (its determinate power to "see" that incomprehensible order within the One). The essence of intellection, the Intellectual-Principle is therefore rooted in the

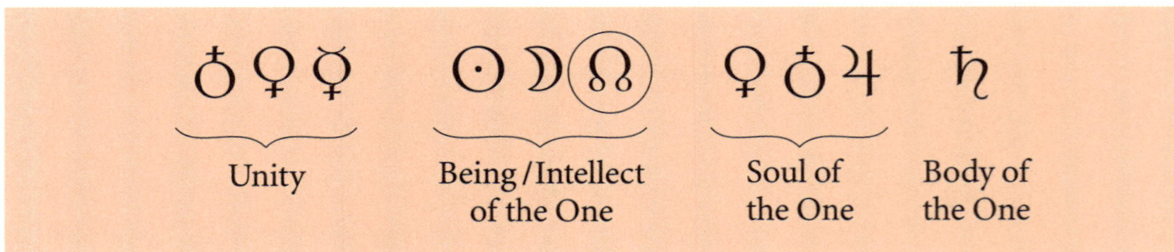

Fig. 40

Head of the Dragon. We turn to the ☊ to understand how when Being "takes lot with multiplicity" it is awakened to manifoldness.

The principle of Number ☊ organizes the order of unfoldment and emanation of the Beings which in their totality take shape as the reflex image of the One, the Intellectual-Principle. Divine numbers are also the very "means" by which the One's power operates in unfolding and regulating the outflow of the Infinite's unlimited possibilities, and a certain [atemporal] series of these possibilities [the Ideas] takes shape as the determined content of and as the Divine Mind. [*fig. 41*]

The unfolding of super-essential Number differentiates Beings from the One and from each other and correlates the Ideas and their powers. An important point brought to our attention here is that Number structures the eternal motion of Intellect.

> Number must be either the substance of Being [and/] or its Activity; the Life-Form as such and the Intellectual-Principle must be Number. (Plotinus, VI.6.9)

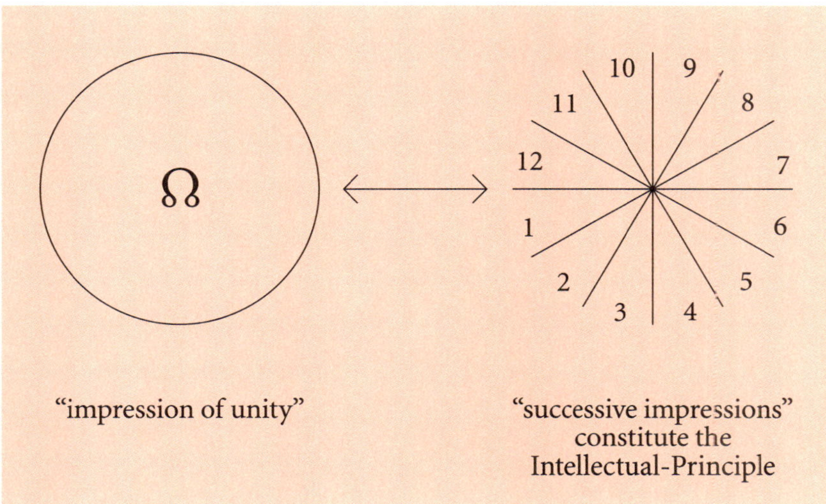

FIG. 41

EDITORS: Plotinus writes about Ideas and the intellective vision:

> The Intellectual-Principle, thus, is informed of itself by the fact of being a multiple organ of vision. . . . in its outgoing to its object it is not (fully realized) Intellectual-Principle; it is an eye that has not yet seen; in its return it is an eye possessed of the multiplicity which it has itself conferred . . . —V.3.10, 11

These metaphors of an "eye that has not yet seen" and an "eye possessed of vision" parallel the earlier (p. 105) description of an "impression . . . of unity" and multiple or "successive impressions" in V.3.11.

EDITORS: Number is the power by which the Intellectual-Principle is determined. It is the power that "measures out" the infinite. By determining a relation among the unities, Number allows them to *be* within the Intellectual-Principle. Unless the beings can be specifically and uniquely characterized, distinguished, and interrelated in the fabric of Being, they cannot *be* at all—for to be is not to be a One only.

The principle of Number determines the relationship between a) Ideas or substance, which are modes of a unitary foundational Consciousness, and b) the activity of the Intelligibles or the gnostic power of the Ideas. [*fig. 42*] ▼ [21, p. 320]

From the essential but undifferenced paradigm of intellection (Dragon's Head ☊) now emanates the Idea of the Good, determined as the whole Intellectual-Principle. The next diagram [*fig. 43*] focuses our attention on the Divine Mind's ordered complexity, a complexity of Ideas and their powers prearranged by the inherent quality of each unific number.

As we attempt to itemize the main constituents of the Intellectual-Principle, the properties of Number prove an invaluable tool, for they reveal to the philosophic seeker the rationale behind the order of Ideas suggested in astrological symbolism.

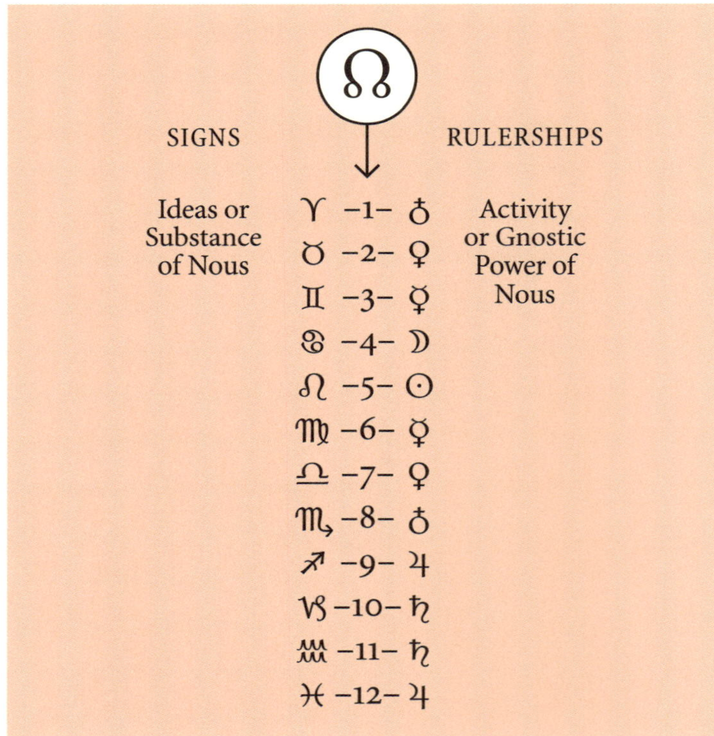

Fig. 42

▼ Editors: When Plotinus says that Number is either the substance of Being or its Activity, he means that it interrelates both.

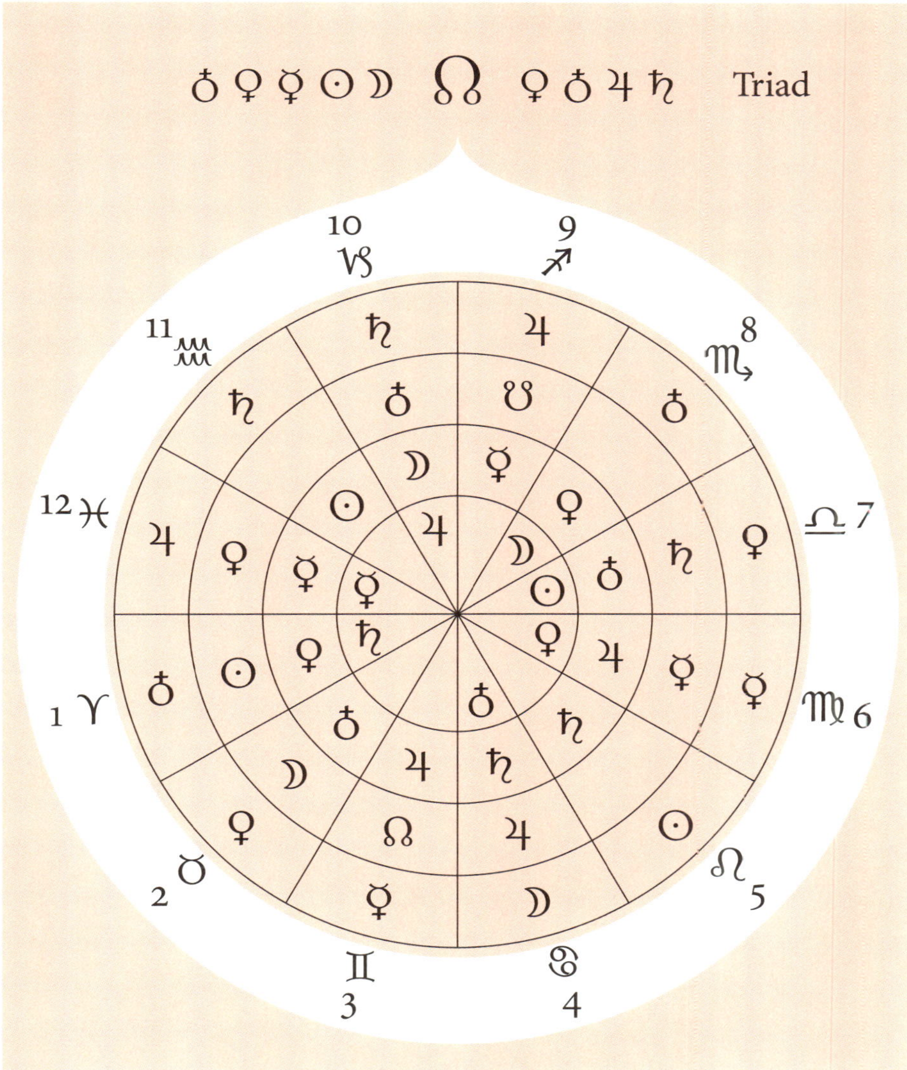

Fig. 43

The Divine numbers are placed around the circumference of the circle to indicate the manner in which they coordinate the unmanifest intelligible substances (signs) with their predominant expressive activities (dignities).

Anthony Damiani,
Notes and Papers, p. O41

The Intellectual-Principle as Revealed through Deific Powers

The undifferenced complexity of the One is made explicit in the interrelation of the various Intelligible Gods, which represent the inseparable manifesting powers of Ideas. "True Intelligence (or Intellect) is Being taken with all its concomitants." (Plotinus, vi.2.18) The different series of Gods are the powers which manifest the divine intelligence of the Intellectual-Principle. [16, p. 314; 17, p. 317] These powers are an imitation and reflex of the One's power, ordered in its very outflow through these various principles, and by the operation of Number.

▼ Editors: The two rulerships (outer ring) of each planet represent Being and Life successively; the exaltations (second ring) represent Intellect; the two detriments (third ring) represent Soul and Nature; the falls (innermost ring) represent Body.

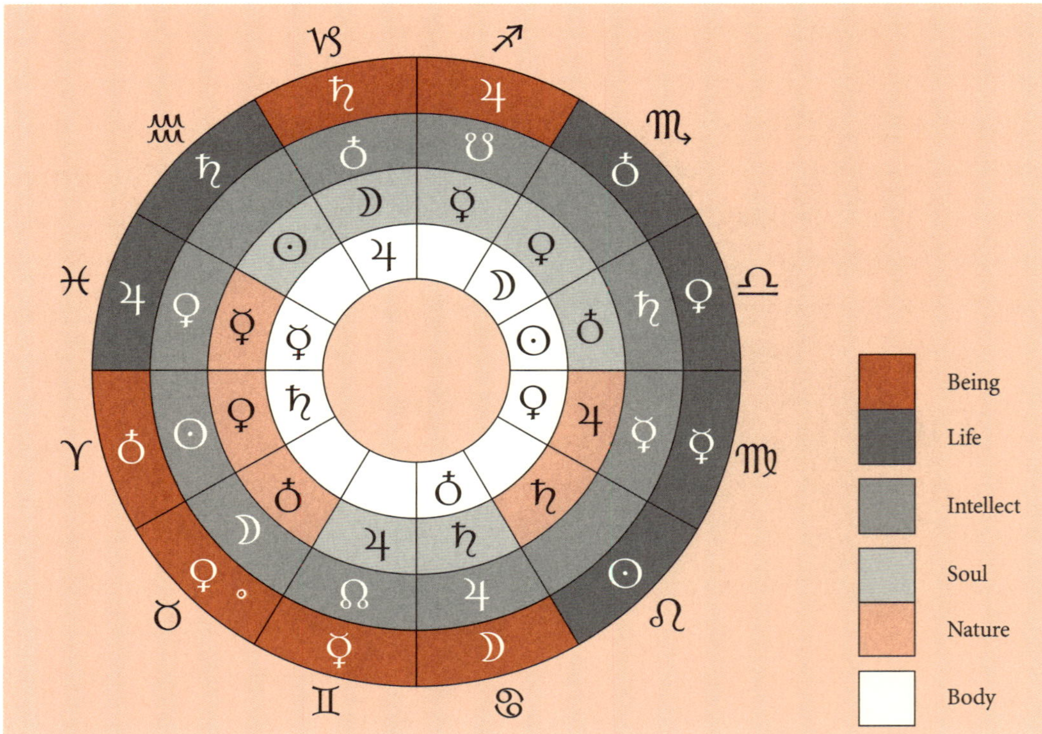

Fig. 44 ▼

. . . from the immense principle of principles, in which all things causally subsist, absorbed in superessential light, and involved in unfathomable depths, a beauteous progeny of principles proceed, all largely partaking of the ineffable, all stamped with the occult characters of deity, all possessing an overflowing fulness of good. From these dazzling summits, these ineffable blossoms, these divine propagations, being, life, intellect, soul, nature, and body depend; *monads* suspended from *unities,* deified natures proceeding from deities. (Thomas Taylor, introduction to *The Six Books of Proclus,* pp. x–xi) [22, p. 325]

The Intellectual-Principle in its unity is also a multiplicity, an atemporal ordered "succession of impressions" of substantial Ideas revealed by a hierarchical succession of deific intelligible powers which Proclus names Being, Life, Intellect, Soul, Nature, and Body. [*fig. 44*]

The super-essential unities are in themselves the principles or paradigms of Being, while their emanations are the series of powers which are the very constituents of Being. Within each series, each intelligible function is associated with a particular substance or Idea, for Being and Intellection are never apart. [*fig. 45*]

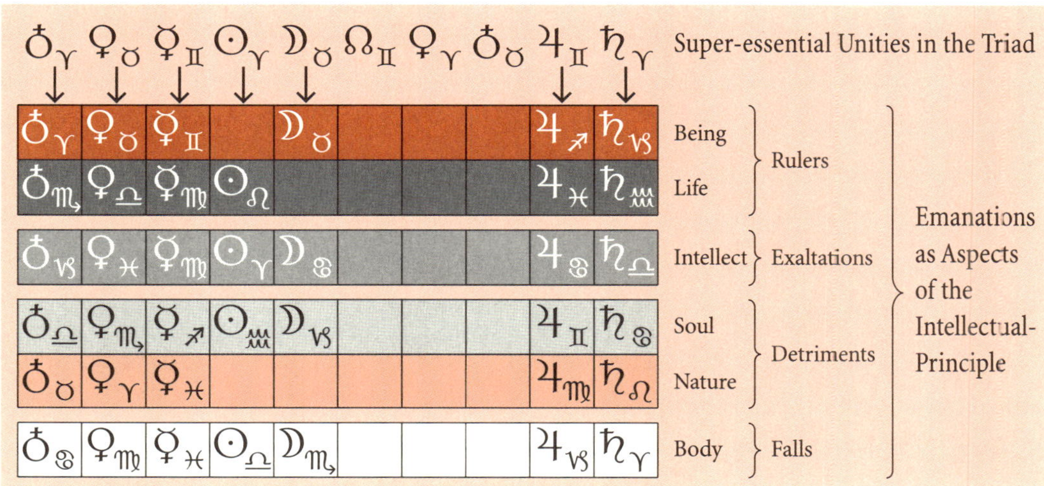

FIG. 45

EDITORS: These diagrams (figs. 44 and 45) illustrate Anthony's remark in the chapter on the One (p. 31) that "the Intellectual-Principle, in its entirety, will exhibit the transcendental paradigms or indivisible aspects of the [One's] super-knowledge from which all the unities within Being are suspended."

EDITORS: In "The One: Symbolization," pp. 49, Anthony first designated the rulerships as *Being* and the exaltations as *Intellect* (Intellectual-Principle). For him, the rulership is the power that establishes the Idea as a part of the whole Intellectual-Principle; the exaltation is the power through which it distinguishes its uniqueness. Here he comments:

> If we apply the analysis to our diagram, we would point out that the foundational consciousness which encircles the twelve houses or Ideas is that unique oneness. The many Ideas, each with their own self-gnostic powers, i.e., the rulers and exaltations, would represent the Intellectual-Principle as a One-Many. These two together constitute a whole of parts, some of which are contrary to others. Furthermore, this element of oneness which gives to each and every one of its parts (Ideas) their Existence and characteristic aseity must be considered together with this whole. . . . The principle of Number will coordinate the specific relations of the Ideas with the appropriate powers or dignities which essentialize (ruler) and singularize (exaltation) the Ideas.
> —Notes and Papers, pp. O184, O86

In unfolding to us the remarkable interpenetration of the real Beings that constitute the intelligible world, Plotinus, as Plato, makes use of the analysis of Being through the five, or sometimes six categories. For Plotinus, these apply to "Real-Being," the Intelligible or Intellectual-Principle, while to the realm of becoming an entirely different analysis is necessary. ▼

▼ Editors: In vi.2.1, Plotinus discusses "Being" in the context of three levels: One, Being, and Becoming. The discussion tries to clarify misconceptions about these levels.

> What is that which, often taken for Being (or the Existent), is in our view Becoming and never really Being?—Note, however, that these concepts are not to be taken as distinguished from each other in the sense of belonging to a genus, Something, divided into Being and Becoming; and we must not suppose that Plato took this view. It would be absurd to assign Being to the same genus as non-Being: this would be to make one genus of Socrates and his portrait. The division here (between what has Being and what is in Becoming) means a definite marking-off, a setting asunder, leading to the assertion that what takes the appearance of Being is not Being and implying that the nature of True Being has been quite misapprehended. Being, we are taught, must have the attribute of eternity, must be so constituted as never to belie its own nature.
>
> This, then, is the Being of which we shall treat, and in our investigation we shall assume that it is not a unity: subsequently we ask leave to say something on the nature of Becoming and on what it is that comes to be, that is, on the nature of the world of Sense.

As in Plato's *Sophist,* to be is not to be a One-Only, but a One-Many. The Manyness is not fortuitous—it is intrinsic to the essence of Intellectual-Principle and is even derived from the One itself, though there is none of this "in" the One. This "logic" of Being is quite different from the logic of A-and-not-A. It is the tool whereby the adept can exercise dialectic—the distinguishing of the Ideas in Real-Being.

> It [Dialectic] is the Method, or Discipline, that brings with it the power of pronouncing with final truth upon the nature and relation of things—what each is, how it differs from others, what common quality all have, to what Kind each belongs and in what rank each stands in its Kind and whether its Being is Real-Being, and how many Beings there are, and how many non-Beings to be distinguished from Beings.
>
> Dialectic treats also of the Good and the not-Good, and of the particulars that fall under each, and of what is the Eternal and what the not-Eternal—and of these, it must be understood, not by seeming-knowledge (opinion) but with authentic science.
>
> All this accomplished, it gives up its touring of the realm of sense and settles down in the Intellectual Cosmos and there plies its own peculiar Act: it has abandoned all the realm of deceit and falsity, and pastures the Soul in the 'Meadows of Truth'... (Plotinus, i.3.4)

The Primals or first categories (genera of Being) are said to be, in *Ennead* v.1.4, Intellectual-Principle, Existence or Isness, Difference, Identity, Motion, and Rest (or Permanence). Of these six, we may take Intellectual-Principle to be the Ideas or forms themselves, and the rest are then the five genera which Plotinus describes most fully in *Ennead* vi.2.

Each Being or Idea in this Intellectual-Principle participates in Real-Being and Intellection, each actuality *is* the Intellectual-Principle. Yet each Idea is distinct from every other, and is different from Intellectual-Principle entire. Universal Sameness and Difference are involved in the constitution of Being. (cf. Plotinus, VI.7.13)

> Similarly, Identity and Difference are not later additions: Being did not grow into plurality; its very unity was a plurality; but plurality implies Difference, and unity-in-plurality involves Identity. (Plotinus, VI.2.15)

All the Ideas, moreover, interpenetrate in the motion of eternity: a "Life—instantaneously entire." (Plotinus, III.7.3) ⊞ This is the motion of the intellective act, each simultaneously involved in the others, and yet each retains its Permanent position in the eternal order of Being, and is a unique form of the whole.

⊞ EDITORS: Anthony often said "instantaneously infinite" when referring to this quote.

> Motion is Being's Act; neither is separated from the other except in thought. . . . Now Motion, thus manifested in conjunction with Being, does not alter Being's nature—unless to complete its essential character . . . at once, then, we are forced to introduce Stability. . . . unalterable condition, unchanging mode, a single rational plan—these are characteristics of the higher sphere. (Plotinus, VI.2.7)

These categories or genera are the criterion by which a real Being is characterized. They are not posterior to Being, but constitute Being. It is not that Being first is, and then takes Motion, and so on, but that these are the very structure of Being.

> Are not Being and Reality . . . distinct; must we not envisage Being as the substance stripped of all else, while Reality is this same thing, Being, accompanied by the others . . . (Plotinus, II.6.1)

The universal fabric, then, is Reality in which Being, Movement, and so on are distinct constituents. The fabric of the levels of reality below the One is the simultaneous operation of the genera. In their interpenetration they retain their functional distinctness, and are at the same time unified and inseparable as the constituents of the Divine Ideation.

We can use the mind thinking thoughts as an analogy to Being and its genera. Analogously to the Ideas we have the actualized thought, and analogously to the genera we have the activity of thinking. The activity of thinking is concomitant with the expression of the thought. Inherent in the very thinking are the motion of time, spreading out the thoughts, the locus and fixity of a place, the difference of one thought from another, and the relative identity of the thought as an object to awareness. All of these must operate as an integral whole, in order for there to be any presentation of a thought. The particular categories required for the thinking act are not important here. What is important is that these activities are simultaneously operative with the manifestation of the thought. The thought is not manifest first and then determined by the categories, but rather it takes its birth in time and space form through the inseparable activity of thinking. ▼

▼ EDITORS: Anthony explored various correlations of the genera with the unities. Earlier in the text (pp. 64, 72–75) he correlated Saturn ♄ with the principle of Difference, for example. Figure 139c, p. 319, unfolds Venus ♀ as Sameness through its dignities which represent particular variations of the principle: beauty, harmony, desire, love, attraction, receptivity. In the following fragment, Anthony suggests using dignities to illustrate genera:

> The One is said to be the transcendental source to the produced genera. The genera are unific principles: the principle of Being is not in being, the principle of Motion or life is above motion, the principle of Difference is above difference. As *principles,* they are inseparable from the super-intelligence of Unity [the Triad]; but as *genera* they are inseparable constituents of the Intellectual-Principle.
>
> Each of the dignities manifests a peculiar function of one of the genera while internally maintaining its own characteristic nature. Similarly, each god may be traced back to its unity in the One through a generic series. There is a complex combined functioning of the genera, interweaving the fabric of Being as a whole, and also combining and interrelating to constitute each distinct level of existence, and each Idea. We see also woven into this array the inner nature of the unities themselves, expressed as various levels of gods manifesting within the Divine Mind. —Notes and Papers, pp. O80A, O82

Each Idea Is a Unique Form of the Whole Intellectual-Principle

We have seen that each of the twelve Idea-principles is determined in its unique character, and assigned a particular hierarchy of Divine intelligible activities according to its particular Number paradigm. We can also now consider each Idea separately as a self-existent principle.

Insofar that this element of unique oneness gives to each and every Idea its characteristic idiom and existence, each stands distinct and cannot be identified with other units within the whole. Plotinus says:

> What, then, is that content [of the Intellectual-Principle]? An Intellectual-Principle and an Intellective Essence, no Idea distinguishable from the Intellectual-Principle, each actually being that Principle. The Intellectual-Principle entire is the total of the Ideas, and each of them is the (entire) Intellectual-Principle in a special form. (v.9.8)

Each and every Idea will reflect and duplicate the organization and structure of the Intellectual-Principle—or again, the Intellectual-Principle exhausts itself in the indefinite variations upon itself. This means that any one Idea can be considered as a unity of Being and situated in the 1st house [of a twelve-Idea circle], and can be unfolded by applying the archetypal meanings revealed by the Intellectual-Principle or the chart of Being.

Insofar that the Ideas which in their totality constitute the Intellectual-Principle are archetypal wholenesses, any one of them can represent the unity of the Intellectual-Principle or

Intellection; and all the Ideas, insofar as they are archetypal wholenesses, can unfold the profound depths of its possibilities. So, in a sense, each Idea is a miniature Intellectual-Principle.

It is not possible to understand one of the Ideas by itself—all the others are implicit in it. The consequence is that each can be taken to be absolute. We also have here an illustration of how a one can be a many and its uniqueness not sundered in the process. Any one of the Ideas could be regarded as an absolute or leading Idea for a particular philosophic tradition and would be that version of the Divine Mind most suited to fill the spiritual needs of the people served by that tradition. ▪

▪ Editors: The 1st house Idea of Self-Existent Brahman, for example, is the central idea for Advaita Vedanta. The Idea of the Five Dhyani Buddhas or Divine Incarnation, however, which T. Subba Row has in the 5th house, is central to Mahayana Buddhism; the hierarchy of principles or Thirty-six Tattvas Unfolding Reality that Row places in the 7th house exemplifies the central teaching of Kashmir Shaivism; the relativity of manifestation in different forms or *lokas,* which Row places in the 11th house, is central to the Idea of Christ-Consciousness on the cross of matter; the Idea of Creative Deific Imagination that Row places in the 8th house is at the core of astrology and some forms of deity yoga. Each of these central Ideas can be understood more deeply through a process of applying all twelve prototype Ideas to it.

The following diagram gives a graphic representation of what Plotinus means when he says that each Idea is the Intellectual-Principle entire, but in a special form. [*fig. 46*]

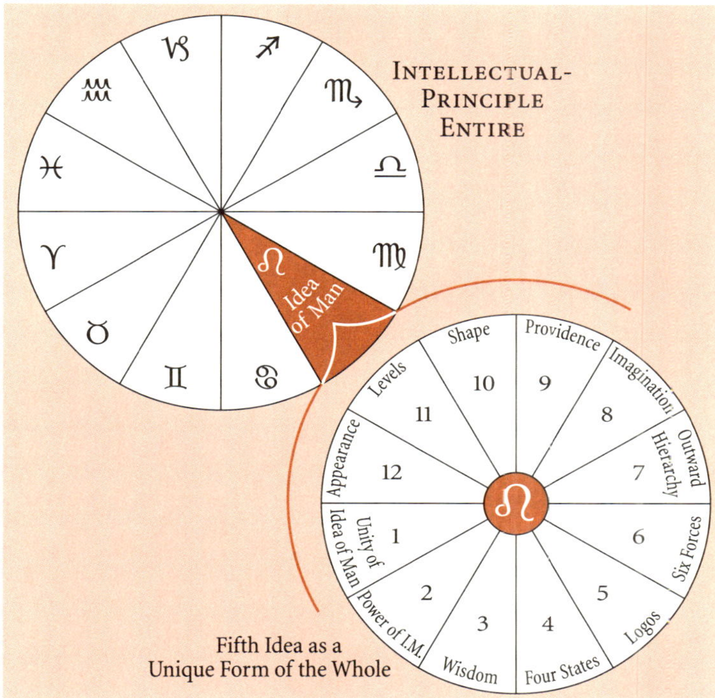

Fig. 46

In other words, we are now applying the same analysis which we have applied to the Intellectual-Principle to one of its parts. Shall we consider the Idea of Man as one of the unities of Being that we shall put in the 1st house?

Every Idea has 1) a unity, 2) a dyadic power, 3) a defined intellectual structure, 4) a fourfold subsistence, and so on. Through this analysis we can obtain a more penetrating insight into a particular Idea, and we are at the same time led to a further understanding of the Ideas in general. ▽

▽ EDITORS: Editors' Extension III, "The Fifth Idea," p. 147, develops the 5th house Idea as a central logos and applies a twelvefold analysis to it.

1ST HOUSE	The unity of the Idea of Man—a permanent possibility
2ND HOUSE	Its active aspect
3RD HOUSE	Its passive aspect
4TH HOUSE	How it will participate in fourfold Being, i.e., four states of consciousness, etc.
5TH HOUSE	Its immanence in every human soul
6TH HOUSE	Its powers
7TH HOUSE	The thirty-six principles that constitute or unfold all its potentiality
8TH HOUSE	The ensuing world-conception of the Idea of Man
9TH HOUSE	The entirety of manifestation as the unity of Universal Man
10TH HOUSE	The shape of the Idea of Man in the Intelligence
11TH HOUSE	The Idea of Man distributed through seven *lokas*
12TH HOUSE	(Manifestation of) the macrocosm in the microcosm

If the Idea of Man is a permanent possibility within the Divine Mind, it is an indivisible principle, self-identical and without parts. This is the key. This Idea has the power to determine itself and this self-determination will introduce the element of multiplicity or otherness. These two constitute a whole of parts which is both the same as and different than the sum of its parts, as we found previously with the Intellectual-Principle. It will include all that is necessary for its manifestation and actualization.

The tremendous and diverse complexity of what the Idea of Man entails is subsumed under its self-determining providential wisdom. It is this Idea of Man—a variation of the Intellectual-Principle itself, or a special version of the Intellectual-Principle imprinted upon or within the unit soul—that determines that soul as human. This is the higher knower within us that Plotinus refers to and which stretches all the way "down" to include the sensible man. We have here the fundamental conception—Mentalism—that the Mind of God or the Intellectual-Principle is present within the macrocosmic soul, the whole of the Intelligible world is within the soul.

Ideas and the System of Nature

After the discussion of the relation of the One to the Intellectual-Principle, we may now examine the relation of Being [Intellectual-Principle] to the sensible appearance. Plotinus speaks of all the three primals being applied to the system of Nature, or universal manifestation.

We would disagree with those who maintain that the appearances of the universe are illusory. For Plotinus says:

> The Intellectual-Principle . . . in its unperturbed serenity has brought the universe into being, by communicating from its own store to Matter: and this gift is the Reason-Form flowing from it. For the Emanation of the Intellectual-Principle is Reason, an emanation unfailing as long as the Intellectual-Principle continues to have place among beings. (III.2.2)

As an image of the intelligible, an image which has taken form in time (the motion of the Soul) the universe is a perpetual image, and an emanation of the authentic realities. The image may have as its base the principle of matter or space or non-Being, and may be filtered and altered through our own individual cognitional apparatus; but in the sense that it images form, our experience is of the Ideas, or the emanation of Ideas. Anything that is so cannot be illusory, although it may be involved with non-Being. How could Real Being ever become anything that is not real? We must therefore be careful to apply "illusory" only to the spurious being of matter in our experience, but not to experience as a reflection of Reason.

> . . . every particular entity is linked to that Divine Being in whose likeness it is made, the divine principle which the Soul contemplated and contained in the act of each creation. (Plotinus, IV.3.11)

Plotinus goes further when he tells us that every Idea is cause in a true sense—it contains within it every form of the particular manifestations of itself: "the Idea itself, unfolded, reveals the cause inherent in it." (VI.7.2) ▼

▼ PLOTINUS, VI.7.2: We see a man—or an eye, if you like—but this is an image or part of an image; what is in that Principle is at once Man and the reason of his being; for There man—or eye—must be, itself, an intellective thing and a cause of its being; it could not exist at all unless it were that cause, whereas here everything partial is separate, and so is the cause of each. In the Intellectual all is at one so that the thing is identical with the cause.

Even here the thing and its cause are often identical—an eclipse furnishes an example—what then is there to prevent other things too being identical with their cause and this cause being the essence of the thing? It must be so; and by this search after the cause the thing's essence is reached, for the essence of a thing is its cause. I am not here saying that the informing Idea is the cause of the thing—though this is true—but that the Idea itself, unfolded, reveals the cause inherent in it.

. . . since nothing There is chance-sprung, and the multiplicity in each comprehends the entire content,

Thus every entity in the empirical world partakes of the Idea to the extent it is able. It is the reason-principle that creates the partial, and through this reason-principle the entity is linked to its prior.

Intellectual-Principle as Revealed to the Sages

We have attempted to penetrate into this incandescent intelligence by using our analysis of the One (the paradigm for any unity) as applied to the primary image of the One—the Intellectual-Principle. We found the Divine Mind to be a unity and also a multiplicity of Ideas, each of which is itself an infinite potentiality to which we can apply the principles of unity, power, and so on. Our ordered investigation and symbolic interpretation of the unfathomable wisdom must of necessity fall short of the Truth which is that realm, but may yet guide us in our reasoned progress toward that holy seat.

The sages who have plumbed the depths of that source of wisdom which lies within their souls and ours if we could attain the vision, have told us, as Plotinus tells, of the blinding beauty of the Ideas, their living Power, and yet the essential mystery which is their core:

To 'live at ease' is There; and to these divine beings verity is mother and nurse, existence and sustenance; all that is not of process but of authentic being they see, and themselves in all: for all is transparent, nothing dark, nothing resistant; every being is lucid to every other, in breadth and depth; light runs through light. And each of them contains all within itself, and at the same time sees all in every other, so that everywhere there is all, and all is all and each all, and infinite the glory. Each of them is great; the small is great; the sun, There, is all the stars; and every star, again, is all the stars and sun. While some one manner of being is dominant in each, all are mirrored in every other.

...There each being is an eternal product of a whole and is at once a whole and an individual manifesting as part but, to the keen vision There, known for the whole it is....

(Perfect wisdom:) for all the Principles of this order, dwelling There, are as it were visible images projected from themselves, so that all becomes 'an object of contemplation to contemplators immeasurably blessed'. The greatness and power of the wisdom There we may know from this, that it embraces all the real Beings, and has made all and all follow it, and yet that it is itself those beings, which sprang into being with it, so that all is one and the essence There is wisdom. (Plotinus, v.8.4)

We affirm again, with Plotinus, that the radiant mind of the World, and of all worlds, is not a blank homogeneity, but is itself the source of the precise wisdom-knowledge of the sage and—according to the Shri-Mala Buddhist text—the abode of Buddha characteristics inconceivable and more numerous than the sands of the Ganges. (Maitreya, *Sublime Science,* pp. 114–115) Yet these Ideas are not discursive conglomerates. This, writes Plotinus in v.8.6, the wise of Egypt indicated by their use of picture symbols to represent truth:

Thus they exhibited the absence of discursiveness in the Intellectual Realm.

For each manifestation of knowledge and wisdom is a distinct image, an object in itself, an immediate unity, not an aggregate of discursive reasoning and detailed willing.

We can at best attempt to point to this wisdom, which is in itself unfathomable to our minds, through our understanding and symbols—unless by the grace and light of Intellect

then the cause of every member can be named; the cause was present from the beginning, inherent, not a cause but a fact of the being; or, rather, cause and manner of being were one. What could an Idea have, as cause, over and above the Intellectual-Principle? It is a thought of that Principle and cannot, at that, be considered as anything but a perfect product. If it is thus perfect we cannot speak of anything in which it is lacking nor cite any reason for such lack. That thing must be present and we can say why. The why is inherent, therefore, in the entity, that is to say in every thought and activity of the Intellectual-Principle. Take for example the Idea of Man; Man entire is found to contribute to it; he is in that Idea in all his fullness including everything that from the beginning belonged to Man. If Man were not complete There, so that there were something to be added to the Idea, that additional must belong to a derivative: but Man exists from eternity and must therefore be complete; the man born is the derivative.

our intuition should be set aglow and in a marvelous sympathy with the inner deity envision those Splendors in the Night.

> For a thing of this kind (says he) cannot be expressed in words, like other disciplines, but by a lasting familiarity and conjunction of life, with this divine object, a bright light on a sudden, as it were leaping from a fire, will illuminate the soul, and there preserve and nourish its splendour. (Plato, Seventh Epistle, in *Shrine of Wisdom,* "Extracts from Thomas Taylor's Dissertation on the Platonic Theory of Ideas," xi, no. 42, p. 176)

Editors' Extension I

What Is an Idea?

An Introduction to a Neoplatonic World-View

To get a feeling for the Neoplatonic understanding of Ideas, it is important to enter sympathetically into a world-view which in many ways inverts contemporary materialistic views. Neoplatonism views the world, or cosmos—including individuals, space, time, matter—as a *living Idea,* a thought within a universal intelligence. Intelligence in this view is not a secretion or by-product of matter. Cosmos and experience are within Mind and for that reason are meaningful. Yet the world cannot express the entirety of infinite wisdom. The world both reveals and conceals something of the intelligence which thinks it. Every moment of experience reveals intelligence as thought, feeling, act, and awareness, while these very forms also limit the infinity of intelligence.

For Plotinus, Ideas are real and powerful, infinitely beyond mere concepts. They are more like myth as a living presence—in the sense made popular by Joseph Campbell and elaborated by Mircea Eliade in his introduction to *Myths, Dreams and Mysteries.* Like a myth, which begins "once upon a time," an Idea unfolds as a mystery and cannot be completely determined. An Idea encompasses various descriptions and understandings as well as the particular events so described and understood. Like myths, Ideas integrate knowing, feeling, and activity and are not static, but include many levels. Although Ideas have meaning for human minds, they are not limited to the human mind. We "participate in" the Ideas. In some sense, the Ideas have us, and not the reverse.

The Neoplatonic view of Ideas resurfaced strongly in the Italian Renaissance. In the 1900s, the notion of Ideas took root in a psychological context in the work of C.G. Jung and his followers, who use the word "archetype" to describe underlying currents of intelligence on which the human psyche depends. Whereas for Jung they are intimately connected with the unconscious, for Neoplatonism the Ideas have metaphysical implications as well: Human beings have an inner reality continuous with the intelligence that manifests the cosmos and is the source of our limited existence. The cosmos is a sympathetic system of living intelligences.

Neoplatonism distinguishes the mind which actively enlivens the world every moment (World-Soul) from the unchanging formless Intelligence upon which this activity is based (Nous, or Divine Mind, or Intellectual-Principle). The latter is too infinite to fit into the finite human conception, yet it is present in every moment, closer than our own thoughts. This Divine Mind expresses absolute reality, the ultimate mystery from which even Being and Intelligence emanate.

Plotinian Context for Ideas in the Divine Mind

For Plotinus, Ideas are living intelligences in the Divine Mind. They have unending omnipresent power and radiant beauty. They are also present at all levels of manifestation, providing meaning in the flux of appearances, and maintaining a continuity between reality and its appearances.

> A real Idea is a unific, dynamic, primary Principle, an origin or fountain from which flows a succession of secondary principles, types and forms, which ultimately find their outermost expression in the material world. An Idea is a universal, eternal reality, an archetypal wholeness, imparting life to all that springs from and depends upon it, by itself formless and undifferentiated, established in unchanging perfection and fullness, unindigent of that which it produces. (Foreword by the editors of the *Shrine of Wisdom* to Thomas Taylor's "Dissertation on the Platonic Theory of Ideas," xi, no. 41, p. 137)

Eidos, the Greek word for Idea, means "the form or figure, the characteristic look that a thing presents to vision . . ." (Findlay, pp. 29–30) From this, we get a sense that each Idea is a way in which Intelligence appears to the vision of the Intellectual-Principle. Ideas are traces of the Good in the Divine Mind, its vision of the ineffable source. "Intellectual-Principle springs from the Supreme, and with it the beauty of the Ideas; at once all these, Life, Intellectual-Principle, Idea, must inevitably have goodness." (Plotinus, VI.7.18) If the Divine Mind is like light, Ideas are the different frequencies or colors of that light. If the Divine Mind is like a diamond, the Ideas are its facets.

According to Plotinus, the Good or the One is the source of the Ideas and their knowledge, but it does not contain or directly produce the Ideas which are the Intellectual-Principle:

> The Good comes to it not as it was in its primal state but in accord with the condition of the Intellectual-Principle. The Good is the source from which the objects of contemplation come to be seen in the Intellectual-Principle; Intellectual-Principle has produced them by its vision of the Good. . . . the Good bestows what itself does not possess. (VI.7.15)

Two of Plato's dialogues provide important background for the meaning of Ideas: the *Parmenides* and the *Republic.*

Plato's *Parmenides:* How Ideas Are Participable

Anthony develops a whole logic of the Ideas and their participation based on his analysis of Plato's *Parmenides.* Here we summarize some of the most important features of his interpretation of this dialogue, as recorded in his unpublished essay "The Eight Hypotheses of the Parmenides."

Plato wants to show that any attempt to understand the problem of participation from the standpoint of empirical logic will not work. The problem of how anything "participates in" reality is directly relevant to the human condition. If we are only a flux of appearance as Heraclitus had argued, then there is no meaning to existence. If there is only the Absolute and it is imparticipable as Parmenides had argued, then individual existence is meaningless. Participation in Ideas preserves meaning from destruction by either of these extreme positions.

In *Parmenides,* Socrates realizes that Zeno's talk showing that manyness is impossible is the same as Parmenides' discourse reducing all to one. Through Socrates' attempted solution in the dialogue, the presence of Ideas would preserve flux from meaninglessness, and provide a multitude of authentic essences necessary for a theory of participation. But Socrates' original views are inadequate, and Parmenides provokes him to investigate possible ways that a multiplicity of Ideas could form a medium between the theory of absolute flux and the theory of absolute unity.

The examples of whole and part (body), concept (mentality) and similitude, or image and original do not suffice for Plato to explain participation in reality. Further, as the dialogue continues, without participation we are left without knowledge of the gods and they are left without knowledge of us or efficacy in the world. We should note first that each of these partial views has some validity at a specific level, and second, that the context of unity in the last part of the dialogue explains satisfactorily how we can participate in reality.

So Plato next has Parmenides set forth the teaching about the logic of the One in order to explain participation. If the One is, what follows for itself, and for the others? If the One is not, what follows for itself and for the others? In Anthony's interpretation, all these hypotheses are meant to apply simultaneously: They do not cancel each other out. They can each and all be applied to the One to provide an inclusive rather than exclusive view of the One. There is a One, and it both includes and does not include the many. Further, this reasoning can be applied to anything which is a one.

The connection between the two parts of the dialogue is crucial to solving the problem of participation in reality. Unless we understand what it means for an Idea to be a one, which is beyond the dualism of Being and non-Being, we cannot understand how Ideas can remain what they are while giving rise to multiplicity; how they can be present in all their instances and be participated without losing their integrity.

J.N. Findlay's descriptions of Ideas in *Plato: The Written and Unwritten Doctrines* help us see how the logic of the One structures the participation in Ideas. If we apply the notion of Unity to Ideas, then each Idea, he says, can contract into a core in which all knowing and naming cease, and can also expand to indefinite expressions. (p. 247) Therefore, no instances can exhaust the mystery of an Idea, and at the same time the Ideas convey some presence of the infinite. An Idea is the central essence that makes knowledge possible. An Idea is all we ever know: The object of knowledge *is* our knowledge. Therefore, amidst the changing flux of becoming, Ideas are the stable content of our experience. Just as each authentic quality of our mind, such as compassion, pervades the entire mind yet retains its unique character, an Idea pervades everything.

Plato's *Republic:* The Analogy of the Sun and the Divided Line

Directly related to the logic of participation in *Parmenides* is Plato's theory of knowledge in his *Republic.* An Idea-intelligence can be participated at different levels coordinate to the level or faculty of knowledge which we can actualize, while retaining its essential integrity. Every grade of life is a form of thought (see Plotinus III.8.8). The deeper one can sink into the heart of Being, the deeper will be one's participation in the Ideas, and the more perfect will be his or her unity with the Idea.

Anthony's epistemological concerns were about knowledge at all levels—noetic knowledge by itself is incomplete, and so is only sensible knowledge. In the following class transcript, Anthony discusses Plato's view of this problem:

> How do I know what I know? That could include delusive knowledge as well as authentic knowledge. If you remember, Plato made this very clear. Plato said, If we take the position of Parmenides, the only knowledge is that in the Nous; and if we take the position of Heraclitus, the only knowledge is that of a barking dog—there is only sensation, how can you speak of

knowledge? He says, Now both of these men are right, but what about that middle domain, in-between, where you have human opinion? . . . Plato divided a line into four and said that there are four levels of knowledge. So knowledge as he meant it was a very broad term.

When some schools of thought use the term "knowledge," they only mean divine knowledge. So then everything else becomes ignorance. Sure, that's great, but what about the first atomic bomb explosion—that was delusive?! So we keep organizing our mind with the scope of the meaning of the word "knowledge" and apply it to what is proper.

You have a sensation and that is an undeniable fact. It may not be authentic knowledge, but it is some kind of knowing. You don't put your finger on the hot stove because your senses will immediately withdraw it. That is knowledge, very primitive, but it's there. Then you have an opinion as to why that wood stove is hot—that's knowledge, too. You know if you take the wood out it will get cold. It is not ultimate intellection, and it is not pure consciousness, that is another grade of knowledge. So Plotinus says this is what the felicity of the sage consists of: intellection, rational discursion, and sensible perception—all three. [cf. I.4.10] Now insofar that the sage has a pure perception, that is, he seizes the intelligible directly . . . he has that, and that is always the inner state of his being. . . . He can add to that sensible knowledge or not, as suits the occasion. (6/17/83)

▼ PLATO, *REPUBLIC,* V. 1, PP. 345-352 (T. Taylor trans.): Conceive then, said I, that this is what I called the offspring of *the good*, which *the good* generates, analogous to itself; and that what this is in the intelligible place, with respect to intellect, and the objects of intellect, that the sun is in the visible place with respect to light and visible things. . . . That therefore which imparts truth to what is known, and dispenses the power to him who knows, you may call the idea of *the good,* being the cause of science and truth, as being known through intellect. And as both these two, knowledge and truth, are so beautiful, when you think that *the good* is something different, and still more beautiful than these, you think aright.

Understand then, said I, that we say these are two; and that the one reigns over the intelligible genus and place, and the other over the visible. . . .

As if you then took a line, cut into two unequal parts, and cut over again each section according to the same ratio, both that of the visible species, and that of the intelligible, you will then have perspicuity and obscurity placed by each other.

Anthony diagrammed the coordinate levels of knower and known using the schema of the Divided Line in *Republic* Book VI. The Divided Line illustrates a twofold source of knowledge, which Plato calls intelligible and manifest (visible). ▼ We have access to the Ideas by direct participation in the intelligible, and through our continuing relation to the system of Nature. Each of these sections is divided into a paradigm and image section, giving us four levels of knower and known. [*fig. 47*]

Plato's analogy of the intelligible Sun and visible Sun suggests an intimate connection between the levels of knowledge and the cosmic framework of epistemology. Anthony makes this correlation explicit by drawing a parallel between the fourfold nature of the Divided Line and the nature of the cosmos represented by four levels of the Sun as a cosmic logos.

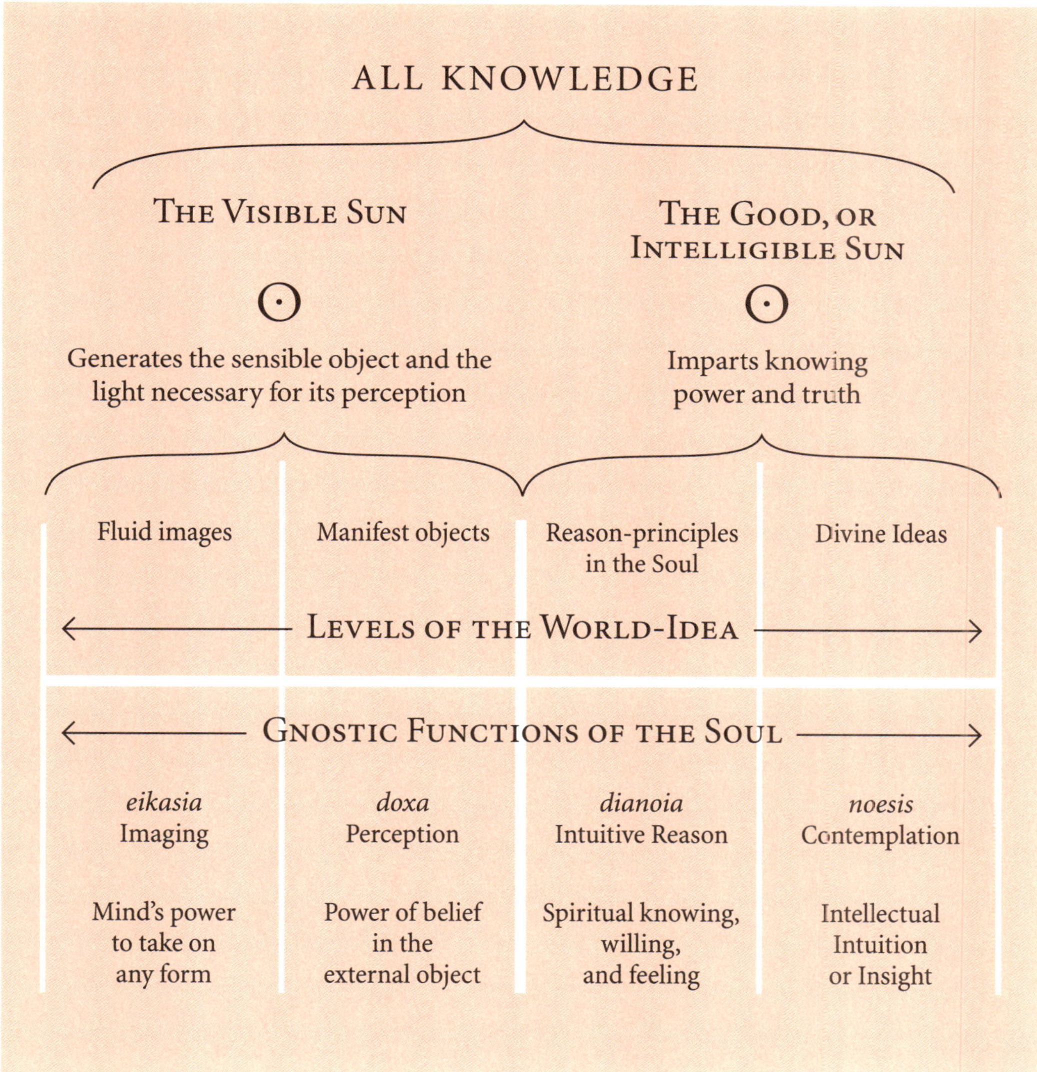

ALL KNOWLEDGE
The Visible Sun
The Good, or Intelligible Sun
Generates the sensible object and the light necessary for its perception
Imparts knowing power and truth
Fluid images
Manifest objects
Reason-principles in the Soul
Divine Ideas
Levels of the World-Idea
Gnostic Functions of the Soul
eikasia
Imaging
doxa
Perception
dianoia
Intuitive Reason
noesis
Contemplation
Mind's power to take on any form
Power of belief in the external object
Spiritual knowing, willing, and feeling
Intellectual Intuition or Insight

Fig. 47

In a fragment from his notes, Anthony gives one possible way to interpret the Divided Line cosmologically:

> Knowledge [A] has its origin at the highest level of the celestial, where this universal consciousness [B] formulates its World-Idea through its absolute Knowing, Willing, and Feeling. This [A & B] is the indivisible whole that must ultimately become reflected in the sensible world (an image). But it is the *reasoning phase of the Soul* which transmits this, moment by moment, and spreads it out in the realm of the Earth's memories which becomes our sensible world. (Notes and Papers, p. S136)

The Stars/Zodiac represent Intelligence. The motion of the planets, trans-Saturnians (B) and Chaldeans (C), translates the Ideas (A) into individual experience (D), reconciling individual understanding with the Ideas. The Earth represents the continual embodiment of Ideas in our experience, and the assimilation of Ideas by the individual. [*fig. 48*]

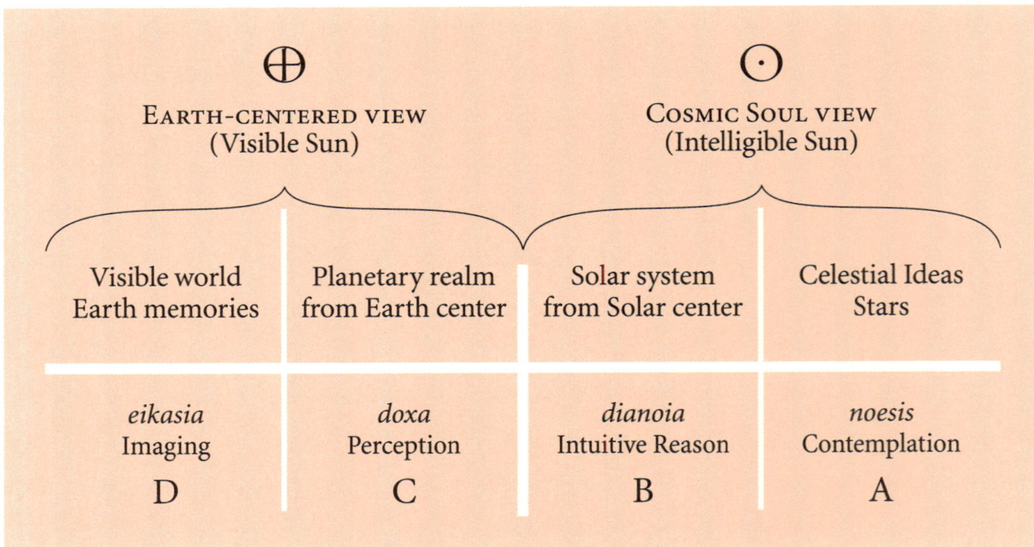

FIG. 48

In each section of the line, there is a correspondence between a soul function and a level of the World-Idea. The proportions in the Divided Line indicate the harmony of the different functions of knowing. The self-duplicating ratios of the upper to lower section, and of the whole to the parts, represent the reproduction of the cosmic wisdom within the individual, as well as the deepening of our knowledge into the cosmos. [*fig. 49*] ▽

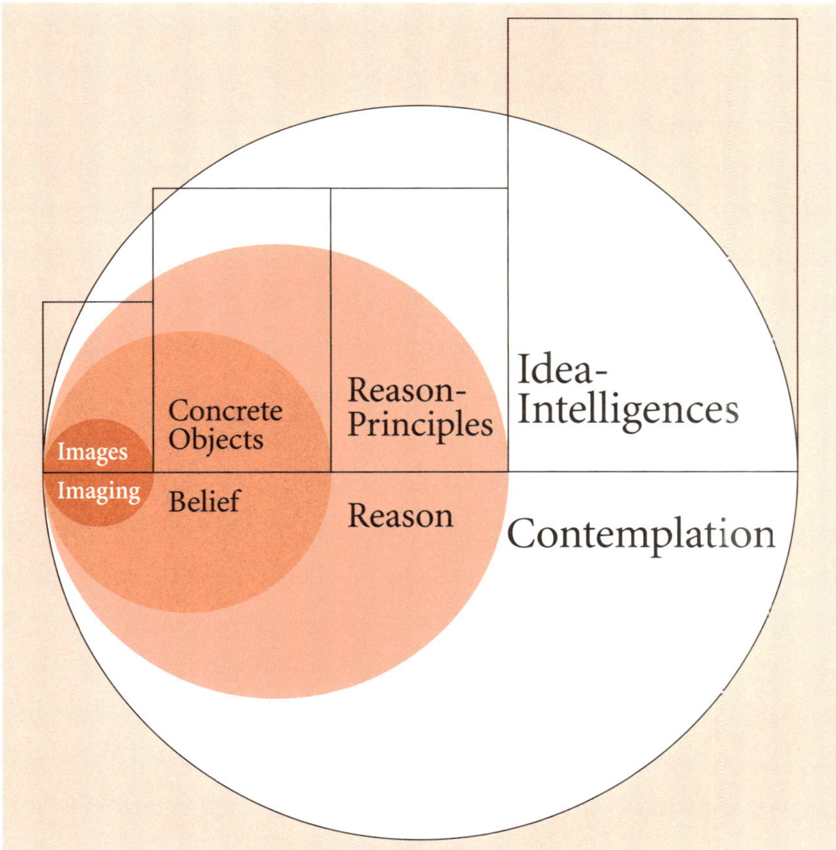

FIG. 49

▽ EDITORS: Plato suggests in his analogy that each section of the line is cut into two unequal portions, with the middle two sections being the same length, as in figure 49. The Platonic esoteric tradition assigns to these divisions the specific self-similar ratio called the golden ratio, derived from the relation found in the pentagon or five-pointed star.

The key point here is not the specific correlation of levels of the cosmos to the Divided Line but the big idea that we can distinguish levels of knowledge within the cosmic framework. There are many possible correlations.

> When you reach the limit of the conception of the Earth's mind, that's a boundary, and it's the beginning of the powers of the Sun soul. You reach the limit of that and then you would be going into the inerratic sphere. What the Platonists are saying is that you have an Idea, and when you reach the limit of that Idea you're at the boundary of the beginning of the next Idea which is going to include the lesser Idea, and this is a process that goes on and on so that the greater Ideas include the lesser Ideas. . . . Evidently the ancients thought it was an absolute necessity, a prerequisite, into order to explain how does knowledge arise. What is knowledge that they have to have all this as a background? Once you begin to grasp the enormity of the conception they work with, the categories of thought are going to seem like a petal thrown into the ocean. (Anthony Damiani, transcript, 1/6/84)

For Anthony, the teaching about Ideas finds its fulfillment in astrology, and the experience of Ideas at various levels is illustrated in the context of astrological analysis.

> The purpose of life, according to Plotinus, is the acquisition of true knowledge and the individual's life in the sensible world is for that purpose. The astrological chart can and does reveal the specific ideas being manifested by that life and assimilated by the individual mind towards that goal. This is the fundamental spiritual nature of astrology. (*Astronoesis,* p. 282)

The corollary to this is that knowledge is not abstract only, but "must be lived." (*Ibid.*)

These statements echo tractates in Plotinus. *Ennead* III.8.6 says that the reason-principle must be made one with the soul of the aspirant. *Ennead* III.5, on love, describes the individual soul as having a lack (*penia,* poverty) which is filled by abundance (*poros*), the reason-principles. Experience is the individual soul or mind transforming itself into the form of the world, touching the cosmic reason-principles and, as a consequence, gradually conforming to and becoming the wisdom of the World-Idea. (This theme is explicated in Anthony's book *Living Wisdom.*) Ultimately, the soul learns about the Intellectual-Principle from the cosmos, and learns to know itself as the Intellectual-Principle knows it.

Editors' Extension II

The Twelve Idea-Principles

In his Invocatory Prelude to *Astronoesis,* Anthony speaks of the zodiac as the "Nous in the heavens." For him, the twelve signs were an image of the Ideas in the Divine Mind. Plotinus provides the primary meaning of these Ideas as "successive impressions" (v.3.11) by which Intellectual-Principle knows the One. Each of the sectors of the circle represents an Idea-principle—an "impression" of unity. As a whole, these sectors give a complete picture of the unfolding of reality.

Plotinus states that number is intrinsic to intellection and Ideas. Following his suggestion, we use the numbers to help organize our understanding of the meanings of the Ideas. For example, intellect first grasps the One as unity, and in that act its own being is determined as a self-existent: This is the "Idea" of the Good. Next, there is an impression of the power of the One, determined as the beauty of the Ideas, etc.

Whereas sequencing the Ideas suggests a hierarchy, placing them in a circle suggests that there is also a simultaneous pattern of knowledge. Although drawn as distinct sections, these Ideas are not to be viewed as separate pieces, but as simultaneous facets of Intellection. No Idea can be separated from the others, any more than Love can reside in a single piece of one's mind. Yet, even though love and justice, for example, pervade the entire mind, they do not get confused with each other. Or we may use a musical analogy and imagine that the Ideas are like the many notes of a musical chord interpenetrating and permeating all of auditory space, and yet each note remains distinct.

Anthony explored philosophic meanings behind the individual signs for many years, trying different systems and arrangements. Ultimately, he found an inspiration in T. Subba Row's essay, "The Twelve Signs of the Zodiac" (pp. 3–19 in T. Subba Row, *Esoteric Writings*).

Row was a key figure in the original Theosophic movement in India in the late nineteenth century. His intuition about the zodiac is elaborated in this short article, which elicits metaphysical meanings from the Sanskrit names of the signs. [*fig. 50*]

In this extension, we sketch philosophic meanings for Anthony's interpretations of these twelve divisions. This is a difficult task, for Anthony expounded many levels of interpretation and instantiation of these Ideas. He pointed out, for example, correspondences with such diverse systems as the twelvefold *nidana* chain of Buddhism, the alchemical

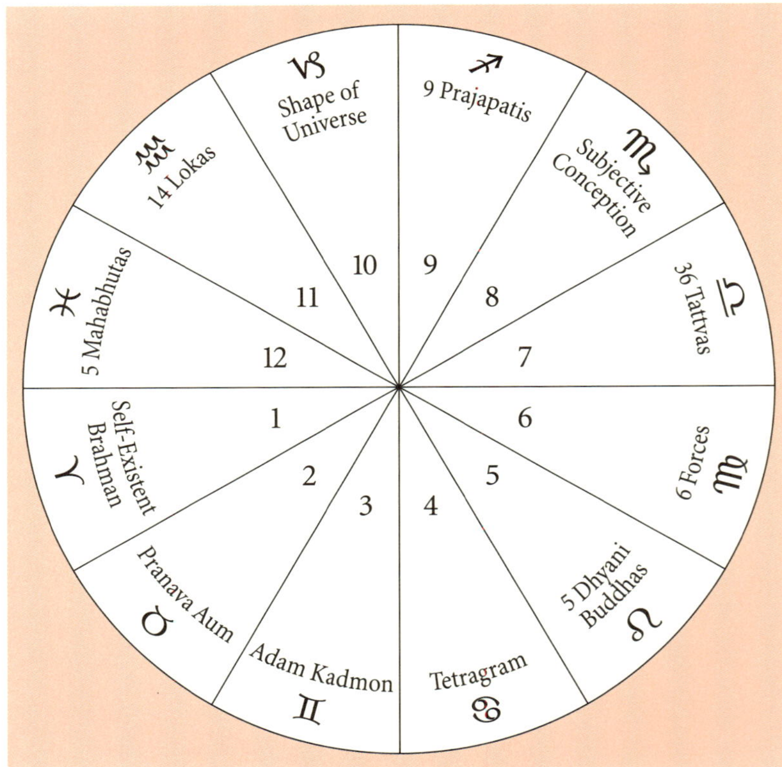

Fig. 50: Chart of Intellectual-Principle, based on T. Subba Row

processes, and the Greek Liberated Gods. Figure 51 shows one of Anthony's most common correlations of Ideas and sections of the chart.

This summary applies principles from Anthony's "interlude" on Number (pp. 86-95) to the inspiration Row provided in that seminal essay. For each sign/house/Idea-principle we feature a key phrase adopted or adapted from Row's essay, and show how his meanings relate to the Idea-intelligences as developed in Neoplatonism. The twelve intelligences described here form a philosophic paradigm for the zodiac.

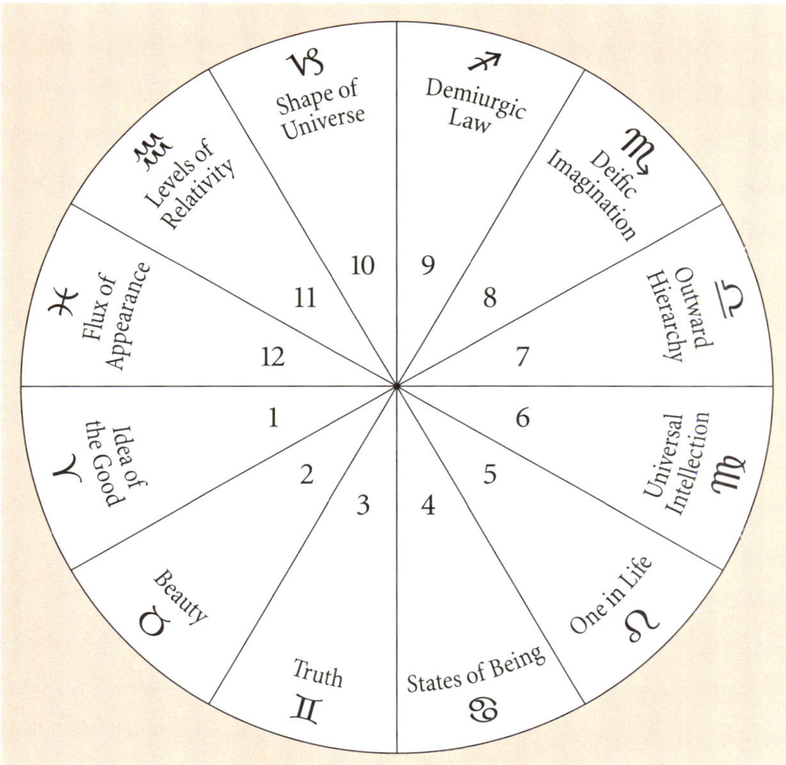

Fig. 51: A chart of Ideas in the Intellectual-Principle

♈

Self-Existent Brahman
Self-Existent Reality

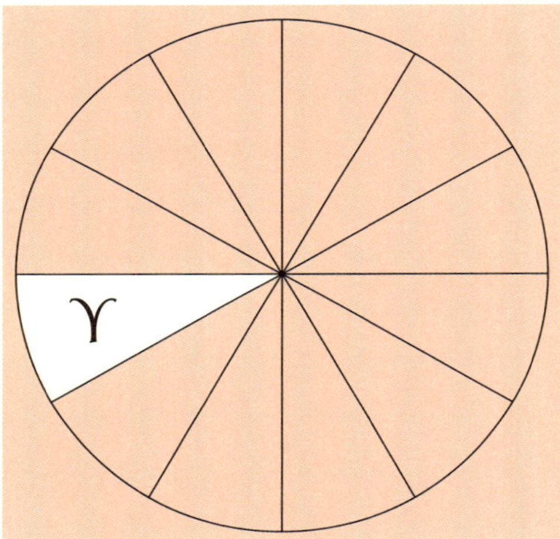

Fig. 52: *Mesham*—Aries

T.S. Row: Aja *literally means that which has no birth and is applied to the Eternal Brahman in certain portions of the Upanishads....* Parabrahman *[is] the self-existent, eternal, self-sufficient cause of all.*

Brahman is the Vedic term for Reality. It is described in Gaudapada's commentary on the Mandukya Upanishad:

> Brahman is birthless, sleepless, dreamless, nameless, and formless. It is ever effulgent and omniscient. (*The Upanishads,* v. ii, p. 304)

Paradoxically, Brahman—itself birthless (*aja*)—is also described as "effulgent." (The Sanskrit root *Brih* means "that from which all things grow.") Our studies in Plotinus find the same paradox in the nature of the One: It is absolute self-sufficient self-existent perfection and it overflows. (cf. v.2.1) Parabrahman means either the highest (*para*) Being or that which is beyond (*para*) Being. Anthony explores a similar distinction in relation to the One's super-knowledge: Is the One ultimate knowledge, or is it rather beyond knowledge?

In his presentation of the One, Anthony uses the 1st house to represent this ultimate nature of Reality in and of itself. When we view the mandala as the Intellectual-Principle, the 1st house indicates that each real Being has a central unity, a core of self-existence.

> It is in virtue of unity that beings are beings. . . . What could exist at all except as one thing? Deprived of unity, a thing ceases to be what it is called . . . (Plotinus, vi.9.1)

At the heart of any being is transcendent and unfathomable mystery. Insofar as the primal good of all life is this participation in Reality, we may call the first Idea, with Plato, the "Idea of the Good."

♉

Pranava Aum
Infinite Power and Beauty

Pranava means word or syllable. *Aum* placed in this house is the unmanifest divine Word in the Absolute before it is spoken (emanated), prefiguring there everything that will become articulated in emanation.

> *Aum,* the word, is all this. . . . As a thing is known through its name so the highest Brahman is known through *Aum* alone. Therefore the highest Brahman is verily *Aum.* (Shankara's commentary on verse I.1, *Mandukya Upanishad,* p. 9)

Swami Nikhilananda comments on this same verse: "The substratum of all sounds is AUM, and the substratum of phenomena is Brahman." (*The Upanishads,* v. II, p. 224)

Anthony refers to this unmanifest divine power by quoting from Eckhart: "God never spoke but one word, and that is still unspoken." (p. 57) The hidden Word is the power of speech or emanation. If we think of sound in relation to vibration and energy we begin to get a feeling for the Word as the power of Reality to "speak itself." [23, p. 325]

The radiant power of the Ideas in the Divine Mind is a trace of the fathomless power of the One. Plato calls this the Idea of the Beautiful. In Neoplatonism, it is the power that shines forth Intelligence.

> There is infinity in Intellectual-Principle. . . . All is within it, all the powers and intellections . . . (Plotinus, VI.7.14)

> The power in that other world has merely Being and Beauty of Being. Beauty without Being could not be, nor Being voided of Beauty. . . . Being is desirable because it is identical with Beauty . . . (Plotinus, V.8.9)

This Beauty, the loveliness of the Divine, is the power that attracts beings to the One.

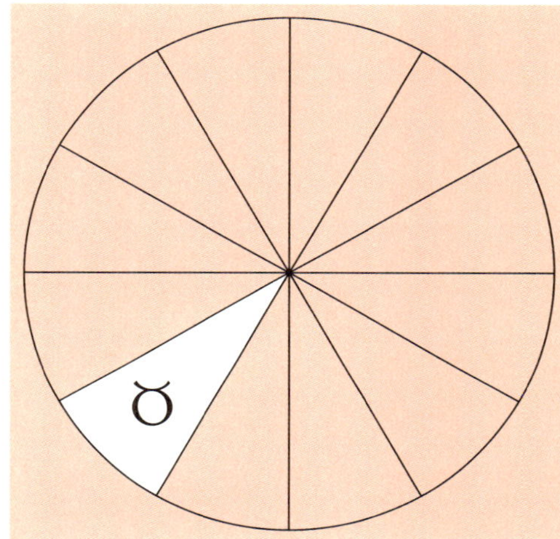

FIG. 53: *RISHABHAM*—TAURUS

T.S. Row: This word is used in several places in the Upanishads and the Veda to mean Pranava *(Aum).*

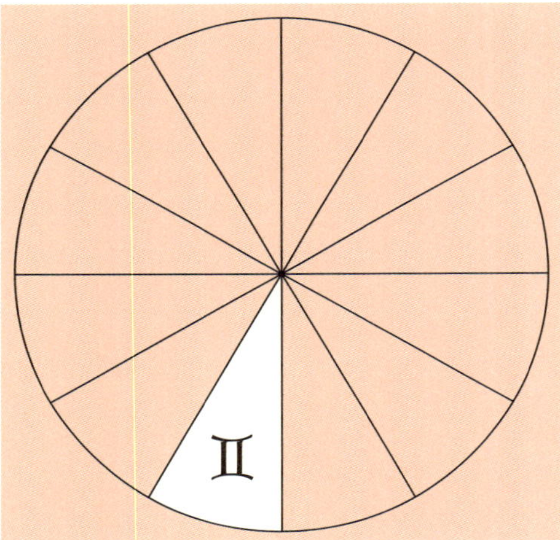

Fig. 54: *Mithunam*—Gemini

T.S. Row: This sign is intended to represent the first androgyne . . . the bi-sexual Sephira—Adam Kadmon.

II

The Bi-Sexual Sephira: Adam Kadmon
The Paradox of Mind

Adam Kadmon, the primordial androgynous Adam, is a term used in the Kabbalistic system of thought to represent the first of the ten unmanifest *sephiroth,* also called the "principle" or the "head," *kether.* This term refers to the nature of Mind, and the 3rd house provides a view of the Absolute as infinite Mind.

Mithunam, the Sanskrit word for Gemini, means union or conjunction, and *androgyne* comes from the root words *andro* (man) and *gynae* (woman), and indicates a union of opposites. (Tantric Buddhism expresses this conjunction as the union of wisdom and compassion, or emptiness and bliss.) Each unity in the super-intelligence of the One is conceived as a whole sphere (*sephira*) or totality uniting complementarities, a complete union of wisdom and power, a transcendental identity which contains a potential multiplicity. This androgynous nature of reality is called Symmetry in the Orphic writings, and the Tao in ancient China.

Interpreting this sign in the context of Intellectual-Principle means that any real Being participates in androgynous wisdom, reflecting the super-intelligence of the One as a dual-unity.

> The true Wisdom, then (found to be identical with the Intellectual-Principle), is Real Being; and Real Being is Wisdom . . . (Plotinus, v.8.5)

The principle of Thought is paradoxical—literally beyond (*para*) the grasp of the finite one-sided operation of empirical opinion (*doxa*)—for it is a knowing-being, both transcendent and immanent. In its role as measuring out reality, thought must both conceal and reveal reality.

♋

The Sacred Tetragram
Matrix of Four States of Being

The Mandukya Upanishad is the primary source for the idea of four states of consciousness:

> All this is verily Brahman. This *Atman* [Self] is Brahman. This *Atman* [Self] has four quarters. (*Mandukya Upanishad,* verse 1.2, p. 11)

The 4th Idea is the overflow of the One as distinct states of Being. To be is not to be a one only, but to reflect the One. Emanation does not mean actually leaving the One spatially—it involves a making distinct or measuring out of the infinite. Thus Row says, "Pranava resolved into four separate entities." We represent this differentiation of fourfold reality, described by Plotinus in v.1.8, 10, by the four quadrants and the four rings of the circle.

The vast intelligence pouring out from the One, reflecting on itself, resolves itself into a multiplicity by means of which it unfolds itself. This vast matrix of possibilities stratified into distinct "states" is intrinsic to the Platonic view of Being, which nourishes and sustains the multiplicity of beings. The 4th house begins the second quadrant, the quadrant of Intellectual-Principle. All is contained here in a universal way.

> Intellectual-Principle . . . is manifested as this universe of Being. It stands over the Beings not as itself requiring base, but that it may serve them as base through its vision of the Form of the Firsts, the Formless Form. . . .
>
> Thus Intellectual-Principle is a vestige of the Supreme; but . . . the vestige is a Form going out into extension . . . (Plotinus, vi.7.17)

Each being or Idea has a substantial existence which participates in all four levels of reality (unity, being, soul, body.) Substance means that which "stands under." In this context, substance is intelligence, or consciousness that supports all beings.

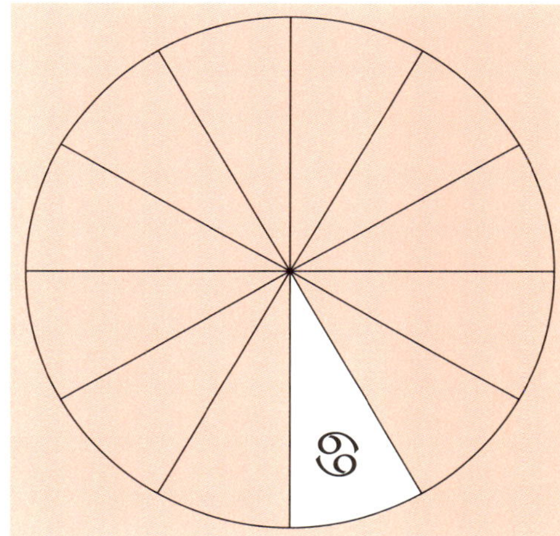

Fig. 55: *Karkatakam*—Cancer

T.S. Row: This sign . . . represent[s] the sacred Tetragram *. . . the Pranava resolved into four separate entities corresponding to its* four Matras *[parts];* the four Avasthas *[states of consciousness]. . . . This sign completes the first quaternary.*

♌

The Five Dhyani Buddhas Representing Jivatma
Life as the Immanence of the One

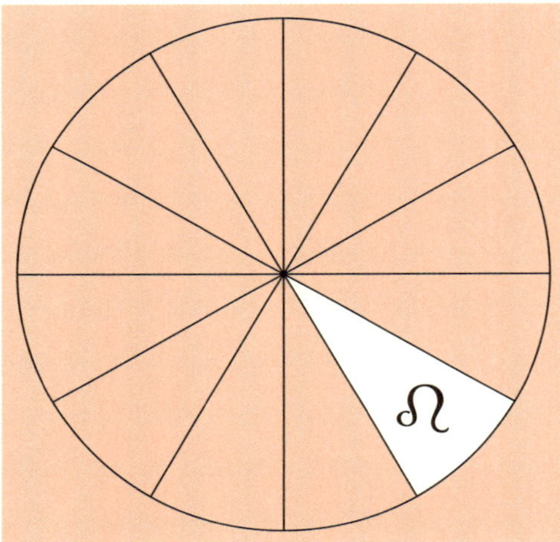

Fig. 56: *Simham*—Leo

T.S. Row: . . . The sign is intended to represent the five Brahmas . . . the five Buddhas. *The second synonym shows it to be . . . the Jivatma or Pratyagatma. . . . The 5th sign is the . . . son's house, according to the rules of Hindu Astrology. . . . I may also add that it represents the real Christ, the anointed pure spirit.*

The five Buddhas (or Dhyani Buddhas as they are called by A. Govinda, H.P. Blavatsky, and other writers on esoteric Buddhism) are various aspects of divine intelligence becoming incarnate in individual souls, called *jivatma.* While the 1st house is *aja,* unborn, in the 5th house we have "birth." *Jivatma* is a self (*atman*) which takes birth (*jiva*). Birth here does not mean physical birth, but rather that reality grants to each individual its authentic and eternal being. In *Notes on the Bhagavad Gita* (pp. 19–20), Row says that an infinite number of logoi pre-exist in the bosom of Parabrahman (the Absolute, or the One). This principle of logos in life is what Row meant by the Christ. In both Western and Hindu astrology, this house is not only the son's house, but it is also the Sun's house: the Solar Logos (the Sun has its rulership in Leo). Each soul is a logos (Plotinus, IV.3.5), a ray of the Divine Sun.

Interpreted as a special form of the whole Intellectual-Principle, the 5th Idea means that there can be unique instances consisting of the entire intelligence of the Ideas, and that each authentic soul is a unique self-realizing intelligence. Out of the matrix of Being (4th Idea) come unique beings, ideas, souls—each a form of intelligence.

> Anything coming from The Good carries the image and type belonging to that original or deriving from it, as anything going back to warmth or sweetness carries the memory of those originals . . . (Plotinus, VI.7.18)

Anthony places the Absolute Soul, the Divine Overself (*Atman*) in this 5th Idea. Soul is the Life of the Nous, and the Nous is the knower within the soul. Each of us is, at heart, a spark of real awareness. This divine birthright is a guarantee of our potential self-realization.

A more detailed discussion of the 5th Idea appears in the next Editors' Extension, p. 147, on the Idea of Man.

♍

The Six Holy Forces of Nature
Universal Intellection

The Idea of the 6th house, the primal power or *Mahashakti,* includes and unifies the six forces or *shaktis* which are the powers of the universal intelligence (*cit-shaktis*). Following Row:

Parashakti is "literally the great or supreme force or power."

Jnanashakti is the power of intellect, wisdom, knowledge.

Icchashakti is the power of will (in the sense of subjective states).

Kriyashakti is "the mysterious power of thought which enables it to produce external, perceptible, phenomenal result by its own inherent energy."

Kundalinishakti is "the power or force which moves in a serpentine or curved path."

Matrikashakti (or *Mantrashakti*) is the force or power of letters, speech, or music.

These six forces are the basis of the internal discrimination and dialectic within universal intellection, before it is broken up or individualized. Plotinus speaks of the higher form of discriminating knowledge as dialectic, "the power of pronouncing with final truth upon the nature and relation of things—what each is, how it differs from others . . . whether its Being is Real-Being," and so on. Dialectic "pastures the Soul in the 'Meadows of Truth.'" (I.3.4) Intrinsic to this inner dialogue of Being is the principle of Number, through which there is a timeless order of knowledge. This simultaneity of unique Ideas in the Intellectual-Principle allows Ideas to interpenetrate without losing their distinct nature.

Virgo includes the mystery of the relation of *noesis* (contemplation) and *poiesis* (production). (Cf. Plotinus, III.8.5–7.) *Kanya,* virgin, means inviolability. The six forces preserve the inviolability of the soul's consciousness (5th house) while presencing it through the rational soul powers.

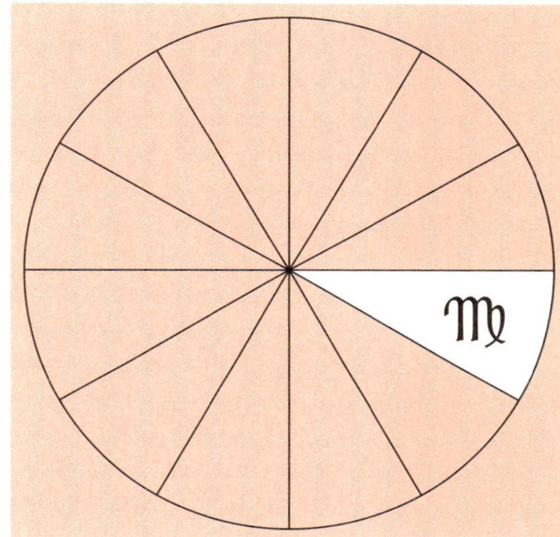

Fig. 57: *Kanya*—Virgo

T.S. Row: Kanya *means a virgin and represents Shakti or Mahamaya. . . . there are six primary forces in nature . . . : Parashakti, Jnanashakti, Icchashakti, Kriyashakti, Kundalinishakti, and Matrikashakti.*

♎

The Thirty-six Tattvas
The Outward-Facing Hierarchy of Reality

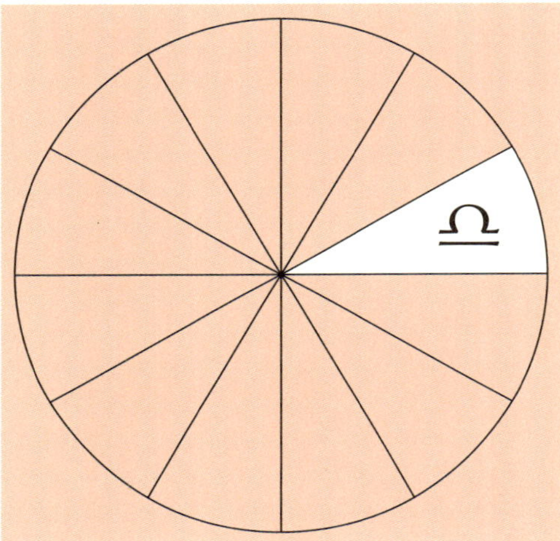

Fig. 58: *Tula*—Libra

T.S. Row: When represented by numbers . . . this word will be converted into 36. . . . the *36* Tattvams. . . . *Jivatma . . . [is] encased as it were within these thirty-six Tattvams . . .*

Tattva means literally "thatness" or principle. Sir John Woodroffe, *The Garland of Letters,* and J.C. Chatterji, *Kashmir Shaivism,* describe the individual *tattvas* from the Shaivite perspective. The thirty-six *tattvas* enable the pure intelligence to emerge from its state of identification with the divine Ideas. Thus they represent the fully differentiated outward or objective display of reality, the production of the many which remain continuous with the One.

> All existences, as long as they retain their character, produce—about themselves, from their essence . . . some necessary, outward-facing hypostasis continuously attached to them . . . (Plotinus, v.1.6)

In the 7th house the reason-principles restrict the infinite consciousness to many distinct individual souls, or many *Purushas,* each with its own set of reason-principles (its own *Prakriti* or vehicles). These reason-principles assure the harmonious unfoldment of meaning from the infinite.

Whereas the second quadrant as a whole is analogous to deep sleep, the third is analogous to dream. An analogy of light and the seven colors displayed by a prism may help us to understand this transition. Resolving the colors back into the white light is analogous to returning to undifferentiated intelligence. Passing light through the prism is analogous to relativizing the universal intelligence, so that we may discriminate above and below, inner and outer, heaven and earth, and so on.

♏

The Universe in Thought
Theophanic Creative Imagination

The vast imagination of Vishnu is infinitely variable. Here we must be careful to distinguish imagination from mere fantasy. This imagination is creative and gnostic, not imitative. Vishnu's dream (*taijasa*) is the transformative power of Theophany—literally a showing forth of deity. [24, p. 325]

Houses 5 and 8 are inverse principles. While the 5th house indicates the focusing of infinite intelligence into each unique individual, the 8th house indicates that every individual has the power to express intelligence through infinite variations.

> For each manifestation of knowledge and wisdom is a distinct image, an object in itself, an immediate unity, not an aggregate of discursive reasoning and detailed willing. (Plotinus, v.8.6)

Scorpio holds the deific image which the soul longs to fulfill. Here soul internalizes the Ideas or reason-principles. The Ideas here are in a dynamic condition: compelling, fluid, and potentially infinite. From here, the individual soul "affines" (is naturally drawn to by affinity) an appropriate place in the cosmos to fulfill its longing: The soul will be where it loves. Every Idea can express infinite variability, division, and transformation within a subjective space. The subtle or seed form of the image in Scorpio has to become actual in and through the agency of the next Idea: the nine Prajapatis.

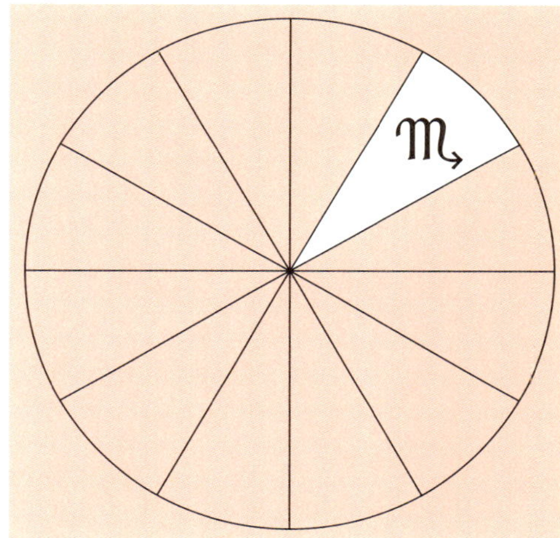

Fig. 59: *Vrishcikam*—Scorpio

T.S. Row: The sign is intended to represent Vishnu. . . . Vishnu literally means that which is expanded—*expanded as* Vishvam *or* Universe*. . . . The sign in question properly signifies the universe in thought or the universe in the divine conception.*

♐

Nine Prajapatis Assist the Demiurgus
Providence: Demiurgic Law

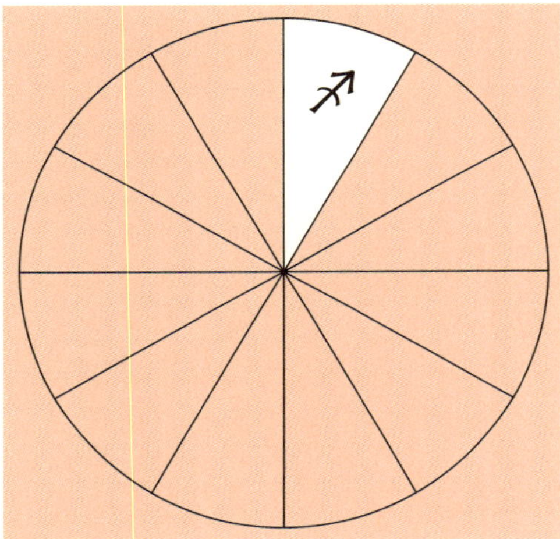

Fig. 60: *Dhanus*—Sagittarius

T.S. Row: The sign . . . indicates the nine Brahmas—the nine Prajapatis who assisted the Demiurgus in constructing the material universe.

In Platonism, the Demiurgus is the Universal Soul or World-Mind that orders and manifests all universes through its "mental" energy and returns all things to the Good. This Demiurgic energy thinks the universe into existence, providing both its germinal unity and its lawfulness. (Cf. especially Plato's *Timaeus.*)

> . . . we . . . explain the providence ruling in the Universe as a universal consonance with the divine Intelligence to which the Cosmos is subsequent not in time but in the fact of derivation, in the fact that the Divine Intelligence, preceding it in kind, is its cause as being the Archetype and Model which it merely images . . . (Plotinus, III.2.1)

The Demiurgic mind contains the divine archetypal pattern of the universe, which Plotinus says provides the "tentative illumining of Matter." (VI.7.7) Demiurgic energy is called "Providence" (III.2.1) because it provides the divine Idea of the universe for all creatures in it.

The Prajapatis, or the Brahmas, are the gods that assist the Demiurge in bringing any universe to birth. These powers are concerned with manifesting the Word, the Ideas, as the material world. They spring from the universal mind (Ishvara). The Prajapatis administer the justice of the world according to the laws of karma, and provide each individual soul with a direction for a series of lives. [25, p. 326]

♑

The Shape of the Material Universe in the Mind of the Demiurgus
Dodecahedron

The "material" universe is not matter as we think of it in modern science, but complex geometric shape. This "shape" is the underlying structure or ground-plan of manifestation—as "tentatively" illumined by the Demiurge (from Sagittarius). A dodecahedron, one of the five Platonic solids, has twelve regular pentagonal faces. It symbolizes the Idea that the universe is based on the twelve living Ideas embedded in five types of subtle space.

The 10th Idea is the world: the Idea that body or embodiment is a divine and eternal Idea, and that the universe is the body of God.

> In the absence of body, soul could not have gone forth, since there is no other place to which its nature would allow it to descend. Since go forth it must, it will generate a place for itself; at once body, also, exists.
>
> While the Soul (as an eternal, a Divine Being) is at rest—in rest firmly based on Repose, the Absolute—yet, as we may put it, that huge illumination of the Supreme pouring outwards comes at last to the extreme bourne of its light and dwindles to darkness; this darkness, now lying there beneath, the Soul sees and by seeing brings to shape; for in the law of things this ultimate depth, neighbouring with soul, may not go void of whatsoever degree of that Reason-Principle it can absorb, the dimmed reason of reality at its faintest. (Plotinus, IV.3.9)

In this view, matter is not some self-standing stuff, but is the pure receptivity, an impassive receptacle for all forms. Matter is the principle which allows the manifestation of a universe to exist, to stand as other than the primal hypostases.

Plato says in the *Timaeus* that the cosmos is constructed according to the most excellent paradigm but is "always becoming and never is." (27d, B. Jowett trans.) Here "becoming" is the Idea of existence *per se,* the Idea that something stands outside of or is other than Being.

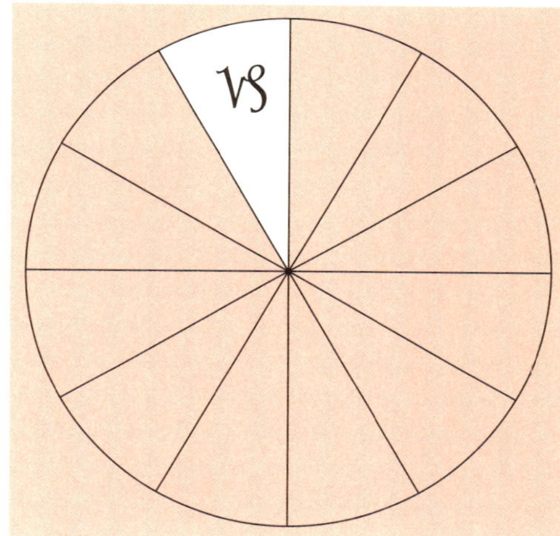

Fig. 61: *Makaram*—Capricorn

T.S. Row: The sign . . . is intended to represent the faces of the universe and indicates that the figure of the universe is bounded by pentagons. If we take the pentagons as regular . . . the figure of the material universe will, of course, be a Dodecahedron, the geometrical model imitated by the Demiurgus in constructing the material universe. . . . From an objective point of view the "microcosm" is represented by the human body. Makaram may be taken to represent both the microcosm and the macrocosm as external objects of perception.

The Fourteen Lokas
Levels of Life and Perception

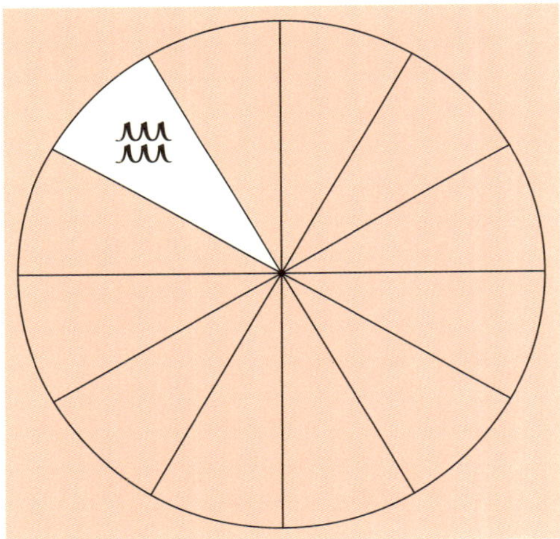

Fig. 62: *Kumbham*—Aquarius

T.S. Row: When represented by numbers, the word is equivalent to 14. It . . . is intended to represent . . . the fourteen lokams . . .

Loka literally means "place," but also has the connotation of a vast "realm" or level of existence. Each *loka* specifies an aspect of living consciousness associated with or determined by an appropriate objective experience (*tala*). There are seven such subjective and objective counterparts, totaling fourteen.

The universal principles of relativity which determine these levels provide the conditions needed for an individual soul's experience. We earlier associated our unconditioned consciousness, *atman*, with the 5th house, which is opposite to the 11th. Now in house 11, Aquarius, we see that cognition as contracted. We have "cognitive isness" as subjected to the conditions and laws of space, time, and causality.

> These firsts, seconds, thirds, of participance are determined by rank, by power . . . by differentiation; and difference is no bar to coexistence . . . (Plotinus, VI.4.11)

All beings who have similar organs of perception will experience, similarly, a particular kind of common world. To develop this idea, Anthony investigated the experience of how consciousness operates differently through each of the different *chakras* or subtle energy centers. The *loka* from which one operates determines how the given (10th house) World-Idea is translated into experience (in house 12). [26, p. 326] Each individual center of consciousness relativizes its experience of the World-Idea, manifesting and distilling its meaning through the appearance of the world in Pisces or house 12.

♓

The Five Mahabhutas Flux of Appearance

The five *mahabhutas* (great elements)—Earth, Water, Fire, Air, and Ether—are the specific forms of universal life. These great elements are the substance of any and every sensible universe, and are similar to what modern physics calls force fields; they should not be confused with the elements found in the periodic table of chemistry. *Mahabhuta* may also be translated as "elemental being," for these five elements are conceived as vast primordial forms of life.

The universal alchemical solvent Row refers to is the life-energy of the World-Soul. The sensible forms emerge from this solvent, and the traces of universal manifestation dissolve into it between the manifestations (and, metaphysically, between moments of manifestation).

Pisces is the Idea of instantiation: the appearance and dissolution of Ideas as the world. To dissolve here means to go back into solution, leaving traces that condition subsequent appearances. Each element retains traces of different types of experience of the living Ideas: physical, emotional, volitional, mental, and intuitional. These vast energy fields provide evolution and apparent continuity for the individual entity and its experienced world. [27, p. 326]

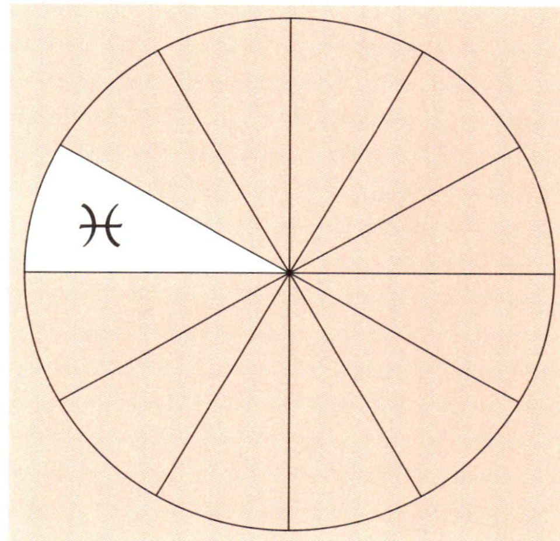

Fig. 63: *Minam*—Pisces

T.S. Row: This word [Minam] *. . . is evidently intended to convey the idea of* pancamahabhutams *[sic] or the five elements. This sign also suggests that water (not the ordinary water but the universal solvent of the ancient alchemists) is the most important amongst the said elements.*

▼ EDITORS: There are many ways of looking at the twelvefold circle of Ideas. It can be divided into thirds, quarters, for example, each of which can be looked at as a mandala in its own right. Figure 64 shows each of the quadrants divided into thirds and looked at separately.

See the last part of the Editors' Extension III, "The Fifth Idea," (p. 157–165) for an example of a threefold division. Note 21, "Transition from the One to the Intellectual-Principle" (p. 320), shows some other ways to view the mandala.

OTHER WAYS OF PICTORIALIZING THE TWELVE IDEAS

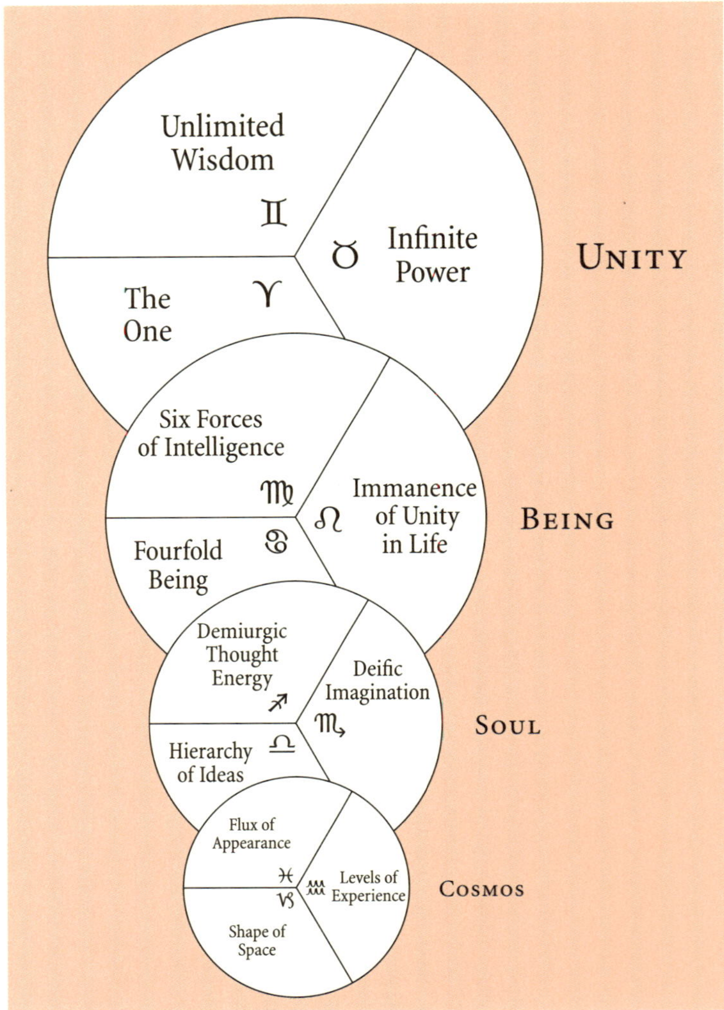

FIG. 64 ▼

Editors' Extension III

The Fifth Idea

An Example of How Each Idea Is a Unique Form of the Whole

We are at a time in world history where there is an intense need for a Principle or principles to provide a unifying background to the rich complexity and variety of philosophic and religious traditions humanity has inherited and is developing. So overwhelming is the proliferation of knowledge that unless we find encompassing principles our mental balance is in jeopardy. If we dare to pursue the glimpses offered by Plotinus, we may arrive at a workable hypothesis as to what that philosophy might be. The pyramiding complexity of knowledge must be boldly faced.

What is the principle, Plotinus asks, that is the giver of Wisdom to the Soul? He responds by saying, "the veritably intellectual, wise without intermission . . ." (v.9.2) It is an Intelligence which is "in actuality and in eternity." (v.9.5) This intelligence—the Intellectual-Principle—he refers to as an aggregate, or the total of the Ideas, and each one of these Ideas is the entire Intellectual-Principle in a special form. It is to this notion that we will direct our attention by positing our initial premise of correlating the zodiacal signs with those Ideas. To attempt to understand this notion in a much more detailed way, we place one Idea at a time in the first house of the chart of Being, then apply the entirety of the other eleven Ideas to it in such a way that the unity of the Being of that Idea is amplified and expanded by the others. If we systematically apply to an Idea the meanings provided by all the other Ideas, we should come out with such an absoluteness that makes that Idea the source and origin of all explanations.

When this process is repeated a dozen times, we should have twelve absolute philosophies—each one legitimately claiming for itself finality. That the World-Idea is involved in such complexity is revealed by the persistence and validity of many schools of thought, which would at the same time indicate the simultaneity of their (the Ideas') presence.

Anthony Damiani, Notes and Papers, pp. C1–4

Our example is the 5th house Idea: the Idea of Incarnation, or the Idea of Man. ▼ First we will present a general description of the 5th Idea, briefly related to how it has been conceived in some Eastern and Western traditions. Subsequently, we will illustrate Anthony's method of applying the twelvefold to this Idea to show how each Idea is a whole mandala.

▼ Editors: The term "Idea of Man," drawn from the classical Greek tradition, may be adapted to contemporary language in a variety of ways: "the Idea of the fully human being" or, more precisely, "the Idea of the completely individuated and integrated human being and the processes involved in reaching that state," "the Idea of the fully enlightened person," "the Idea of the sage," "the Idea of Self-realization," "the incarnation of Wisdom," and so on.

Participation in Divine Mind: The Birth of Intelligence in Soul

The 5th Idea—incarnation and participation—is especially relevant to us as human beings. Through this Idea, divine intelligence instantiates itself in individual souls so that we as souls participate in the divine. This presence of the Intellectual-Principle in the soul or mind is the logos, which Plotinus says (III.2.16) is not a separate principle but a union of wisdom and Life. Logos is the divine Idea that God has of the individual soul, and which that individual is seeking to realize. Row speaks of this logos when he describes the Idea of Leo. Each soul is a unique logos, a center of intelligence, a ray of the divine Sun. In Neoplatonism, the seed or potentiality of enlightenment present in the individual is called the logos principle of that individual—the spark of the divine intelligence in the soul. According to Plotinus in IV.3.5, each soul is a logos of the Intellectual-Principle.

The *bodhicitta* teaching central to Mahayana Buddhism parallels this conception of the logos. *Bodhicitta* means enlightenment-mind or the heart of intelligence. In Mahayana, *bodhicitta* is both the potentiality for enlightenment (relative *bodhicitta*) and the actual enlightened mind itself (absolute *bodhicitta*). [28, p. 326] The notion of a seed or potentiality in the deeper mind, which develops into the fully enlightened mind of the Buddha, is akin to Plotinus' notion of the intelligence shining in the deeper part of the soul.

The Pythagorean teachings about the number 5 relate to logos and incarnation. In a class on the meaning of the twelve signs, Anthony remarked:

> The ancient Pythagoreans spoke of the 5th number principle as the number of divine incarnation of the One in the Tetrad. Five represents the presence of unity in Being, and also signifies

> the essential meaning of participation, that there can be authentic existences or Ones which are not the absolute One. Self-reproductive intelligence is also called logos, the union of life and wisdom, taking determination as centers of intelligence. (transcript, 9/27/76)

Meister Eckhart writes about the logos as the eternal birth that takes place in the stillness of the ground of the soul:

> St. Augustine says this birth is always happening. But if it happen not in me what does it profit me? What matters is that it shall happen in me.
>
> We intend therefore to speak of this birth as taking place in us: as being consummated in the virtuous soul; for it is in the perfect soul that God speaks his Word. (p. 3)

Self-realization, the birth of intelligence within us, is our greatest good and the source of our deepest happiness. This realization means both the Divine Mind realizing itself in us, and our re-birth into the divine. The 5th house Idea is therefore the ideal of the fully human being, the sage.

To explain the 5th house Idea, T. Subba Row, in "The Twelve Signs of the Zodiac," combines teachings at the heart of the Buddhist and Vedantic realizations—the five Dhyani Buddhas and the *jivatma*. These are two ways of describing the core of our humanity.

> This word contains a world of occult meaning within itself . . . [one of its synonyms] shows that the sign is intended to represent the *five Brahmas* . . . [or] *the five Buddhas*. The second synonym shows it to be Narayana, the Jivatma, or Pratyagatma. . . . I look upon Jivatma as identical with Paramatma in its real essence when stripped of its fictitious attributes created by Ajnanam or Avidya—ignorance. The Jivatma is correctly placed in the 5th sign counting from Mesham [Aries], as the 5th sign is the . . . son's house, according to the rules of Hindu Astrology. The sign in question represents Jivatma—the son of Paramatma as it were. (*Esoteric Writings*, pp. 8–9)

Apparently Row realized that these two great traditions present complementary sides of our nature. Whereas *atman* refers to the eternal and unchanging consciousness within the individual, the teachings about the five Dhyani Buddhas refer to the transformation of consciousness, or the drawing out of the potential within the individual. Paul Brunton pointed out the need for both Buddhist and Vedantic understanding: "The defect of all the Vedantic authorities in India today is that they have lost the Buddhist esoteric tradition and even despise it; for only in the combination of both can be realized that restoration of the genuine, archaic Indian wisdom." (*Notebooks,* v. 10, p. 121)

Anthony comments on the meaning of Leo as the *jivatma,* relating it to the nature of the human soul in Platonism:

> *Jivatma* means living self, or soul: a one which lives, or a self-reproduction of unity in life. However, it may also signify the *jivanmukta,* the fully accomplished human—the sage—within whom the universal intelligence is realized, and who thus is truly an incarnation of wisdom. In this view realization involves a transformation of the *jiva,* the unit of life, into *atman,* an embodiment of authentic intelligence.
>
> The essence of the individual human (*jivatma* or individual soul, the microcosm) and the essence of the Solar Cosmos we live in (*Paramatma*) is the same. Or to expand this view, the spiritual essence of the entire universe and the essence of the individual human being is one and the same. Essentially, the soul of the cosmos and the individual soul are the same. Thus this sign indicates the intimate relation of the macrocosmic principles and the microcosm (individual human being). (transcript, 9/27/76)

Next, Anthony relates the transformation of the *jivatma* into *jivanmukta* to the meaning of the five Buddhas. In their totality, the five Dhyani Buddhas may be thought of as the self-reproductive wisdom present in the living reality of the soul or mind.

> The connection of the five Dhyani Buddhas to the *jivatma,* is that the Buddhas represent five potential aspects of a human being, and preside over the process of transformation of the ordinary human entity into enlightenment. Each of the five Buddhas symbolizes the transformation of one of the individual mind's constituents into wisdom. Thus the constituents of universal Being are also present as the five Buddhas or wisdoms, principles of individuated intelligence which constitute a being—*jivatma.* Each of the five Buddhas represents the unity or the intellect of a god, and is to be thought of as including a host or a hierarchy. (transcript, 9/27/76)

In Mahayana Buddhism, the transformation of potential wisdom into realization—presided over by the five Dhyani Buddhas—is the central purpose of life and takes place over many lifetimes. Whatever the Divine Mind thinks of us will be actualized through this process of transformation. Anthony provides details through examples of two of the Dhyani Buddhas, Ratnasambhava and Akshobhya:

> We begin to see a kind of transformation process indicated by these gods. Each god is in charge of a certain element which has to be transformed from its lowest to its highest aspect. Each one of these gods ultimately universalizes our experience—turning us, most of the time against our own wills, into universal beings. Although the *jivatma* of an individual and the world are identical in essence, if we make this transformation from the individual to the universal, we would still retain an individual aspect of our vehicles even though we now recognize its [the individual aspect's] identity with the universe.
>
> To use psychological terminology, these human constituents are physical/corporeal (*rupa*), feeling (*vedana*), conception (*samjna*), habituation (*samskara*), and consciousness (*vijnana*). What would we have to do in order to transform these? To achieve *moksha,* or liberation, the Buddhists recommend that we would have to take the bodhisattva path.
>
> For example, the Buddha Ratnasambhava rules *vedana* (feeling). Initially, as an aspect of individual existence, feeling is self-centered. But when feeling gets universalized, it becomes universal compassion. We must start out being concerned with our own individual existence. Before we can have compassion for others, we must first get some for ourselves. We're not going to find out anything about universal compassion until we've had enough feeling for our own self-centeredness to transfer it to something better. So, when we've had enough lives, for example, as mothers and have learned to love children, then there comes a time when that love for a child can't be restricted anymore to just the biological child. We begin to feel love for all children. Given enough time, lives, and experience, this feeling for ourselves gets transformed into universal compassion for all other beings. We might believe in universal compassion, a feeling of compassion for everyone. But if we reach that kind of a level, we would have gone beyond the ego-centered functions. To take the Diamond Sutra to heart is really not to believe in the ego anymore. Then we would have difficulty saying, "I believe

myself to be a self-centered individual who exists in, by, and for myself." Finally, this universal compassion is recognized as an aspect of *Prajna*—the Wisdom of equality which sees that all beings are essentially identical.

Once a person has a feeling for universal compassion or has recognized the universal body, that person recognizes all these individual constituents as cosmic principles which ultimately come from these wisdoms. Sages haven't given up the physical body or all the psychical essences a sage works through. The sage transcends them but hasn't destroyed them.

For another example, the Buddha Akshobhya is in charge of the material form (*nama rupa*)—not only in terms of any individual but the entire universe as universe. An individual human being has an individual *rupa,* the earth has an individual *rupa,* the cosmos we live in has an individual *rupa*—anything that has any kind of existence would have to have some sort of individual *rupa* or corporeal form.

Concomitant with the form element is the principle of individual corporeality which must be transformed into a universal body, "in which forms of all things are potentially present and are recognized, according to their true nature, as exponents of the great Void . . ." (A. Govinda, p. 108)

The statement by Govinda applies to a cosmos or to an individual entity. For an individual, it means the individual recognizes that the body is a microcosm—a combination of the cosmic principles organized to accommodate him or her. Through this body they become aware of the universe; and as they begin to understand how the body represents a universe for them, it gets universalized. For instance, K.C. Bhattacharyya pointed out [in "Studies in Samkhya Philosophy," paragraph 38] "The body represents in itself the entire process of nature for the self." In other words, the body represents the entire system of Nature for each individual consciousness. This organized body is not for personal use alone but also represents the principles that are governing the universe. When this process reaches the *Prajna* aspect, it becomes Mirror-like Wisdom. (transcript, 9/27/76)

The following table [*fig. 65*] shows the relation of the individual constituents and their transformation into enlightenment components. Each one of the Dhyani Buddhas listed in the first column rules a certain constituent (*skandha*) listed in the second column. The third column is the corresponding aspect of individual existence (the psychological constituent based on that element). This individual function has to be transformed into a universal aspect, given in the fourth column. Lastly, the fifth column lists the wisdom (*prajna*) appropriate to each universal aspect.

DHYANI BUDDHA	SKANDHA	ASPECT OF INDIVIDUAL	UNIVERSAL ASPECT (with its true nature as Void)	WISDOM (*PRAJNA*)
Akshobhya	*RUPA* (form or sensation)	Individual body	Body recognized as cosmic process	Mirror-like Wisdom
Ratnasambhava	*VEDANA* (feeling)	Self-centered feeling	Universal compassion for all that lives	Wisdom of equality
Amitabha	*SAMJNA* (perception / conception)	Individual thoughts and preconceptions	Faculties of inner vision and outer discernment	Wisdom of discrimination
Amoghasiddhi	*SAMSKARA* (volition)	Self-centered will and activity	Altruistic activity	All-accomplishing Wisdom
Vairocana	*VIJNANA* (consciousness)	Individual ego-consciousness	Universal consciousness	Wisdom cognizing essence

FIG. 65

Sagehood and the Symbol of the Star

The realization of eternal intelligence is aptly expressed in Paul Brunton's description of the philosopher-sage, the counterpart of the Buddha:

> The esoteric meaning of the star is "Philosophic Man," that is, one who has travelled the complete fivefold path and brought its results into proper balance. This path consists of religious veneration, mystical meditation, rational reflection, moral re-education, and altruistic service. The esoteric meaning of the circle, when situated within the very centre of the star, is the Divine Overself-atom within the human heart. (*Perspectives,* p. 260)

The star represents the five wisdom aspects of the philosopher as fully discriminated and fully developed—distinct yet interwoven. The circle in the center represents the Overself as the basis of all of them. [*fig. 66*]

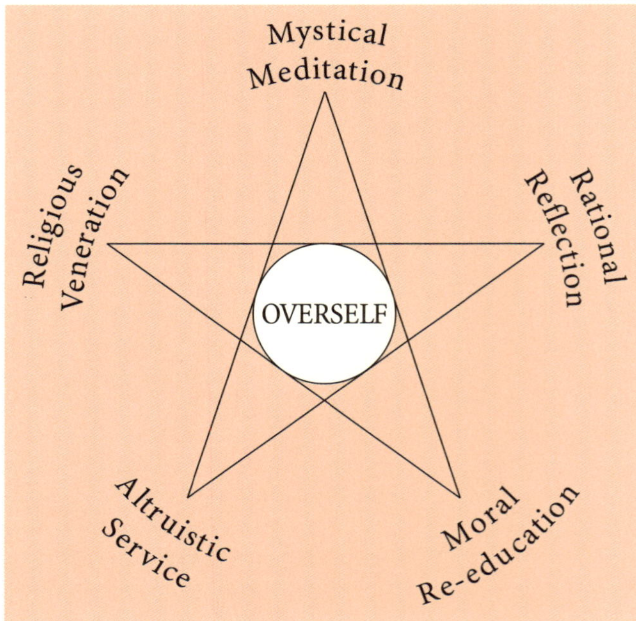

Fig. 66

The points of the star represent the full unfoldment or transformation of the ego functions into the five wisdom consciousnesses, while the circle symbolizes the pervasive permanent self-cognitive insight.

Thus the star symbolizes the perfected human as self-knowing—the logos realizing itself in human soul and body. This principle of logos in life is what Row means when he refers to Leo as expressing the true Christ, the anointed one. This state is the full and permanent philosophic realization, which includes and permeates all aspects of the being. [29, p. 326]

These five constituents of the philosopher can be amplified and correlated with aspects of the Neoplatonic path expressed in Plotinus' *Enneads*, and the teaching of "five wisdoms" of Buddhism:

- Through *opening the heart* in reverence, love, and compassion, self-centered feeling is transformed into the recognition that "each separate life lives by the Soul entire." (v.1.2) This *wisdom of equality* sees all beings as divinely interrelated, recognizes a sympathy with all life, maintains equanimity, and lives in harmony with all.
- Through *reason and intellection,* limited understanding is transformed into the power of *dialectic:* that which discerns reality from appearance. Dialectic brings with it "the power of pronouncing with final truth upon the nature and relation of things." (1.3.4) This is equivalent to *discriminative wisdom.*
- Through *mystical meditation,* relative consciousness is transformed into *universal consciousness.* The soul sinks all its being into the Intellectual-Principle, and thus participates in Intellectual-Principle's vast contemplation. "All its content, thus, is perfect, that itself may be perfect throughout, as holding nothing that is less than the divine, nothing that is less than intellective. . . . this is pure being in eternal actuality; nowhere is there any future, for every then is a now; nor is there any past, for nothing there has ever ceased to be; everything has taken its stand for ever, an identity well pleased, we might say, to be as it is." (v.1.4) Through this participation, soul attains the *wisdom which cognizes the essential nature of all things.* (This wisdom is referred to in Buddhism as the *Dharmadhatu* Wisdom.)
- Through *moral re-education and purification* our finite sense of embodiment is transformed into a sacred place to reflect all things as they are. Soul becomes a *mirror* of the Divine: "When there is peace in that within us which is capable of reflecting the images

of the Rational and Intellectual-Principles these images appear. Then side by side with the primal knowledge of the activity of the Rational and the Intellectual-Principles, we have also as it were a sense-perception of their operation." (I.4.10) [30, p. 326]

- Through continuous remembrance of the divine in the midst of *altruistic service,* egoistic volition is brought into conformity with the divine will. "The Reason within it [soul] becomes Act in virtue of its contemplation of that prior; for its thought and act are its own intimate possession when it looks to the Supreme Intelligence; those only are soul-acts which are of this intellective nature and are determined by its own character . . ." (V.1.3) Thus the soul attains to *all-accomplishing wisdom.*

These constituents of the fully human being in their totality are the wisdom of the Idea of Man present in the soul (*jivatma*). The sage is the complete and balanced actualization of the Idea of Man. Anthony writes (in a draft of a paper on the Idea of Man):

> Through the very process of manifesting the Idea of Man in generation, the end result will be the attainment of sagehood—or Buddhahood. The sage has become a fully actualized unique version of the Idea of Man which his soul has contemplated, or the Idea of Man has actualized itself in the sage. He is the essential man—the Intellectual Man, realized of the Nous in him. Through this highest he has access to the infinity of divine wisdom. The means to actualize this wisdom are inherent in the Idea, and in the soul's essence. This is the true self-knowledge which the sage has attained—he becomes the True Knower, the Nous. (Notes and Papers, p. D6/38)

The Sufis speak of the individual who has attained this state as a Man-God, or Perfect Man:

> Man alone knows God both as the Real (*Haqq*) and the Phenomenal (*Khalq*), for Man himself (the Perfect Man) is the Real and the Phenomenal, the internal and the external, the eternal and the temporal. The Perfect Man's heart is the seat of the manifestation of the universal Logos (the Reality of Realities or Reality of Mohammed, etc.), and in it alone the activities of this Logos find their fullest expression. (A.E. Affifi, p. 80)

Twelve Facets of the Fifth Idea

To complete the exploration of the Idea of Man, we may now examine the 5th Idea in relation to the other eleven Idea-principles. We start with the circle of twelve Ideas, each intrinsic to the Divine Mind. [*fig. 67*]

Using the whole circle to represent the Idea of Man as an Idea in its own right, we apply the twelvefold scheme. The unity of the Idea of Man becomes its 1st house, its power goes to the 2nd, and so on. [*fig. 68*]

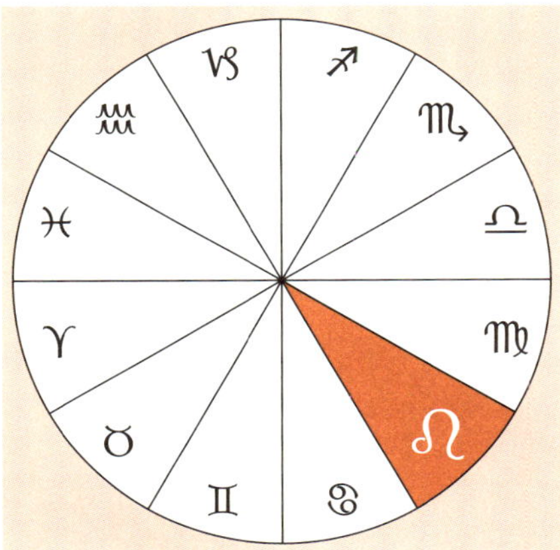

Fig. 67: Fifth Idea as a constituent of the Intellectual-Principle entire

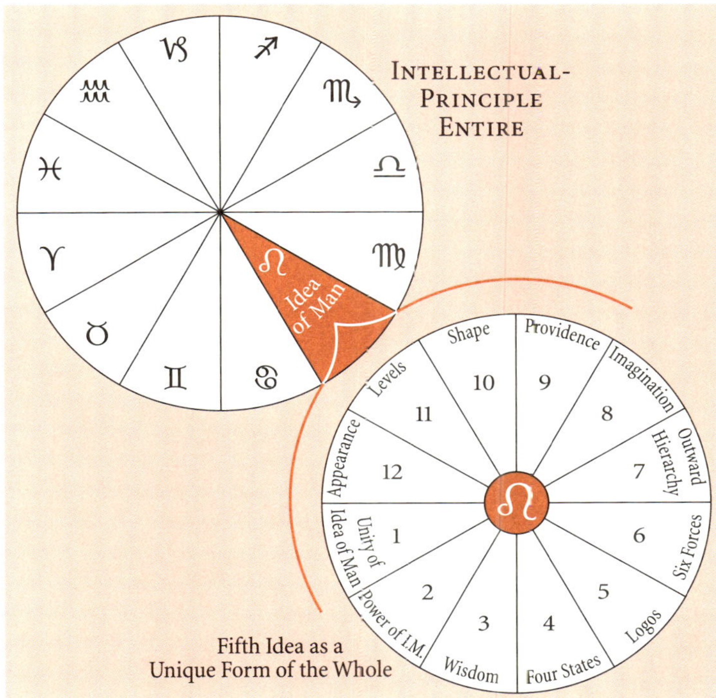

Fig. 68: Fifth Idea as a unique form of the whole

Astrology represents each of these root-Ideas by a certain animal intelligence:

> We can view the animals in the zodiac as representing, in their totality, an image of the Idea of Man—each animal developing one particular function to the fullest. The particular organs that these animals as living species have developed are said to go into the physical constitution of Man. All these warm-blooded animals are then parts of the Idea of Man, not ontogenetic priors. They are all contained in the Idea of Man. Also, there is the unfolding of the gnostic powers of Soul from sense to the highest intuition guided by the Idea of Man, and there is the almost endless series of archetypal images which are expressing the reason-principles in Soul.
>
> In this expression of Reason, even empirically we can see the vastness of what is involved in the Idea of Man. The reason-principles of Man have ordered the totality of all previous civilizations, root races, and so on, stretching back to the beginnings of our aeon—and even before this to the beings who laid down the basis of our humanity from previous aeons. Aeons of evolution are the working out of the reason-principle of Man. This whole sea of humanity is but a small fraction of the Idea of Man which the soul is unfolding. (Anthony Damiani, Notes and Papers, p. D1/80)

We can also correlate the 5th Idea with the three aspects of Man Plotinus describes in tractate VI.7.6. This approach lends itself to a natural division of the circle into thirds. [*fig. 69*] [31, p. 327]

> If material things existed There, the soul would perceive them; Man in the Intellectual, Man as Intellectual soul, would be aware of the terrestrial. This is how the secondary Man, copy of Man in the Intellectual, contains the Reason-Principles in copy; and Man in the Intellectual-Principle contained the Man that existed before any man. The diviner shines out upon the secondary and the secondary upon the tertiary; and even the latest possesses them all—not in the sense of identifying itself with them all but as standing in under-parallel to them. Some of us act by this lowest; in another rank there is a double activity, a trace of the next higher being included; in yet another there is a blending of the third (i.e., highest) grade with the others: each is that Man by which he acts while each too contains all the grades, though in some sense not so.

To present more details of how to apply the twelvefold mandala to this Idea, we will focus on the sage as the ideal of a fully human being. Each soul will act according to the level of the Idea it can actualize—natural, psychical, and Wisdom Universal. All grades are integral and simultaneous to the sage, and the sage incorporates and consciously accesses the three levels and all twelve Ideas. Through the experience of many lifetimes an individual becomes a sage, a fully conscious embodiment of divine wisdom.

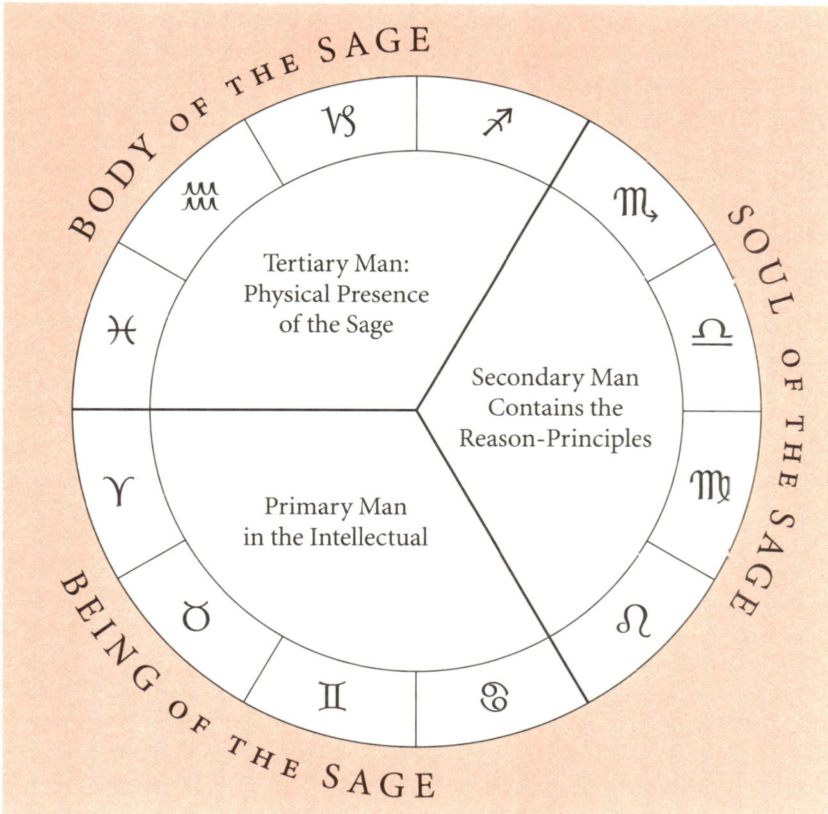

FIG. 69: THREE ASPECTS OF THE SAGE AS THE COMPLETE HUMAN BEING

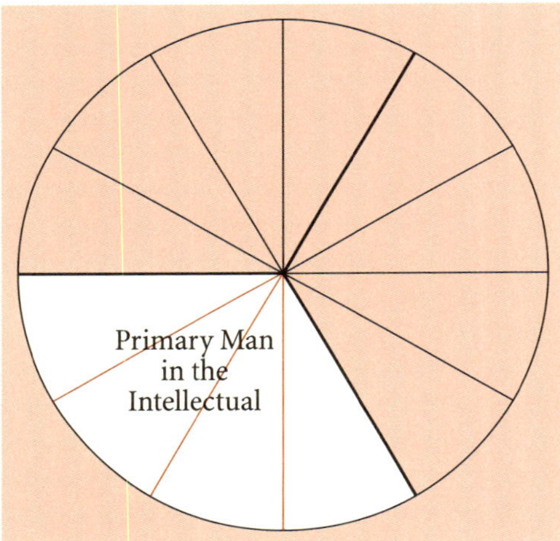

FIG. 70: UNIVERSALITY OF THE SAGE

Idea of Man in Being

The first four houses represent the Being of the Idea of Man, the Idea as a universal possibility. "Man in the Intellectual-Principle contained the Man that existed before any man." (Plotinus, VI.7.6) Here the Idea is in the mind of God, universally pervasive, everywhere and nowhere. The first sector, houses 1–4 of this Idea, represents the participation of Soul in the Intellectual-Principle. Through this relation, something of the individual sage is also non-individual, partaking of the universality of Ideas. [*fig. 70*]

1 ARIES THE UNITY OF THE IDEA OF MAN—A PERMANENT POSSIBILITY

The 1st house of the Idea of Man indicates the eternal self-existent mystery which remains identical in all expressions of the Idea of Man. This is the deepest holy fount of our being, where we participate in the source of our being which is beyond Being. To participate in this Divine Mind is "the Good" for human beings. Plotinus says that the term Good is used to "assert identity without the affirmation of Being." (VI.7.38) If we can find that identity, we have the foundation of all levels of self-realization. This identity for the Idea of Man is beyond the conditioning factors that make one individual different from other individuals, but still serves as the ground of individuality *per se.*

2 TAURUS THE ACTIVE ASPECT OF THE IDEA OF MAN

In the 2nd house we locate the power of the 5th Idea. *Pranava Aum,* Reality's sacred name, represents the infinite radiating power of the Word which God speaks into the ground of the soul. This Word contains an inexhaustible supply of essential living possibilities. This is the source of the omni-potency of the Buddha nature through which each sage is a fountain of overflowing compassion.

3 GEMINI THE PASSIVE ASPECT OF THE IDEA OF MAN

When applied to the Idea of Man, the 3rd house represents the universal mind of the sage: *sahaja,* the nature of the Mind. Gemini is described as androgynous, and just as the androgyne combines polarities, the sage's mind also encompasses a double nature. The sage's insight "knows" reality and appearance, eternal and perpetual consciousness simultaneously. As Paul Brunton observed: "When duality is blended with, *and within,* unity it is the true *jivanmukta* realization. The One is then experienced as the Two but *known* to be really the One." (*Perspectives,* p. 346) This Idea represents the access of the sage to the truth-consciousness of the Intellectual-Principle. [32, p. 330]

4 Cancer How the Idea of Man will participate in fourfold Being, four states of consciousness, etc.

The 4th house of the Idea of Man reveals the four states available to the human: waking, dream, sleep, and *turiya.* The fourth state reveals mind itself, the mystery of the first three. Universal good, power, and intelligence are here differentiated into distinct states so they can be available to all beings.

For the sage, the three states of being-consciousness—waking, dream, and deep sleep—are available simultaneously within the continuum of *turiya* intelligence. The sage's mind is itself a continuum, at every moment capable of all the differentiated states.

This fourfold Being is the universal ground and substance of the *jivatma* (or the Buddhas). This universal ground or matrix is called *tathagatagarbha* in Buddhist terminology. It affords nourishment for all beings, and is the place in which the unique intelligence that matures into sagehood gestates and grows. *Tathagata* means *thus-gone,* an epithet of the Buddha, and *garbha* indicates both the womb and the seed of the *tathagata. Tathagatagarbha* is thus matrix in the Buddhist context, providing the spiritual components of all beings, available at every level, and spread throughout the states of consciousness.

Idea of Man in Soul

In the second sector, houses 5–8, we conceive of the Idea of Man as the guiding wisdom and power shining into the soul. While the first four Ideas describe the universal constituents of wisdom and Being, these next four describe how intelligence is uniquely present in each soul. Houses 5–8 can be viewed as four aspects of living wisdom: a combination of the five wisdoms discussed in the introduction to this Idea. This creative power propels the soul's infinite journey into the manifestation of the cosmos in houses 9–12. From the state of essential Vision unified in the 5th house, Intelligence becomes assimilated to subjective centers of consciousness (in house 8). [*fig. 71*]

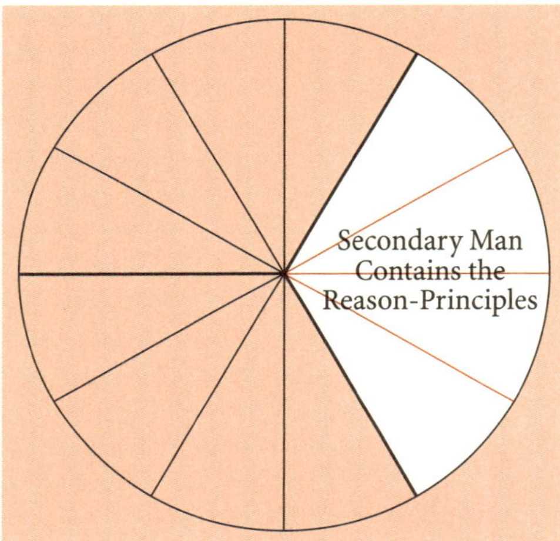

FIG. 71: IDEA OF MAN IN SOUL OR IN THE SOUL OF THE SAGE

5 LEO THE IMMANENCE OF THE IDEA OF MAN IN SOUL

The 5th principle applied to the Idea of Man represents the unique form of Ideas identified with the soul, the intrinsic power of intelligence to individuate itself in and through soul. Each soul is a special living intelligence—a "Logos of the Intellectual-Principle" (Plotinus, IV.3.5). The Idea by which soul is known to the divine Mind is also the vision by which soul uniquely is and knows itself. The 5th house applied to the 5th house Idea reveals the principle of the five Dhyani Buddha intelligences which enable a characteristically unique self-realization, based on each soul's vision of the Intellectual-Principle entire. Each sage ultimately becomes a special expression of the Nous.

6 VIRGO THE POWERS OF THE IDEA OF MAN

The Idea of the six forces (*shaktis*) indicates the internal differentiation of intelligence before it bifurcates into separated components. Each intelligence that contributes to the constitution of the Idea of Man has a transformative power. One sage, for example, will be as a force of nature with a great transformative power of healing, another will have a great transformative power in *jnana,* another a great transformative power in *kundalini,* etc. Through these powers the sage has access to the "infinity of divine wisdom." (*Astronoesis,* p. 156) Looked at another way, these powers together constitute the sage's inner intelligence or integral act. While the 5th house is the subjective wisdom consciousness of the soul, the 6th house is the intelligible content of the soul. Plotinus says that all act (*poiesis*) arises from inner contemplation (*noesis*), and that all action and reason must lead the soul to vision. ▼

7 Libra The thirty-six principles that constitute or unfold its potentiality

The 7th house applied to the Idea of Man indicates the hierarchy of reason-principles in the soul, through which the soul's integral interior vision achieves outward expression. Platonic ratio and reason allow the harmonious and full articulation of each soul's wisdom and power (found in houses 5 and 6). Through the working of the 7th house Idea, the Ideas in the soul are brought into a perspective and relation appropriate for that soul. Proportion allows for unique perspectives on universal knowledge, as well as the appropriate evaluation of Ideas.

A true valuation and appreciation of the reason-principles that constitute all individuals is the foundation of the bodhisattva path. The sage sees the universals present in unique individuals who are similar but not identical, always aware that all forms of life are in essence divine and participate in a universal harmony.

▼ Plotinus, iii.8.7: All the forms of Authentic Existence spring from vision and are a vision. . . . All act aims at this knowing; all impulse is towards knowledge, all that springs from vision exists to produce Ideal-Form, that is a fresh object of vision, so that universally, as images of their engendering principles, they all produce objects of vision, Ideal-Forms. In the engendering of these existences, imitations of the Authentic, it is made manifest that the creating powers operate not for the sake of creation and action but in order to produce an object of vision.

8 Scorpio The creative world conception of the Idea of Man

The 8th Idea is the imaginal power in the soul. Imagination here is fluid, creative, and gnostic, transcending the separate subject-and-object relation. It is the universal divine substance of what T. Subba Row calls "Vishnu's dream." Through this creative power, souls participate in the presence of the gods as a theophany of Idea-images. This imagination encapsulates the unity, wisdom, and power of the soul in forms appropriate for creatively unfolding that soul's wisdom and power. Each soul contains the entire power of the Divine Form in image. This creative Idea is the germinal seed of the soul's unique perfectibility in relation to the universe. The sage's participation in the universal imagination is the basis of tantric deity yoga and other esoteric practices aimed at actualizing one's own divine image.

Idea of Man in an evolving unit of life

In the third sector, we see the Idea of Man evolving within the unit of life—the phase of the soul that seeks embodiment. Here we locate the human as actualizing the Ideas and soul powers in space and time in relation to the World-Idea. Within the system of Nature, one evolves and unfolds unique causal, subtle, and gross bodies that reflect the cosmic intelligence. For the sage, the manifested world in houses 9–12 is known as the instantiation of divine immaterial Ideas. The sage's own embodiment in its most inclusive meaning operates as an organ of perception for the soul, and a vehicle that benefits all living beings. [*fig. 72*] [33, p. 330]

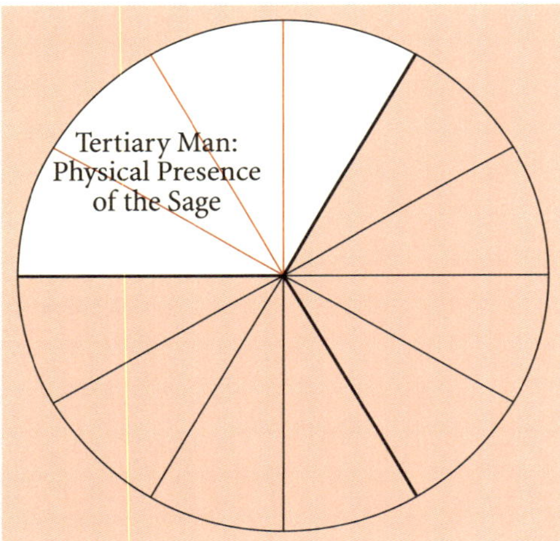

Fig. 72: Idea of Man embodied

9 Sagittarius The entirety of manifestation as the unity of Universal Man

Here the powers of the Demiurge, which Row calls Prajapatis, infuse the emanating ray of soul with the universal energy needed to fulfill its vision. This unity of the soul's embodying phase is in contact with the providential intention of the gods, who assist the sage's purpose in life. Each evolving individual is a union of the soul's intention and a version of the World-Idea determined for it by the gods. In Hindu creation myths, the Prajapatis are archetypal progenitors of humanity who provide specific essences and potentials that order the entire sequence of lives for an individual. In this way, the soul is set under the necessity of karmic law from its first entry into the universe and is purposively directed into embodiment.

10 Capricorn The shape of the Idea of Man in the Demiurgic intelligence

The 10th house Idea applied to the Idea of Man represents universal space as the common ground for all individual embodiment. The "shape of the material universe" here means the congealed or bounded mind-space in which manifestation can occur. This "space" is the causal or intelligible repository of the laws of universal manifestation. Entering into relation with this vast receptacle of Forms, each soul is given a causal vehicle appropriate for its evolution. This individual causal body represents the Ideas potentially available for lifetimes of cosmic activity. There is a unique "shape" to how the universal intelligence is available to a given soul.

11 *Aquarius* *The Idea of Man distributed through seven lokas*

The 11th house of the Idea of Man represents the way the intelligence in the soul is relativized through specific conditions in individual centers of experience. Each individual's experience is determined by the conditions of the World-Idea in which it is located. These are broadly characterized according to seven grades of existence called *lokas,* and their corresponding seven *talas.* Applying the 11th house to the evolving unit of life provides each individual with a power to form and transform itself. The individual consciousness can function according to various modes of space, time, and causality that give rise to different kinds of epistemology and different world-views. The sage has knowledge of and access to a variety of world-views or *lokas,* through which he or she may be able to perform seemingly "supernatural" activities.

12 *Pisces* *The Macrocosm in the Microcosm*

The 12th Idea represents the Idea of Man as embodied in the ever-becoming flux of the five Great Elements. Our microcosmic vehicles are constituted of these cosmic elements, and structure our participation in the cosmic soul. Through these vehicles, which provide the locus for an ego in relation to a world, we participate in the sensible experience of an external world. Here we can place both the genetic and the historical development of individuals, including the cultural and historical forms of collective humanity. Sages will express their compassion in particular forms appropriate to the time and place in which they live. The sage's body is a radiation of the divine, permeated with the perfume of divinity. The sage lives consciously divine in each moment of experience.

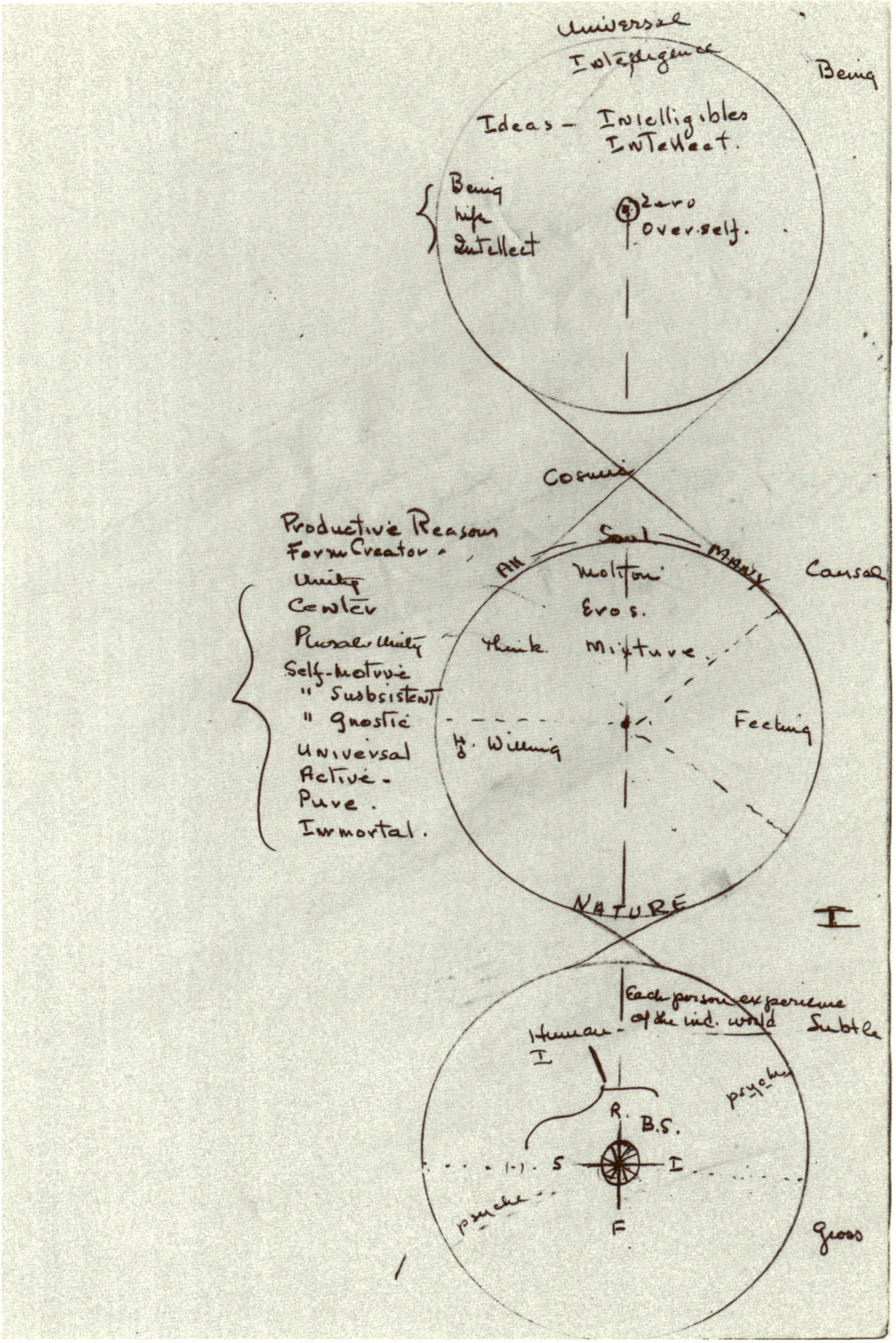

Universal
Intelligence
Being
Ideas - Intelligibles
Intellect.
Being
Life
Intellect
Zero
Over-self.
Cosmic
Productive Reason
Form Creator.
Unity
Center
Plural Unity
Self-motive
" Subsistent
" gnostic
Universal
Active.
Pure.
Immortal.
Soul
All
Many
Causal
Motion
Eros.
Think
Mixture
Feeling
Willing
NATURE
I
Each person experience of the ind. world
Human
I
Subtle
psyche
R.
B.S.
S
I
F
psyche
Gross

Chapter Three

SOUL

Preamble to Soul as Authentic Existence

Soul is a principle of Awareness, but not self-awareness. Nous is intelligible and is self-Aware. Soul, in knowing Nous, is knowing at once itself as well; for it is the Nous' presence in Soul that makes Soul self-aware. Each unit soul is of such a nature—a permanent, indivisible Mind, a living Awareness. In this way also, Soul participates in the omnipresence of the Intellectual Realm. In this inner tendance toward Nous, Soul is also receiving the imprint of the Intelligible Cosmos—which is its World-Idea, its Logos. Because Nous is a self-knowing awareness, which knows itself as intelligible or Idea, when the Soul contemplates Nous it too is possessed of the intelligible and knows it within itself as its guiding Idea. As Plotinus writes in III.9.6: *"Since, then, the activities of self-intellection are intellective-forms, We, the Authentic We, are the Intelligibles and self-intellection conveys the Image of the Intellectual Sphere."*

The contemplation of Nous is the source of Soul's knowing itself, and is source of the World-Idea it will manifest. This is direct knowing by being—without image or reasoning. The life activity inherent in Soul becomes active in virtue of this intelligible presence within it.

Inwardly the soul is intent upon the Intellective Act of the Intellectual-Principle. This we have called in-sight, the conjunction of the soul's inherent receptivity "tendance" or seeing power with the intellection of the Nous. In this seeing "we know the reality of a knowing nature." Here, soul and intellect are unified, yet are distinct. Soul is self-intent, at rest, and its life power is intent upon the receptivity to the intellective light which is its (soul's) in-forming principle. "The Soul itself must exist as Seeing—with the Intellectual-Principle as the object of its vision—it is undetermined before it sees but is naturally apt for intellection . . ." (Plotinus, III.9.5*)*

Any principle of Awareness is traced directly to the One. The One is *"in the way it is awake." (Plotinus,* vi*.8.16, John Deck trans.) It is Pure Awareness or All-Awareness. Intellectual-Principle and Soul are secondary and tertiary "modes" of Awareness. Intellectual-Principle is a self-knowing, intellective Awareness. Soul is an active, living or productive Awareness, and aware of other. The Soul's own Awareness conjoined to the Intellectual Awareness makes soul "intellective." The common thread is Awareness—which is pervasive. In this Contemplation, Plotinus says, "We know, and it is ourselves that we know; therefore we know the reality of a knowing nature . . ." (*iii*.9.6)*

An authentic existence is a principle of Awareness—or Mind. What is authentic remains itself—cannot become other or be objectified. What can be objectified, can become other and so is inauthentic. To be permanently itself, indivisible, is to be authentic.

Anthony Damiani, Notes and Papers, pp. S163–164

An Outline

▼ EDITORS: All references in parentheses within this outline are to Plotinus' *Enneads.*

The third hypostasis [Soul] is an authentic essence, real Being, and characterized primarily by life. (V.1.2, V.1.3, V.1.8, VI.2.6) Soul is said to be both a one and many. (VI.2.5–6, VI.2.1–2) ▼

In Plotinus, Soul as a generic term is regarded in a variety of distinct ways. Nonetheless in all of these its essence or principle is said to be Life—self-motion. (V.9.14, VI.2.6) He distinguishes:

- Soul has its source in the unlived life of the One. We conceive of that aspect of its infinite Life, this unpolluted fountain of all animation, the rootless root, the source from which flows forth Absolute Soul. (V.2.1, III.8.5) We point to this indivisible aspect of the One, the unvarying fountain of life excluding all process and change. This Soul essence is the thread, then, upon which we must focus. This essence flows unhindered from the infinite Nature of the One, through Absolute Soul, Universal Soul, cosmic souls, and individual human souls—remaining self-identical, imparting Grace upon all. (IV.4.10)
- Absolute Soul is the unity of Soul in the Nous, and in this are rooted both Universal and Individual Soul. Within the Intellectual-Principle it is the living matter which simultaneously reflects all the Forms. (IV.3.5, V.1.3)
- The Universal Soul is spoken of as twofold. It is the Abiding Wisdom in eternal contemplation, and secondarily the one life of the universe which provides the reason-principles that "illumine matter." Thus Plotinus tells of the Royal Soul which contemplates the Ideas and Royal Intellect, the reason-principles of the Universe which belong to Zeus. (IV.4.9–10)
- The Individual Souls also are rooted in the infinite Life of the One (VI.9.9) but reach down to association with bodies. (IV.3.4–5) In this last category, Plotinus places planetary souls, star souls, cosmic souls, and individual human souls.

- Each Soul is aprioristically a unity. The omnipresence of Soul does not fragment its identity, for it is at once both divisible and indivisible. (IV.1.1, IV.2.1) Plotinus describes each individual "I AM" as omnipresent, indivisible in its unity; and as indivisible, it manifests the entire heavenly system as a unit within itself. It is also an outflow of graded powers and life, to unfold what it contemplates in the Intellectual-Principle. (II.9.2, III.4.6, IV.3.5) It is a logos or transmitter of the Ideas to the cosmos through its multiple powers.

The indeterminate life of Soul is eternally conjoined with a vision of the Idea of Man. This is the Authentic Man, a true essence, a permanent possibility in the Intellectual-Principle. (VI.7.6, VI.7.5) The tendency of souls toward secession must be with a desire toward a fuller self-realizing vision of this Idea. (IV.1.1)

Because of the prior activity of Universal Soul, the ground-plan, it now becomes possible for an Individual Soul which can be cosmic in its sweep, to elaborate what is to become the objective manifestation. It is also now possible for the Individual Souls that are conjoined with various bodies within that cosmos to participate and evolve therein. (II.9.8, III.3.3, III.8.5, IV.3.2, IV.8.4)

Through their association with bodies, they are individuated, and eventually taught to be self-conscious of their divine origin in the Nous. (V.8.4, V.3.2–3, V.3.6)

Summing up: For Plotinus, Individual Soul must be of such a nature: a unity and a multiplicity, divisible and indivisible. At its highest—in Divine Mind—it is a unity, but it has phases and powers which unfold through various levels down to the sensible world. But even in its association with body, with the universe which has been illumined by Universal Soul and elaborated by an individual cosmic soul, it remains impassive. Soul is entire and indivisible even where its distinct powers are operative. (IV.2.1–2, IV.3.19, IV.3.8, IV.3.11, IV.3.7, V.1.2, IV.8.8)

Essence of Soul as Indivisible and Divisible

▽ Editors: The psychosomatic organism is the mind-body complex, and is approximately equivalent to the ego.

⊞ Plotinus, iv.1.1: In the Intellectual Cosmos dwells Authentic Essence, with the Intellectual-Principle (Divine Mind) as the noblest of its content, but containing also souls, since every soul in this lower sphere has come thence: that is the world of unembodied souls while to our world belong those that have entered body and undergone bodily division.

There the Intellectual-Principle is a concentrated all—nothing of it distinguished or divided—and in that Cosmos of unity all souls are concentrated also, with no spatial discrimination.

But there is a difference:

The Intellectual-Principle is for ever repugnant to distinction and to partition. Soul, there without distinction and partition, has yet a nature lending itself to divisional existence: its division is secession, entry into body.

In view of this seceding and the ensuing partition we may legitimately speak of it as 'divided among bodies.'

continues on p. 173

Our Venerable Guide reveals that he has penetrated into the mysteries of Soul, its greatness and wonder, a Unity-Being integrally present in all its graded powers. This is not knowledge by mere acquaintance—that is, the ego's capacity for creating concepts from the remembrance of previous bodily modifications and manipulating them to our heart's desire. No, this is a knowledge that goes beyond such formulation; for this is that principle which illuminates the thoughts and is itself not one of the thoughts. This unobjectifiable and immaterial principle, which is illuminating the ego's functioning, is to be differentiated from the ego's functioning and isolated in the meditational practices. Only after that is done can the investigation or inquiry into the nature of this principle, the Soul, begin. We are now speaking of Plotinus' intimacy with this innermost core of his being, which is an offspring from the Absolute Soul.

In general, Plotinus is very guarded in his mystical utterances. Understanding them will require a very careful reading of the text, even by those who have some proficiency in these matters. Reading mystical literature can not be accepted as a substitute for such inner experience. We must make clear to ourselves why this is so. The functioning of the psychosomatic organism ▽ is accompanied by bodily modifications, and these appearances—duly noted, registered, classified, stored away in the memory—are the means whereby the individual ego is creating its world, both subject and object. In its conjectures as to the meanings of these appearances, the ego is dominated by the instinct for survival and biological adaptation. This realm of existence must be sharply distinguished from the immaterial principle of Illumination to which we have referred and which is altogether on a different plane.

We cannot unreservedly apply the logic of the sensible to this immaterial principle. What is required for us to have accurate knowledge about this pure principle is an intimacy with it and a recognition of it as our "Isness," in contradistinction to identification with a psychosomatic organism. If we allow ourselves to disregard those persons who have had intimate glimpses of this higher Self, and deal rather with that rare person who has profoundly established residence in egoless Being—in other words, with one who *knows*—we may be able to reorganize our conceptual understanding so as to evoke the intuition within ourselves.

In IV.1.1 Plotinus points out that Soul is a content of the Divine Mind included with other Divine Beings, permanently stationed and resident there. ⊞

Here, Soul or Mind is included in an undifferentiated unity and has a tendency or desire towards secession. We may also rephrase this in accordance with many previous statements: What is authentically real has an outward-facing hypostasis; what is perfect engenders, and the eternally perfect engenders eternally. (cf. V.1.6) We may take this to mean that Absolute Soul eternally engenders souls. Each and every soul, offspring of Absolute Soul, is similar to but not identical with its prior.

This emanation, this thought of the Mind of God—our Absolute Self ⊕—is what we have to investigate, for this is the key to the mysteries. [34, p. 330] This unit of Being-Life-Consciousness goes forth, "formed from the undivided essence and the essence divided among bodies." (Plotinus, IV.1.1) We may refer to this as the individual soul or mind; this point, which is the combination of the mind or soul with the World-Idea, is our absolute individuality or Overself, and is our concern here. We now must try to understand something about the nature of this individual mind or soul. This will require that we review some passages from Plotinus and in great detail itemize our understanding of his statements.

In IV.2.1, Plotinus speaks of this offspring as characterized by unity which it derived from its prior. He says that each soul is distinct from every other, and that each is an intermediate between Absolute Soul and the Cosmic Circuit or World-Idea. Furthermore, regardless of what it is occupied with—either the Idea of the universe or a body within it—Soul gives itself to either or both without abdicating its unity. What, in effect, he is saying

continued from p. 172

But if so, how can it still be described as indivisible?

In that the secession is not of the Soul entire; something of it holds its ground, that in it which recoils from separate existence.

'Formed from the undivided essence and the essence divided among bodies': this description of Soul must therefore mean that it has phases above and below, that it is attached to the Supreme and yet reaches down to this sphere, like a radius from a centre.

Thus it is that, entering this realm, it possesses still the vision inherent to that superior phase in virtue of which it unchangingly maintains its integral nature. Even here it is not exclusively the partible soul: it is still the impartible as well: what in it knows partition is parted without partibility; undivided as giving itself to the entire body, a whole to a whole, it is divided as being effective in every part.

⊕ EDITORS: Here the Absolute Self means the absolute individual—that is, the Overself or unit soul—and should not be confused with the Absolute Soul, which is the non-individual source of Soul in Divine Mind.

is that if the mind thinks the thought of a pinhead or of the continent of Australia, it is totally present in each of the items without its unity being sundered. In other words, the omnipresence of Soul does not fragment its identity, for it is at once both divisible and indivisible: "Its divisibility lies in its presence at every point of the recipient, but it is indivisible as dwelling entire in the total and entire in any part." (IV.2.1)

The recitation of these words is not a guarantee that the meaning accompanies their repetition. Plotinus is referring here to the mystical experience of this illuminating principle as egoless Being-Consciousness. In other passages he even refers to it as the "All." (cf. III.9.3 and III.4.3) We must bear in mind that the ego's intellect cannot—by the very nature of its own makeup—understand, let alone feel, the reality of such a principle (which is illuminating its activities) and, in all honesty, cannot accept such a position.

If we examine these passages with the attention they deserve and deduce the staggering implications that arise as to who and what we are, a shift in the orientation of a culture may take place. However, there is no cause for undue alarm—the Philosophy of Truth will not become a popular "ism" in the near future. The inner erudition and outer perspicuity required is a call to heroic souls to reclaim their own.

Overview of Levels of Soul

We may refer to the three primal hypostases as the Absolute in order to emphasize the transcendent aspect as well as their remoteness and inconceivability. ▽ ⊞ If we do so, then Plotinus refers to the third degree of this mysterious Void as Soul. In some regards Soul is the most complex and difficult of all notions to articulate, because of its scope and diversity of application. So we are required at the very outset to try and organize our understanding of the manifold of meanings it is invested with, so that whenever the term is used the appropriate referent is brought into view and the proper context of connotations applied. ⊕

Throughout Plotinus' text, statements about the Soul are articulated with a planned obscurity. We must remember that these were esoteric teachings and not intended for public or mass consumption. The *Enneads* are formulations of Plotinus' intuitive realizations of ultimate truths to which only our inner being may respond. We must let the Logos in our soul absorb the impact and assimilate the meaning of his intuitions prior to allowing our critical and egotistical intellect to pounce upon them. In other words our mental activity must be stilled, so that the Silence within can receive the passage in question without coloration. This may help us to understand the ordered classifications of the various levels of Soul and how they fit in with the overall scheme of reality that is implied.

The issue is crucial, for its ramifications extend from the unitary life of the universe to the least of creatures within it. For within the universe are to be found all these various grades of life or Soul, suspended from above. (cf. IV.8.4) Each is fixed at that grade of reality appropriate to it.

There cannot be any ambiguity in our understanding if we are to properly appreciate this wisdom-knowledge Plotinus is delivering. The profundity of his understanding makes

▽ Anthony Damiani, Notes and Papers, p. S79: Only in analysis can we logically evaluate and isolate one hypostasis from another. In actuality the view of simultaneity will prevail, for from our view of manifestation the three hypostases (One, Nous, Soul) can be considered together as Absolute.

⊞ Editors: *Astronoesis* pages 41–42, 47, and 71 in chapter I discuss the three primal hypostases.

⊕ Editors: A key point of this essay is that "appropriate" knowledge, i.e., *correct* metaphysical understanding, enhances mystical experience. As Anthony quotes from Paul Brunton: "Accurate ideas about the nature of the soul he is seeking to unite with—that is, right thinking—will not only not hinder his venture in meditation but actually promote its success." (*Notebooks,* v. 11, p. 51)

it possible for us to distinguish the definition of Soul, the differentia that exist among souls, and the consequent levels of reality that they are being co-ordinated with. This complex of problems associated with Soul must be comprehended and not explained away by reducing the many souls to one soul, or confusing them with one another.

We will attempt a brief resumé or outline of the term Soul and try to reveal the hierarchy implicit in his exposition.

At the highest level we must understand Soul as an indivisible aspect of the One itself. Here the formless form of the unsullied Life may be recognized as the transcendent paradigm or immaculate version of this principle from which flows a whole hierarchy.

From Its [The One's] perfection there radiates the life of the Intellectual-Principle or Absolute Soul, which is the recipient of all the Ideas. Just as the life of the Intellectual-Principle is determined by all the Ideas, so reciprocally are all the Ideas made living and permanent possibilities or presences within the Intellectual by Soul.

The next hierarchical level of Soul, and not to be confused with Absolute Soul, is the Soul of the Universe or the Demiurge—Zeus. Here Soul is the principle of universal manifestation, and it is referred to as twofold. On the one hand, it eternally contemplates the Ideas or the wisdom of God, a timeless and abiding wisdom. On the other hand, it organizes the *samskaras* or past traces of universal manifestation (energy = matter) into a basic or primitive plan of the universe to be. Plotinus refers to this vast and cosmic sweep of its intellectual fabrication as the "tentative illumining of matter" (VI.7.7), or the ground-plan of the universe to be.

Next follows that order of Souls which we may for convenience call Star Souls. Each Sun is such a soul. (We are ignoring for the moment each Sun's intellect and unity, that resides in the Intellectual-Principle.) Each of these Star Souls is a part of universal manifestation, and in their totality these souls constitute the inerratic or immutable sphere of Ideas. Also each Sun or Star is unfolding or manifesting within itself, through its own powers, the content of its own Logos, i.e., a world or a cosmos of its own. So for instance our Sun is unfolding or manifesting its own world through its own powers. These powers are symbolically represented by the planets, which are both the agents and the vehicles that manifest that Sun's intellectuality.

Each of those planets is also a god in its own right, manifesting or unfolding the master plan inherent in the Sun's intellectuality. Each provides in its respective sphere habitations for the next degree or level of Soul. (The vehicles or sheaths are provided by the planetary gods to individual souls, which will then experience that particular world by the bodies thus supplied.)

Ignoring for the present the herd or tribe of *daimons* (angels) [35, p. 331] associated with each mundane god (introduced by later Neoplatonists), we come to human souls. Here we must remind ourselves that Plotinus points out that there is a part of the soul that remains in the Nous, but it emanates a reflex of itself, a principle of intellectual Being that is similar to itself. It is this unit soul, the God within us, that may be referred to as our soul, our ultimate individuality, and the true knower within us. It is "man made in the image of God." We hope to explain later on how this formless and infinite consciousness gets contracted and starts on a journey of self-realization. In order not to lose the thread of continuity, we will remark now only that this soul or mind is "reduced" ▽ and will ultimately become identified with bodily vehicles that are being fabricated for its use in order to experience the sensible, i.e., complete its contemplation.

We may refer to this contracted consciousness as impersonal or witnessing consciousness, associated with the planetary mind from which it receives the image of the World-Idea. At this level the world image is in a pristine and subtle form, an alogical unity that includes all the features that will be found later on in the reflective consciousness. Included in this world image is the particular body which is being evolved, with which the human soul identifies.

The entrance of the soul into the world, to associate with a particular body, is determined by the root tendencies inherent in the soul. These compel the soul to identify with a body with a certain phylogenetic and ontogenetic history produced at a certain moment of time, which is indicated by aspectual relationships among the planets. The human soul is now a fated member of the "Circuit." ⊞ Here the experiences provided through the body will develop the soul's faculty of understanding and reason through an indefinite series of incarnations. Soul incarnates or participates in the Ideas of the planetary mind by perceiving the world and experiencing, through a body, what the gods think. Thus Earth—footstool of the Gods—nurses our soul to heavenly maturity and humanity.

▽ Editors: "Reduced" may mean contracted, limited in scope and power, or concentrated, or put into equivalent lowest terms. The ray of soul is associated with the Earth-centered perspective. It must have within itself a distillation of the karmic tendencies, personality, and intelligence that make it unique and determine the moment of birth.

How is soul reduced or contracted? Part of the answer here is attention. Whatever the soul puts its awareness on exists for it at the time. Soul is entire in each perception. The vehicles of the soul focus its attention on certain aspects of the World-Idea. Our human soul, for example, associates with the reason-principles of the Earth mind and the traces of life activity and elements. The paper "Omnipresence of Soul to the Cosmos" (p. 205) explores how Soul associates with the cosmos.

⊞ Editors: For Anthony and Plotinus the cosmic circuit is the solar system together with its planetary powers.

Diagrams of Levels of Soul

SOUL MUST FIRST BE UNDERSTOOD as an aspect of the Indivisible, as one of the forms of the formless First; then as Absolute, i.e., as having an intellect of its own; then as hypostasis, or as principle in its own right. Finally we must see Soul's relation to the effects it produces through its own activity; that is, we have to understand that Soul is an eternal principle *sui generis* in juxtaposition to its temporal activity, which latter is not properly speaking hypostasis.

A summary with accompanying diagrams may help us to make clear the variety of contexts in which Plotinus uses the term Soul. [*fig. 73*]

EDITORS: Soul as the outflowing of life intelligence, one essence with divisible aspects, can be symbolized by the detriment (Soul) ring. In each quadrant, Soul expresses its life and is reciprocally determined in accordance with that level of reality.

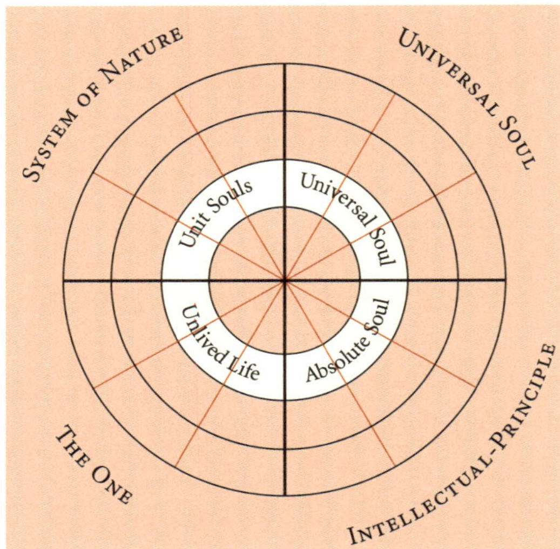

FIG. 73

Soul in the One

The meaning of Soul must first be understood in the context of the One. Plotinus indicates that "it has its source in the First and thence, along an unhindered path, it [Soul] flows into a total of things, conferring grace." (IV. 4.10) This indivisible aspect of the One is the unvarying fountain of life excluding all process and change. [*fig. 74*] ▽

Conceiving of the One as having a fourfold nature (and of course this is an obviously incorrect usage of words, for there is no heterogeneity in the One), its infinite and operational nature is referred to as Soul, which is Unity, the root principle. The continuing identity of this essence can be identified regardless of the usage the term is put to. The entire ring suspended from the Absolute Life of the One represents one view of this multiple unity. Figure 74 illustrates our foregoing remarks on [how] the essence of Soul, as it traverses the hierarchy of Being, provides for the possibility of conceiving the Supreme Identity. That is, the life of any individual creature is ultimately rooted in the One. The life

▽ Plotinus, VI.9.8–9: Thus the Supreme as containing no otherness is ever present with us; we with it when we put otherness away. It is not that the Supreme reaches out to us seeking our communion: we reach towards the Supreme; it is we that become present. We are always before it: but we do not always look: thus a choir, singing set in due order about the conductor, may turn away from that centre to which all should attend; let it but face aright and it sings with beauty, present effectively. We are ever before the Supreme—cut off is utter dissolution; we can no longer be—but we do not always attend: when we look, our Term is attained; this is rest; this is the end of singing ill; effectively before Him, we lift a choral song full of God.

In this choiring, the soul looks upon the wellspring of Life, wellspring also of Intellect, beginning of Being, fount of Good, root of Soul. It is not that these are poured out from the Supreme, lessening it as if it were a thing of mass. At that the emanants would be perishable; but they are eternal; they spring from an eternal principle, which produces them not by its fragmentation but in virtue of its intact identity: therefore they too hold firm; so long as the sun shines, so long there will be light.

Editors: Soul is rooted in the One by virtue of the unities symbolized by ♀♈ ♁♉ ♃♊.

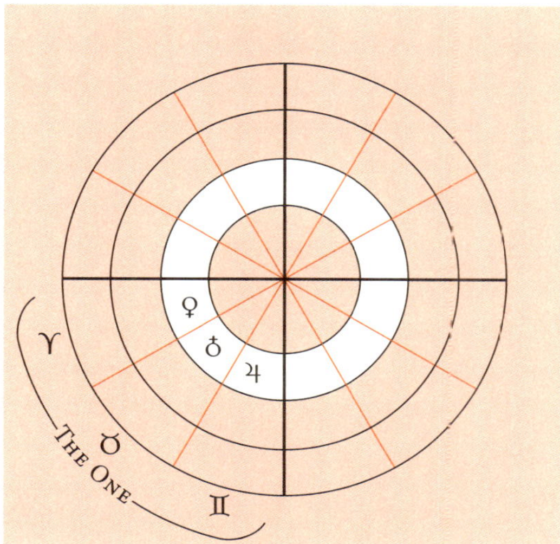

Fig. 74

of the One is unsullied, uncontaminated, pure, "unlived life." Plotinus speaks about this life as the undivided power of the One which radiates forth. That is, it is to be conceived of as the double act itself, and the principle of divisibility which remains indivisible regardless of its station. [36, p. 331]

Absolute Soul

Flowing from the Life Supreme—the Source and Principle—is Absolute Soul, life emanated and simultaneously determined by and as Intellectual-Principle. We have this important distinction to consider at the onset—Soul as an aspect of the transcendent Nature of the One, and Soul as the Absolute Life of the Intellectual-Principle. Life essence is thus required by the One, and goes forth from it just as the Intellectual-Principle does. There is a reciprocal and mutual determination on the part of each.

In the Nous, Soul essence is the recipient of the eternal Ideas and is the Life Absolute of the Divine Mind. In the Intellectual-Principle this essence receives into itself the Ideas, which makes of it a living and abiding wisdom. ▼ Considered most precisely as the Intellect of the Soul, Absolute Soul is an eternal Presence—totally and unextendedly present in the Now—Life instantaneously infinite. It does not descend, for as descending it would annul its own Perfection. Yet the super-abundance of Soul gives of itself ceaselessly. It is "the entire stream of life sent forth by that Principle." (Plotinus, v.1.3) ⊞ Intellectual-Principle remains abiding as the Kingdom of Being; Soul proceeds. [*fig. 75*]

▼ Editors: Anthony writes further of the inseparability of Soul and Nous: "We treat Soul as essence, that is, isolating its definition from the Ideas in order to conceive its Being as other than the Being of the Nous—to trace it from its origin as it winds its way though the various levels of reality, carrying with it the paradigm of the double act which we will find operative at each level. Of course this isolation cannot in fact be achieved because of the eternal interrelationship between Nous and Soul." (Notes and Papers, p. S76)

⊞ Plotinus, v.1.3: Soul . . . is yet a secondary, an image of the Intellectual-Principle: reason uttered is an image of the reason stored within the Soul, and in the same way soul is an utterance of the Intellectual-Principle: it is even the total of its activity, the entire stream of life sent forth by that Principle to the production of further being . . .

The principle of Life or Soul is self-motive; that is, it springs into activity of itself. The paradox is that this essence is the one principle of divisibility—a single essence, a self-identical nature, even though it divides itself endlessly. This brief amplification of Plotinus' terse definition of Soul as One-and-Many can be contrasted to his reference to Intellectual-Principle as a One-Many.

This principle, Life Essence, *is*—eternally self-begotten. Its determination by Intellectual-Principle is to be kept distinct from it in order for us to understand this essence in itself.

Being is not added to it to comprise it as Life Essence. Being, so fundamentally comprehensive and all-encompassing, prevents the necessary refinement required to distinguish its complementary—Life—so that we are always mistakenly reducing Soul's distinctive aseity to this type of isness [i.e., that of Intellectual-Principle]. Nevertheless, Plotinus clearly indicates the distinction between Soul Being and Intellectual Being. (cf. VI.2.5)

Self-identical in each manifestation, the Intellectual-Principle contains all possible manifestations of itself within itself. The Ideas remain indivisible, eternally abiding and immutable. They are participated entire and cannot be divided without losing their proper aseity. Soul is the very means by which the Ideas in the Mind of God, which are permanent possibilities, become actualized and expressed. Consequently, Soul or Life divides or duplicates itself endlessly, remaining one self-identical essence, a principle *sui generis* in each of its offspring.

Soul in the Nous is recipient of all the Ideas simultaneously, so that the whole of its life is totally present and complete in the Now. [36B, p. 333; 37, p. 333]

EDITORS: Soul as the absolute receptor of all the Idea-intelligences in the Intellectual-Principle is termed Absolute Soul and represented as ♄♋ ♄♌ ♃♍. As the life of the Intellectual-Principle, Soul is an abiding Contemplation. Anthony says of this relation that Soul is in the Nous and Nous is in the Soul. Based on its source in the Nous (Intellectual-Principle), every individual soul is an inviolable unit of consciousness and power.

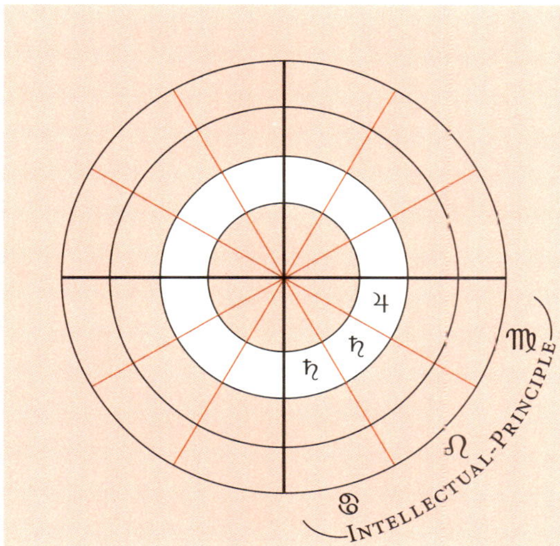

FIG. 75

Universal Soul

The Universal Soul is the Providence presiding over the universe, the principle of universal manifestation. It is the one life of the cosmos, the logos or transmitter of Forms to the universe. The name Zeus, used in Greek theology, signifies that through which life is imparted. Zeus is linked in his inner vision to an unchangeable abiding wisdom.

Two aspects of Demiurgic activity have to be considered in producing the groundplan or sketch of universal manifestation: the Ideas and the power necessary to actualize them. The former [the Ideas] are associated with the third quadrant, the Ideas which will be developed and evolved into the system of Nature or the fourth quadrant. The latter [the power to actualize them] is associated with the third ring or dynamic aspect of Soul.

In the fourfold diagram, the third quadrant represents one view of the Universal Soul as a self-subsisting hypostasis. The third-ring view represents another. [*fig. 76*]

Editors: Anthony identifies the Universal Soul with the third quadrant, and particularly associates the Demiurge with Jupiter in Sagittarius ♃♐. The Demiurge tentatively illumines the universal matter, providing the ground and receptacle for the self-actuating system of Nature. The Demiurge is Providence. *By receiving (or envisioning) the Ideas in the Intellectual-Principle, and transforming them into an Idea of the universe, the Demiurge* provides *what all souls need to be nourished in manifestation.*

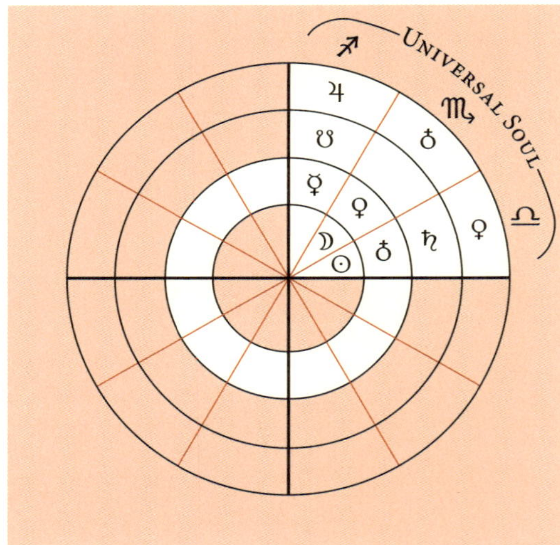

Fig. 76

The perfection inherent in Absolute Soul—Soul in Intellect—has as its out-going act the Universal Soul, or the Demiurge. The more revealing appellation of twofold Zeus refers again to a version of the double act at a lower level of reality. It must be clearly stated that the term "twofold Zeus" does not apply to a combination of Absolute Soul and Universal Soul. Rather the term applies to the double nature of Soul as hypostasis, that is, both to its inner contemplation of the Ideas* intellect in Soul—and its perpetual creativity, the Royal Intellect productive of the ground-plan of universal manifestation (fourth quadrant). These two aspects are the Demiurge, Soul considered as a distinct hypostasis. [*fig. 77*] [38, p. 333]

> Zeus—ordering all, governor, guardian, and disposer, possessor for ever of the 'kingly soul' and the 'kingly intellect', bringing all into being by his providence . . . (Plotinus, IV. 4.9)

> The ordering principle is twofold; there is the principle known to us as the Demiurge and there is the Soul of the All [universal manifestation]; we apply the appellation Zeus sometimes to the Demiurge [♃♐] and sometimes to the principle conducting the universe [♃♓].
>
> When under the name of Zeus we are considering the Demiurge we must leave out all notions of stage and progress, and recognize one unchanging and timeless life.
>
> But the life in the Cosmos, the life which carries the leading principle of the universe, still needs elucidation; does it operate without calculation, without searching into what ought to be done?
>
> Yes: for what must be stands shaped before the Cosmos, and is ordered without any setting in order: the ordered things are merely the things that come to be; and the principle that brings them into being is Order itself; this production is an act of a soul linked with an unchangeably established wisdom whose reflection in that soul is Order. It is an unchanging wisdom, and there can therefore be no changing in the soul which mirrors it, not sometimes turned towards it and sometimes away from it—and in doubt because it has turned away—but an unremitting soul performing an unvarying task. (Plotinus, IV. 4.10)

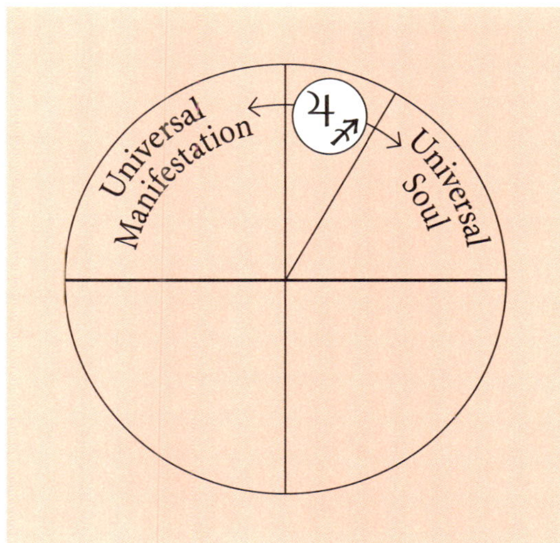

FIG. 77

The highest phase of the individual soul has its eternal residence in and vision of the Divine Mind. ▼ From that indivisible Soul remaining in the Nous is emanated an eternal offspring. This, its "issuing phase," is the Mercurial or *Asmita* aspect of individual soul, whether cosmic, planetary, or human.

As this principle exists in the Nous, it gives rise to the many I AMs which are of the essence of Soul. [*fig. 78*] Thus each soul retains eternally its aseity as an indivisible and individual emanant which includes the principle of divisibility. This individual unit of life is the principle of manifestation for every individual soul. ■

> No doubt the task of the Soul, in its more emphatically reasoning phase, is intellection: but it must have another as well, or it would be undistinguishable from the Intellectual-Principle. To its quality of being intellective it adds the quality by which it attains its particular manner of

▼ Editors: The highest phase of individual soul is represented by Saturn in Leo ♄♌. Here, each soul has a unique vision of the Nous; "the whole of its life is totally present and complete in the Now." (*Astronoesis*, p. 181) The following papers "Powers Intrinsic to Soul" and "Cosmic Soul" develop the relationship between individual soul and its root in the Nous.

■ Editors: Individual souls enjoy all of the characteristics of soul essence. For most of us these are "merely" potential, but they all are *inherent* in what Soul is. Individual characteristics are emphasized only in soul's relations to and within the evolving World-Idea. In the system of Nature, souls determine and exhibit different levels and powers; yet the unity of Soul is the inviolable unlived life of the One, its intellect is the presence of Ideas, and its rational powers participate fully in the manifesting energy of the Demiurge.

Editors: Individual soul is rooted in the Nous, in ♄♌ and ♃♍, and has a continuity through all the Soul ring planets. Finally, in Pisces are the many individual principles of manifestation, ☿♓, which produce their own worlds; each is "a self-identical, undivided individual mind without beginning or end." (Anthony Damiani, Notes and Papers, p. S81)

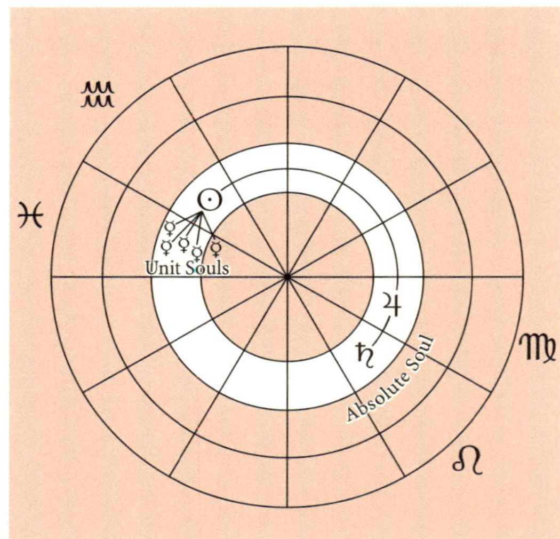

Fig. 78

being: it ceases to be an Intellectual-Principle, and has thenceforth its own task, as everything must that exists in the Intellectual Realm. (Plotinus, IV.8.3)

Whether we consider cosmic soul or individual human souls, ⊕ the I AM principle—remaining an indivisible unit of soul essence—is in inward contemplation of the Ideas, the logos principle within it. As the principle of manifestation, its outward act is the projection and production of its universe.

> Each soul is permanently a unity (a self) . . .
>
> Thus the gist of the matter is established: one soul the source of all; the many founded in that one, on the analogy of the Intellectual-Principle; those many are at once divided and undivided; that Soul which abides in the Supreme is the one expression or Logos of the Intellectual-Principle, and from it spring other Reason-Principles, partial but immaterial, exactly as in the differentiation of the Supreme. (Plotinus, IV.3.5)

The universality of the "I" is qualified by an immaterial "AM," and each one of these "I AMs" is linked to a reason-principle in the Intellectual which it will manifest. [*fig. 79*]

In tractate IV.2.2, Plotinus establishes the canons of a higher logic of Soul. He has already pointed out what is required in order to understand Soul; his words will appear paradoxical if we have decided

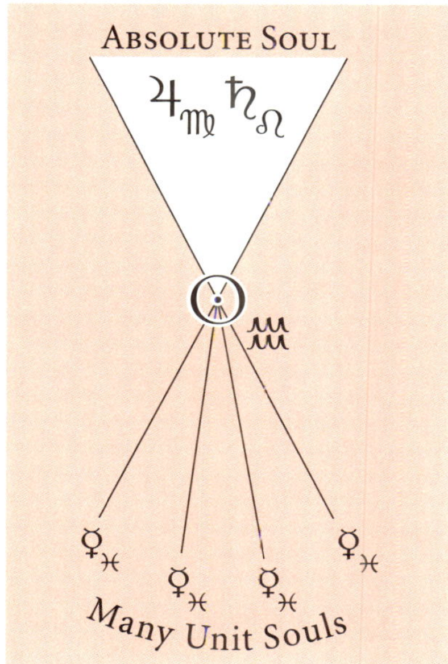

FIG. 79

⊕ EDITORS: Anthony uses the term "individual soul" for a wide variety of souls that emanate from Absolute Soul:

> When you speak about Absolute Soul, it's not the soul *of* anything. But as soon as you start speaking about life in the sense that it's in manifestation, then you're speaking about either the soul of the cosmos or the soul of this plant, this animal, whatever. You're speaking about a unit soul. And if you have a unit soul, the World-Idea is inevitably associated with it; it has to be. There isn't an undetermined life anywhere. —transcript, 2/15/84

Individual or unit souls are not "parts" of soul, but ways of seeing (experiencing) the whole of Soul as a principle of individuation. We may think of souls as rays which continually pour out from the Supreme. We cannot completely separate individual unit soul from Absolute Soul, nor separate the notion of individuality from the system of Nature.

"Cosmic soul" refers to a variety of "individual" souls, but usually means the Sun soul. Cosmic means that it is participant with the Universal Soul in manifesting a cosmos, providing a habitation for other individual souls. There is an uncountable nested hierarchy of cosmic souls, when considered in terms of their manifestations. However, all cosmic souls are also essentially of the nature of individual soul, having an individuality connected with the cosmos, and an inviolable essence.

beforehand that Soul cannot be both a one or unity and "multipresent" in its graded powers. This descriptive definition of Soul is an absolute prerequisite to his theory of knowledge.

> It can be demonstrated that soul must necessarily be of just this nature, and that there can be no other soul than such a being, one neither wholly impartible nor wholly partible but both at once. (IV.2.2)

In the extraordinary revelatory passage, VI.2.6, Plotinus refers to the soul's tendency to be divisible about body, while retaining its individual essence. He is here referring to individual soul as an eternal offspring of the Absolute Soul.

> A single Existent, it [Soul] makes itself many by what we may call its motion: it is one entire, but by its striving, so to speak, to contemplate itself, it is a plurality. . . . The cause of its appearing as many is this contemplation, and its purpose is the Act of the Intellect; if it were manifested as a bare unity, it could have no intellection, since in that simplicity it would already be identical with the object of its thought. (VI.2.6)

The quality and the grandeur of each individual unit of life is specifically related in tractate V.1.2.

> Let every soul recall, then, at the outset the truth that soul is the author of all living things . . . but soul, since it can never abandon itself, is of eternal being.
>
> . . . each separate life lives by the Soul entire, omnipresent in the likeness of the engendering father, entire in unity and entire in diffused variety.

Exceeding the bounds of belief, he describes each individual "I AM" as omnipresent and as indivisible in its unity. As indivisible, it manifests the entire heavenly system as a unit within itself.

If this point is not seized, any interpretation of Plotinus' epistemology will go astray. The life of each unit soul is the medium in which this reality is being manifested. The grandeur and sublimity of this principle cannot be retold but must be experienced.

Powers Intrinsic to Soul

THE SCOPE AND AMPLITUDE of the powers of the soul are so immense that they are overwhelming when we begin to get an insight into them. Only those which have the intellective nature are true soul powers.

> Thus its [Soul's] substantial existence comes from the Intellectual-Principle, and the Reason within it becomes Act in virtue of its contemplation of that prior; for its thought and act are its own intimate possession when it looks to the Supreme Intelligence; those only are soul-acts which are of this intellective nature and are determined by its own character . . . (Plotinus, v.1.3)

All the reasoning powers of the Soul are prefigured in the Jupiter principle in Virgo (♃♍). ▼ Here reason embodies the six forces of nature in and through the soul's life essence. [*fig. 80*] [39, p. 335]

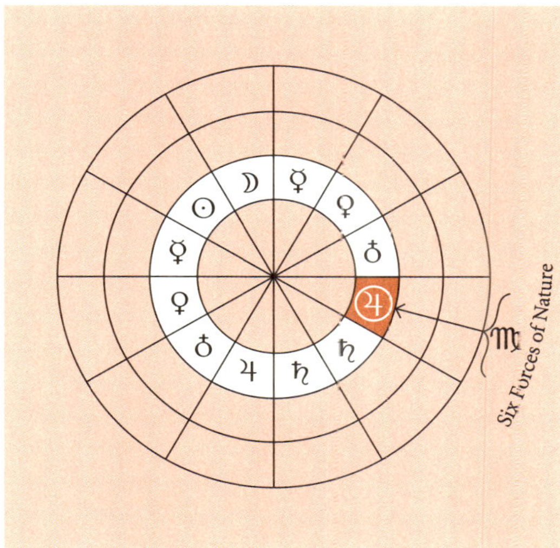

FIG. 80

▼ EDITORS: "Reasoning" in this context means a fundamental act common to all individual souls. Through this vital act, Soul expresses its manifoldness—coordinating the creative powers of each manifesting soul with the intelligible form of its own highest phase (which remains in the Nous). As such, it is the middle phase of individual soul, mediating that soul's intelligible and sensible phases.

These powers of the soul, with their content of living reason essences (Ideas in the soul), are operative as the six forces T.S. Row mentions in his article "The Twelve Signs of the Zodiac." These six are *Parashakti,* the highest and unifying power; *Kriyashakti,* the power of action and will; *Icchashakti,* the power of devotion, feeling, and desire; *Jnanashakti,* the power of knowledge and thinking; *Kundalinishakti,* the power of the continuity of life and embodiment; and *Mantrashakti,* the power of the word, or *manas,* and productive imagination. [*fig. 81*] ▼

Through these powers, soul can develop the pure thought of the highest phase, Saturn in Leo ♄♌, into sensible image Mercury in Pisces ☿♓. These universal powers represented in the Soul ring will be imparted to the Mercurial phase of an individual soul concomitant with the cosmic soul's manifestation of the cosmos. [*fig. 82*] ⊞

▼ Editors: From the source or principle of reason, Jupiter in Virgo ♃♍, the reasoning of Soul flows through the third quadrant into the system of Nature. Each dignity in the ring represents a Soul power articulating the vision of Soul and absorbing the Ideas of the particular house in which it is located.

Soul's vision in the Nous, the realm of pure Thought, is represented by Saturn in Leo ♄♌. That vision is in turn transformed through these powers and present within Mercury in Pisces ☿♓, the manifesting principle of cosmic soul.

⊞ Editors: Here Sun in Aquarius ☉♒ represents the principle of all the Solar Logoi, while Mercury in Pisces ☿♓ in the Soul ring represents the many individual souls. Every unit soul, represented by this Nature Mercury in Pisces ☿♓, participates in the universal powers represented in the Soul ring. These are what enable it to manifest and know a cosmos for itself.

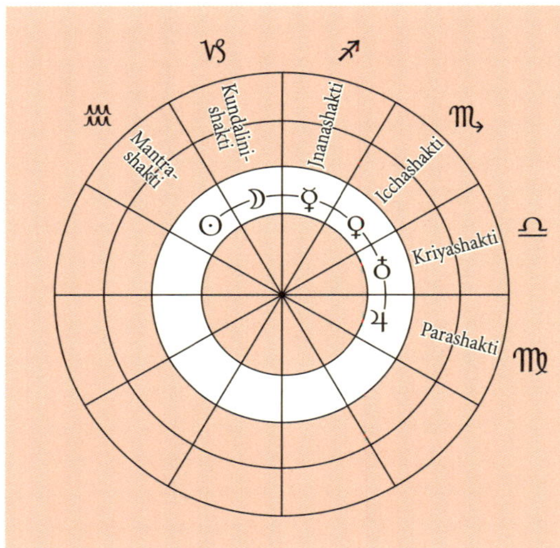

Fig. 81: One pairing of *shaktis* and Soul powers

Through these powers, the reasoning phase of the Sun soul combines the forms according to the soul's vision in the Nous. It must produce a world according to the ground-plan of the universe. [*fig. 83*]

The rational [reasoning] soul powers of the cosmic soul are imaged as the planetary spheres. These are the instantiation of the activity of the six *shaktis* or forces of nature.

> The World Soul is distributed so as to produce the fixed sphere and the planetary circuits corresponding to its graded powers: so with our souls; they must have their provinces according to their different powers, parallel to those of the World Soul . . . (Plotinus, III.4.6)

The activity of these rational powers will translate and combine the reason-principles to produce all the manifest creatures and their psychosomatic existences. We have to

EDITORS: Soul expresses its inner vision in the Nous (Saturn in Leo ♄♌) as it elaborates for itself the illumined matter provided by the Demiurge (Jupiter in Sagittarius ♃♐). Both these sources combine in ☿♓, the manifesting phase of the soul of the Sun. Further, these two sources help to understand why the Emperor Julian in his "Oration to the Sovereign Sun" (p. 59) describes the Sun's emanating light as "the secretion of forms and the concretion of matter."

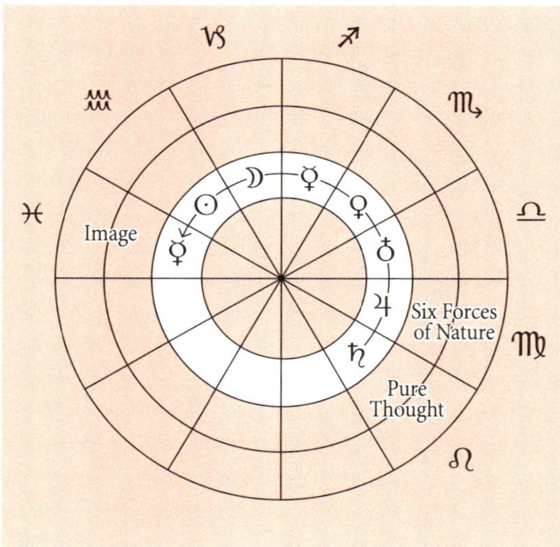

FIG. 82

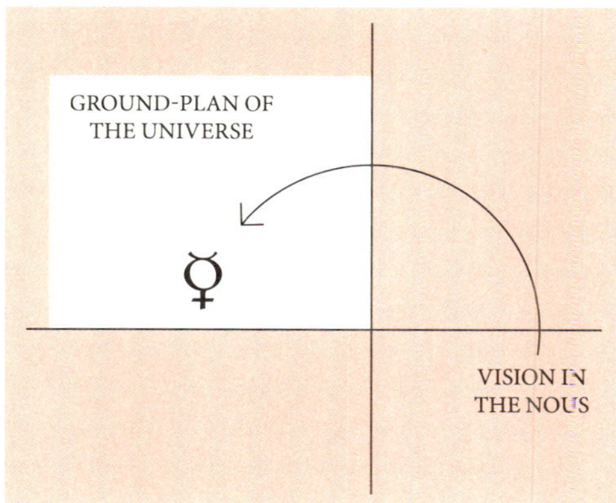

FIG. 83

conceive of this picture as the subtle topography of both the cosmic (or Sun) soul, and all the individual souls in the solar system that will evolve within it. These powers are working to fabricate bodies. The light of the Sun, of which ninety-nine percent of the solar system is composed, is the lowest bodily phase of our Sun soul. What we see as its light emission is an image of the Sun's own bodily phase.

The state of Thought has passed to that of Image. All the prior principles are imaged in the emanated light of the Sun as the manifest cosmos. The Ideas and rational soul powers are imaged in this imaginative projection. This means that within the cosmos of the Sun we find an image of the Intellect, an image of its rational powers as the Chaldean planetary spheres, and of its body as the elements composing the bodies of the planets and all the creatures on these. [*fig. 84*] ▼ ⊞

▼ Editors: Three "levels" of the sensible universe imaged by the Sun correspond with the levels of the tripartite Soul: The inerratic sphere or stars is an image of Saturn in Leo ♄♌ (Intellect), the planetary spheres (the Sun's power(s) of reason) are an image of Jupiter in Virgo ♃♍ (Reason), and the Earth's elements and storehouse of imaginal essences are the third phase of Soul, (Nature) Mercury in Pisces ☿♓ (embodying phase).

⊞ Plotinus, iv.3.30 (K.S. Guthrie trans.): Perhaps reason, whose actualization always accompanies thought, has the function of receiving it and transmitting it to imagination. Indeed, thought is indivisible, and so long as it is not evoked from the depths of intelligence, it remains as it were hidden within it. Reason develops it, and making it pass from the state of thought to that of image, spreads it out as it were in a mirror, for our imagination.

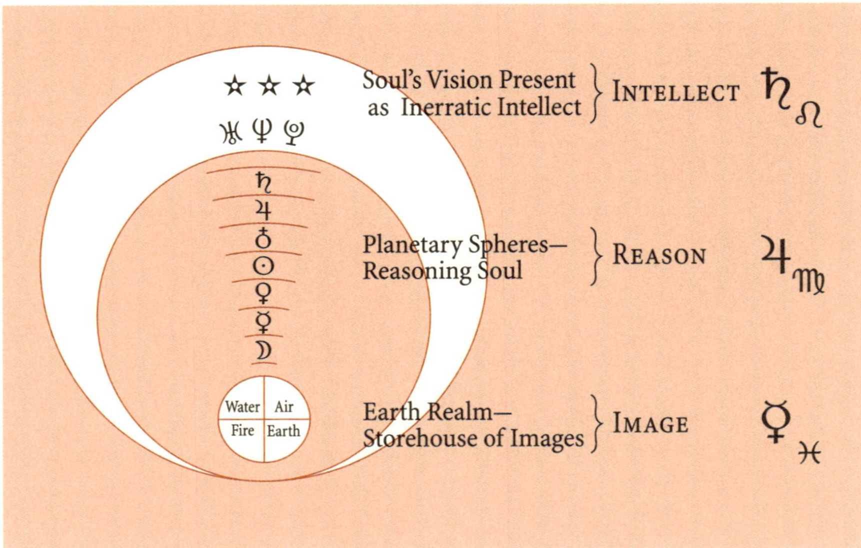

Fig. 84

Editors' Extension IV

Seven Soul Powers

Whereas most astrological writers emphasize psychological meanings of the planets, Anthony also emphasized their metaphysical, cosmological, and epistemological significances as powers of the soul *per se*. He saw this spectrum of meanings as necessary to provide an adequate basis for interpreting a natal chart. His work traces these powers from their source in the Absolute Soul, through the principles of the Soul ring, to the cosmic soul and other individual souls. The current paper explores seven key powers ranging from Saturn in Leo ♄♌ through the Sun in Aquarius ☉♒. [*fig. 85*] Other Soul ring dignities and the psychological level of all these powers are explored elsewhere. (See Note 54, p. 351, "Planets as Rational and Instinctual Functions.")

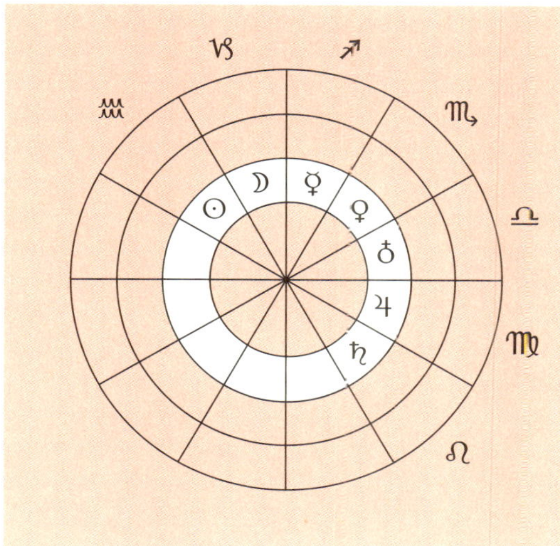

Fig. 85

Soul Powers in the Second Quadrant: Contemplation

Soul's contemplation as rooted in the Intellectual-Principle is represented by Saturn in Leo ♄♌ and Jupiter in Virgo ♃♍—the paradigms, respectively, for self-cognition and appropriation of the noetic vision. [*fig. 86*]

♄♌—*Self-cognition as paradigm for the self-reflection that accompanies reason*

Saturn in its Leo dignity (♄♌) represents Soul's access to self-cognition. Plotinus distinguishes two self-knowers.

> Thus the self-knower is a double person: there is the one that takes cognizance of the principle in virtue of which understanding occurs in the Soul or mind; and there is the higher, knowing himself by the Intellectual-Principle with which he becomes identical: this latter knows the self

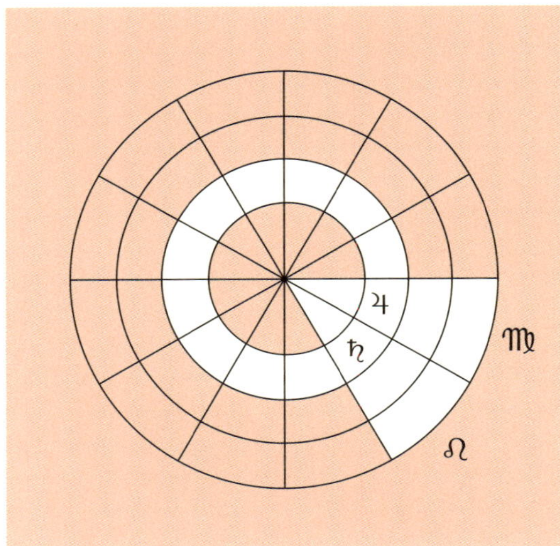

Fig. 86

> as no longer man but as a being that has become something other through and through: he has thrown himself as one thing over into the superior order, taking with him only that better part of the Soul which alone is winged for the Intellectual Act and gives the man, once established There, the power to appropriate what he has seen. (v.3.4)

We can see the function of self-cognition in two realms. Soul knowing itself "by the Intellectual-Principle with which he becomes identical" is represented by ♄♌. Saturn in the Soul ring of the second quadrant is the soul's access to the light of the Nous. There is also another self-knowing which accompanies all reasoning, "the one that takes cognizance of the principle in virtue of which understanding occurs in the Soul or mind." This refers to Saturn as the soul's own reflective cognizance of itself as a "subject," which accompanies its other functions at every level.

♃♍—*Soul's appropriation of vision as paradigm for its growth in relation to cosmic ideation*

As Plotinus points out, accompanying the noetic vision is the phase of soul "winged for the Intellectual Act." This power to "appropriate" what is experienced in the Divine Mind, and to assimilate—through reasoning—what is appropriate for each stage of an individual soul's development is represented by Jupiter in Virgo ♃♍. Plotinus also speaks of this as the inner dialectic of the soul, integrating the complexity of the Ideas without deliberation—that is, by a simple act of its own essence.

This ♃♍ also represents the outgoing activity of the soul in relation to the cosmos. The twofold activity of reason conjoins the intelligible vision (in ♄♌) and its creative expression in manifestation (houses 10–12). Because it provides the presence of the divine in manifestation, Jupiter is called *Providence.*

In the solar system, Jupiter represents the rational soul's growth of understanding through its assimilation of the World-Idea, while Saturn represents the self-cognizing that results from this understanding. Looking at the relation in the reverse way as well, Saturn is the self-cognition which Jupiter manifests in life.

In the individual soul, Jupiter represents the power of sophrosyny—temperance, prudence, and balance—which mediates the wisdom appropriate for the embodying soul.

> Every soul, authentically a soul, has some form of rightness and moral wisdom; in the souls within ourselves there is true knowing . . . (Plotinus, v.9.13)

Hence Jupiter often appears as the hierophant, the guide, because it has an intuitive view of the interconnections of Ideas that the soul needs.

Soul Powers in the Third Quadrant: Creative Ideation

Mars in Libra, Venus in Scorpio, and Mercury in Sagittarius ♂♎ ♀♏ ☿♐ represent the metaphysical principles of those powers through which Soul participates in creative Ideation, and incorporates the reason-principles into its life. [*fig. 87*]

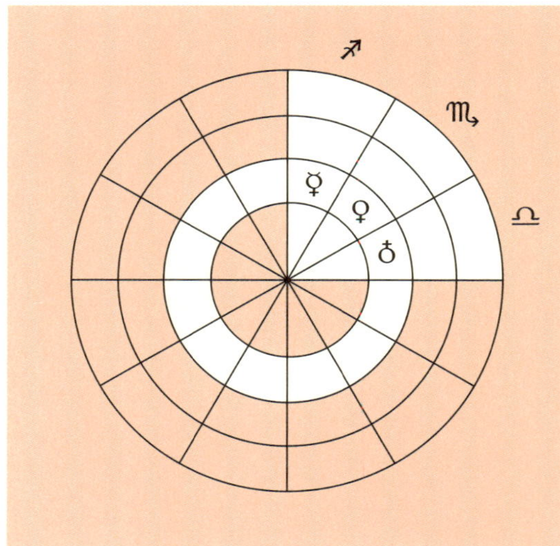

Fig. 87

♂♎—*as paradigm for Soul's choice of reason-principles*

Mars in Libra ♂♎ selects the objective reason-principles, or Forms, which go into the constitution of individual souls. Through these particular principles, Soul is aroused from the infinite consciousness to a state of distinct individuality.

As a rational soul power, Mars also represents what Anthony calls (Notes and Papers, p. S54) "the courage and fortitude needed by a given soul in its journey" toward freedom. This will is what actualizes the soul's potential. Mars not only propels the reason-principles into life, but provides the energy needed to purify experience so as to distill and reclaim these reason-principles for the soul's return journey to itself.

> ... purification is its awakening from the baseless visions which beset it, the refusal to see them; its separation consists in limiting its descent towards the lower and accepting no picture thence, and of course in the banning of all that it ignores when the pneuma (finer-body or spirit) on which it is poised is not turbid from gluttony and surfeit of impure flesh, but is a vehicle so slender that the Soul may ride upon it in tranquillity. (Plotinus, III.6.5)

Anthony writes, "The ancients speak about the Mars amputating base matter—the base matter being the appearances taking place in the dragon.... It is a particular requirement to bring about the purification and separation of the soul essences from their association with impure images.... It requires fortitude to bring a halt to the compulsion of the phantasy, or—to put it conversely—to choose which essences you will experience and will yourself not to be overcome by." (Notes and Papers, p. S54) The embodied individual experiences the presence of Mars as craving and will.

♀♏—*as paradigm for Soul's attraction to its ideal state*

The soul power of Venus in Scorpio ♀♏ binds together the subtle images which are called forth by Soul's love of its intelligible vision, and "congeals" a center of subjectivity. ♀♏ enables the individuating soul to participate in the Idea of the Subjective Conception of the Universe in the Mind of God.

Soul "is where it loves," and desires to fulfill its longing for the divine. Venus ♀♏ is that love, that force through which Soul seeks to fulfill its longing. Anthony writes: "This evaluation (♀♏) provides the subjective feeling states needed to accompany the objective principles (♁♎) in the soul." (Notes and Papers, p. S55)

> And the Soul includes a faculty peculiarly addressed to Beauty—one incomparably sure in the appreciation of its own, when Soul entire is enlisted to support its judgement. (Plotinus, I.6.3)

In the rational soul, Venus represents the power to fulfill Soul's vision through union with its inner Idea.

☿♐—*as paradigm for Soul's power of articulated expression*

Mercury in Sagittarius ☿♐ is the power of the word or thought—*logos.* Through this power Soul participates in the fabricative and gnostic power of the Demiurge. ☿♐ provides the proper language of expression for each kind of soul. We can think of this power as establishing in the soul the "Celestial Name" by which God uniquely knows each soul.

As a rational soul power, Mercury makes available the inner meaning though which we can understand experience, and also communicate our inner knowledge outwardly. Mercury in Sagittarius ☿♐ weaves together the objective reason-principles (♁♎) and subjective feeling states (♀♏) at each level of experience. This communion of inner and outer reason is the soul's language.

> For Understanding, the true, is the Act of the Intellections: in many of its manifestations it is the assimilation and reconciliation of the outer to the inner. (Plotinus, I.1.9)

The individual participates in this function through the power of naming and communicating. Mercury reconciles the outer forms to the inner meaning by relating, naming, and recognizing the reason-principles of Nature.

Fourth Quadrant: Soul in Manifestation

In the Soul ring of the fourth quadrant, Moon in Capricorn ☽♑ and Sun in Aquarius ☉♒ represent the powers through which Soul is located in the universe. [*fig. 88*]

☽♑ – *as paradigm for Soul's continuous location in the cosmos*

Moon in Capricorn ☽♑ represents the "shape" of the individual mind as retained in the universal mind. The ☽♑ positions the soul in the universe, and represents the portion of divine matter allotted to that soul. It is the principle of continuity for a series of incarnations. On the soul entering the cosmic body, Plotinus writes:

> In the absence of body, soul could not have gone forth, since there is no other place to which its nature would allow it to descend. Since go forth it must, it will generate a place for itself; at once body, also, exists. (IV.3.9)

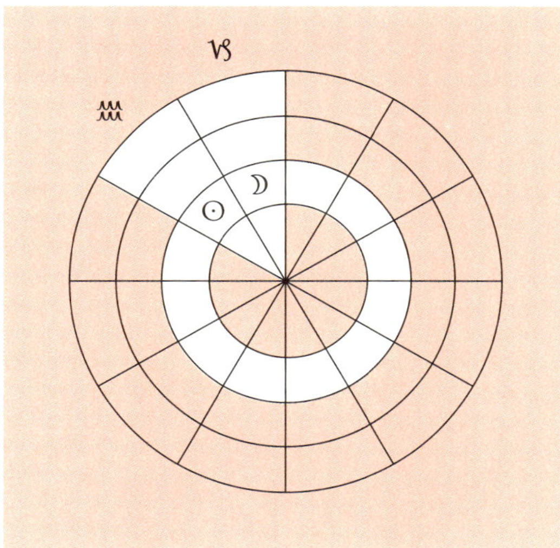

Fig. 88

> Thus, in sum, the Soul, a divine being and a dweller in the loftier realms, has entered body: it is a god, a later phase of the divine: but, under stress of its powers and of its tendency to bring order to its next lower, it penetrates to this sphere in a voluntary plunge . . . exhibiting those activities and productions which, remaining merely potential in the unembodied, might as well never have been even there, if destined never to come into actuality, so that the Soul itself would never have known that suppressed and inhibited total. (IV.8.5; see also IV.8.3–4)

The ☽♑ is the World-Mind's power of memory, which provides a karmic continuity for the cosmos and all individuals within it. This cosmic memory holds together the impressions as a vehicle in which the soul can incarnate.

> . . . aspiration will belong to soul, and so, as a consequence, will that memory of the aspiration and of its attainment or frustration, without which the Soul's nature would fall into the category of the unstable. . . . Deny this character of the Soul and at once we refuse it perception, consciousness, any power of comparison, almost any understanding. . . . it possesses certain activities to be expressed in various functions whose accomplishment demands bodily organs; at its entry it brings with it . . . the powers necessary for some of these functions, while in the case of others it brings the very activities themselves. (Plotinus, IV.3.26)

Through the power of memory at this level, the rational soul a) assimilates the experience of the World-Idea, and b) provides a continuity through the discontinuous or moment-to-moment manifestation of the world. Memory here is primarily a power, and not mere passivity. (cf. Plotinus, IV.6.3) Plotinus points out that memory is even "impeded by the body." (IV.3.26) [40, p. 335]

This higher memory is the rational soul power of "Recollection" (the Platonic *anamnesis*) which recalls the soul to its true place in the Divine Mind. In contrast, the lower memory is that "associated with the psychosomatic organism, the convolutions or engrams left in the brain—mythologically Minerva's snakes." (Anthony Damiani, Notes and Papers, p. S57)

☉♒ – *as paradigm for Soul's living of its good*

The Sun in Aquarius ☉♒ is the centrality and pervasiveness of the Life of the Soul. Through the Sun's act of visioning, Divine Mind imparts the power of perception and the truth to what is known. The ☉♒ represents productive imagination. It is the universal soul-power that brings Ideas into life/consciousness, creating Forms for the purpose of manifesting vision.

> All act aims at this knowing; all impulse is towards knowledge, all that springs from vision exists to produce Ideal-Form, that is a fresh object of vision, so that universally, as images of their engendering principles, they all produce objects of vision, Ideal-Forms. In the engendering of these existences, imitations of the Authentic, it is made manifest that the creating powers operate not for the sake of creation and action but in order to produce an object of vision. This same vision is the ultimate purpose of all the acts of the mind and, even further downward, of all sensation, since sensation also is an effort towards knowledge; lower still, Nature, producing similarly its subsequent principle, brings into being the vision and Reason-Principle that we know in it. It is certain, also, that as the Firsts exist in vision all other things must be straining towards the same condition; the starting-point is, universally, the goal. (Plotinus, III. 8.7)

Anthony speaks of the Sun as the "higher purpose for ego. . . . Soul is free when it moves toward the Good, or lower soul can become a true individual acting as a channel for the higher." (Notes and Papers, p. 1085)

As a rational soul power, the Sun essentializes life according to the Good, and is the manifest symbol of the Good. The Sun is not bare life, but *meaningful* life—the good for the individual. It is the fluid power of imagination to manifest and perceive the creative Idea of any life.

Cosmic Soul

Within the framework of Plotinus, the God of our universe is a cosmic soul (or the lowest phase of an Intellectual-Principle). [41, p. 336] Cosmic soul refers to the fact that each is the author of a cosmos, and within the infinitude of universal manifestation each star is such a "One." Thus there are many such unities, or cosmic souls, each of which includes individual souls for whom that star soul is their God—the Ishvara of their world.

We have symbolized their inner intellectual essence—the intellectual and dianoetic phases, or pure cognitive isness and its power of intellectual act—as Saturn in Leo ♄♌ and Jupiter in Virgo ♃♍. But to this we must add that Soul has also a principle of manifestness, and through this principle it has the quality by which it attains a peculiar manner of being. It has a tendency toward secession [i.e., individuation] and this is equivalent to its outward-facing hypostasis. It is not pure Intellectual-Principle, but exercises its various powers in and through the cosmic manifestation. Soul's *asmita* function, with an inherent logos or unique version of the Ideas, is like a celestial ray, a principle of manifestness within which a cosmos will appear. [This phase is represented by Mercury in the Soul ring in the fourth quadrant, symbolized as the Nature Mercury in Pisces ☿♓N.] [*fig. 89*]

In IV.3.11, Plotinus discusses the nature of the Sun behind the Sun, which in Western terminology may properly be called God, in Hinduism Ishvara, and in Buddhism the World Constructor. A Himalayan panorama is brought into view, which must be made explicit. We find an application and example of the distinctions he makes concerning Soul in an instance of philosophic astrology.

> The sun of that sphere—let us return to it as our example—is an Intellectual-Principle, and immediately upon it follows the Soul depending from it, stationary Soul from stationary Intelligence. But the Soul borders also upon the sun of this sphere, and becomes the medium by which it is linked to the over-world; it plays the part of an interpreter between what emanates from that sun down to this lower, and what rises—as far as, through Soul, anything can—from the lower to the higher. (IV.3.11)

"The sun of that sphere" refers to the intellect of this particular sensible Sun (our star soul), its highest phase, which we represented as its ♄♌ phase. It would be correct to refer to this phase as an Intellectual-Principle, for as Plotinus points out, it is only logically distinct from the Nous. Suspended from this is the rational soul, the abiding reason of this Intellectual-Principle, represented by the ♃♍ phase of the tripartite division of soul. This reasoning phase of cosmic soul borders on the sensible and is the life streaming into that cosmos through the intermediary of the *asmita* principle (☿♓N).▼ This continuity of reason links

▼ EDITORS: "The Soul borders also upon the sun of this sphere . . ." Through the emanating phase (☿♓N) of the Sun soul, the soul behind the (sensible) Sun is present to the cosmos. This presence is the *asmita* or I AM.

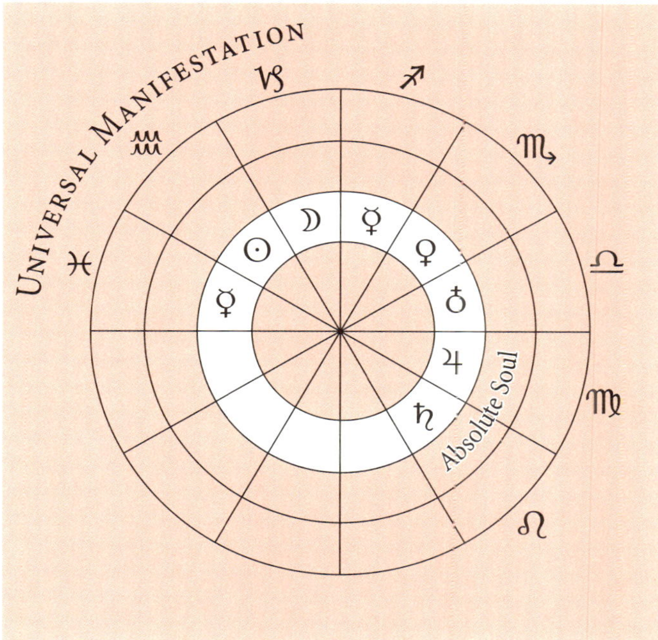

FIG. 89

the sensible Sun to its highest phase in the Nous through its intermediary. The reasoning phase (♃♍) translates and transposes the intellectual vision (♄♌), and spreads it out in living forms by means of and through the intermediary—the *asmita* (☿♓N).[*fig. 90*] ▼

The I AM principle of the cosmic soul includes the reason-principles as the logos within it, the light that is bearing the images of the world to be, and results in the production of the world. This lowest phase of Soul, the *asmita* principle of the Intellectual Sun, is being referred to as the "producing soul," for within this principle manifestation occurs. This power of the cosmic soul [☿♓N] which is linked to its highest phase [♄♌] through the reason-principles it contains, interprets and elaborates the universal reason-principles which went into constituting its body, into a cosmos. ◘

Thus the light of the [sensible] Sun is truly representative of the principle of manifestness. Its interior intellectual essence [from ♄♌] pours forth a representation of the Ideas through its light, appearing as this world-thought. ●

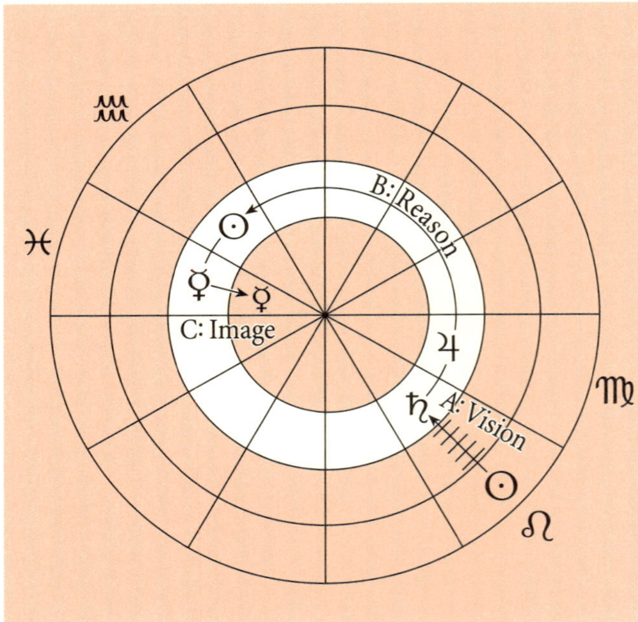

Fig. 90

▼ Editors: The relation of A) the Sun in Leo ☉♌ to Saturn in Leo ♄♌ represents Soul's vision of the Intelligible Ideas, its intrinsic Self-cognition. But that the "Soul borders also upon the sun of this sphere" can also be represented by following B) the Soul ring (Soul's reasoning phase) from Jupiter in Virgo ♃♍ to Sun in Aquarius ☉♒ and illumining the Nature Mercury in Pisces ☿♓N. So there are three steps in the transmission of logos to cosmos: vision, reason as soul powers, and image. The embodying ray of soul, the Nature Mercury in Pisces ☿♓N, experiences these three phases as intellect, reason, and C) image within the cosmos.

◘ Plotinus, iv.3.30 (K.S. Guthrie trans.): Reason [takes Thought and] spreads it out as it were in a mirror, for our imagination.

● Editors: The Sun unfolds the content of its own logos or Idea, manifesting a World-Idea, which the individual human soul in turn projects as its own world.

It is important to consider the correspondence of the inner nature of the Sun soul and the levels of the World-Idea which its emanating phase manifests. ♄♌ ♃♍ ☿♓ represent three phases of the subject consciousness, corresponding to their counterparts in the World-Idea: inerratic, planetary, and sublunar. See figure 91.

To further explain the fourfold nature of a cosmic soul, we must consider the manifested cosmos itself, which is indeed the fourth aspect or "body" of God. The cosmic soul has projected this vast and varied Thought-form, portions of which are organized from reason-principles as suitable vehicles for use by secondary and tertiary souls. The appearance of reason-principles in souls as an apparitional being or sensible world is the consequent result for the individual, who thereby participates in the cosmic soul's thinking of its universe. This point is critical, for the so-called "illusory" world that is being produced is the result of a highly organized efficiency and wisdom of our Ishvara (cosmic soul), or the Imagining which will then be *reproduced* by the individual mind. Prior to its reproduction in the imaginations of all creatures therein, the manifested cosmos which is our cosmic soul's "body" is a world of light, or the World-Idea. [42, p. 336] △

△ Editors: "Notes on the Solar Logos" (p. 336) gives more on the cosmic soul or Solar Logos (Sun in Aquarius ☉♒) as a world of imaginal Light. A cosmic soul draws upon the prior principles in the Soul ring, including the Moon in Capricorn ☽♑ (the portion of divine memory or continuity), to produce, as we shall see, the worlds imaged by the Mercury in Pisces ☿♓ emanant unit souls.

This World-Idea or cosmic thought-form has grades or divisions within it, reaching from the heavens down to the sublunary realm. These grades in their totality are the manifested world for the partial souls (including human souls) that inhabit bodies therein.

> The World Soul is distributed so as to produce the fixed sphere and the planetary circuits corresponding to its graded powers: so with our souls; they must have their provinces according to their different powers, parallel to those of the World Soul . . . (Plotinus, III.4.6) [*fig. 91*]

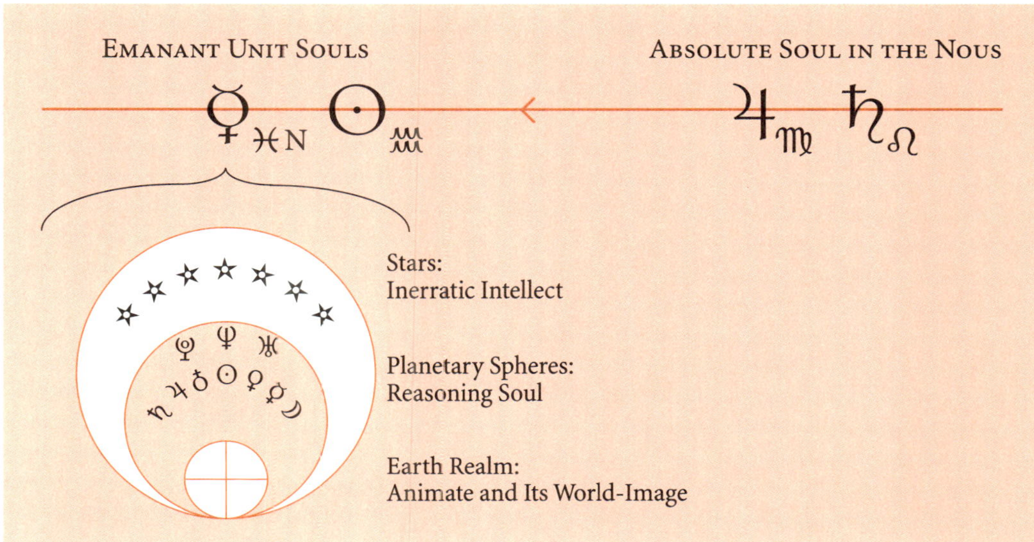

Fig. 91

Cosmic soul is able to maintain the Double Vision—the state of being in continuous contemplation of the Intellectual while it is manifesting the lower through its rational energy. Cosmic soul does not descend to identification with any partial body. It remains resting in the Void of intellect even while it produces. It has the vision—the continuous establishment in reality while it produces the appearance.

Thus for Ishvara (our God the Sun) there is no identification with its body, no brooding over it. Ishvara knows it as an apparent manifestation of these reason-principles which it knows in very essence, i.e., as Idea. In contrast, the individual human soul becomes identified with a body which it takes to be matter—that is, something other than the cosmic soul's Idea—and in so doing it flounders about in that awareness. The mystery of Ishvara's universe—the World-Idea—is duplicated within each human body as a microcosm, though it requires the individual soul's intelligence in order to penetrate and comprehend the mysteries locked up in its vehicles.

In summary, our background in Plotinus reveals to us the existence of primary, secondary, and tertiary soul. The significance of these distinctions can be highlighted by referring them to their respective roles in the universe. Primary is the Universal Soul, for its outgoing act is fundamentally concerned with the primary organization of the disordered motion by the reason-principles, and this becomes the ground-plan for the universe. He then points out that the secondary soul—the *asmita* phase of any cosmic soul—through its contemplation, interprets the World-Idea for Itself. This elaboration, i.e., the cosmos, then becomes a habitation for the tertiary or partial souls, including human souls, to gain experience.

Omnipresence of Soul to the Cosmos

In their inner union with the soul of the Sun, or the All-Soul, all Overselves are interpenetrating or associated as one Soul, yet each soul remains a distinct unit. This necessarily follows from our definition of Soul essence as a One-and-Many. As souls are integrally omnipresent, and their power is "infinite to the core" (Plotinus, VI.5.12), what could distinguish them? The term "the All" can thus refer to the Overself (unit soul) as well as to the cosmic soul or to both together as one. As unique logoi or expressions of the Supreme, souls remain with the cosmic soul—not produced by it, and not set apart from it. They are conjoined as whole soul to whole, or "beings by the side of Being." (Plotinus, VI.4.4)

In an analogy, the Overselves are like radii of a circle, attached at the center in inseparable union with the All-Soul, but manifesting their distinct worlds at the circumference. As abiding in the center they are inseparable, and yet the center remains the endpoint of all the distinct radiations. (Only as they are active and manifesting can these radii be distinguished from their union at the center.)

That phase of our Sun we are calling the soul of the Sun, itself an emanant of an Intellectual-Principle, is what is being referred to as interpenetrating with our innermost essence. Behind the visible Sun is this cosmic soul—an inconceivable immaterial boundless light and life, the substratum of the cosmos. This infinite light is a principle of consciousness and intelligence which will think a world into existence. ▽

▽ Editors: Note 42 "Notes on the Solar Logos" (p. 336) provides a context that is helpful for understanding this essay.

Associated with that light of the Sun are individual souls, and in this undivided awareness the souls hold together, not separated by any boundaries, and indifferently may be referred to as one or many.

Identification with this immaterial presence which is within as an infinite light, everywhere present but contained in no place, is the mystery of our being wherein is lodged the

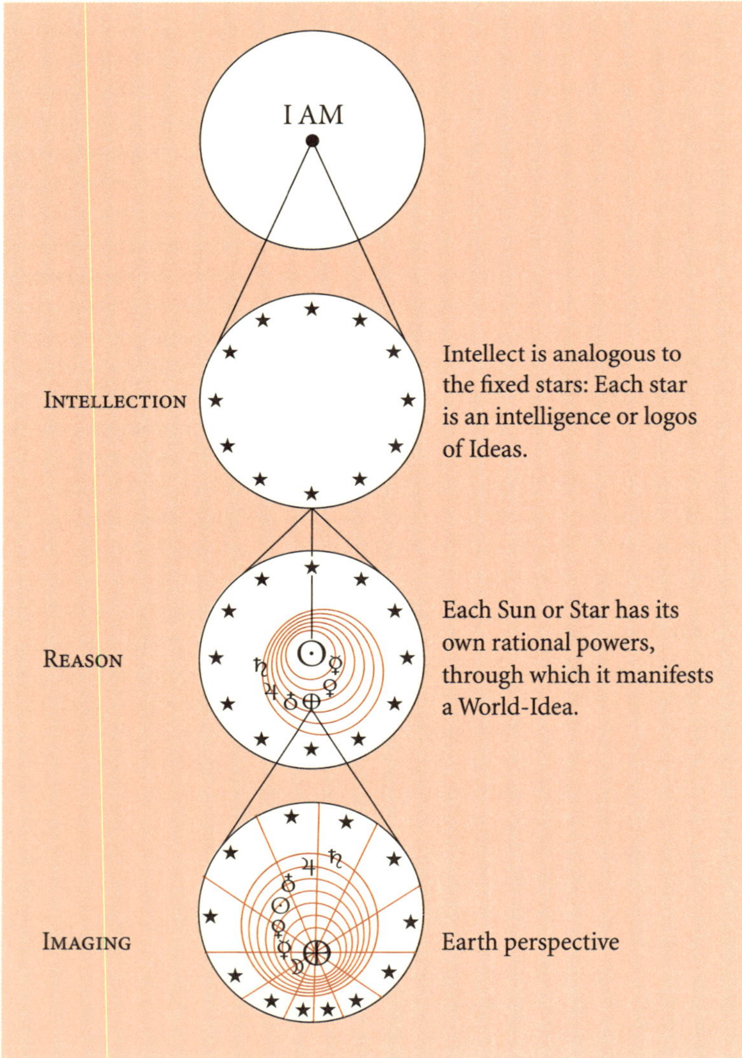

Fig. 92

glory of the archetypal cosmos. And thus our "souls are described as receptacles of the divine beings." (Plotinus, IV.3.1; MacKenna trans., 1st ed.)

Thus, also within the formless light—the soul of the Sun—is hidden the potentiality of the universe to be, the unmanifested Idea of the world. This principle of consciousness is a substance remaining the same regardless of the appearances. (The notion of substance is never used by Plotinus as matter, but as Being-Life-Consciousness.) When this light is focused at a certain point, we can speak of the body of the Sun. Now, since everything in our solar system comes from this principle, every body, insofar as it is a body, is an appearance within that principle of manifestness. More precisely, we can say with Emperor Julian (p. 59) that the principle of manifestation in the Sun is the source of "the secretion of forms and the concretion of matter"—the cosmos precipitates out of its form-bearing light. But the Sun behind the Sun, the infinite power without location, is what we are referring to as present to the entire cosmos, and present entire to each Overself.

Because of its essential nature as we have described it, our soul too is both still and active, and characterized by unlimited power. When its attention is upon the World-Idea present to it, then Soul and Idea form a union, a unique logos expressing the intelligence of the Divine Mind. That is, each soul is an incarnation of the divine Nous. Thus, as concerned with manifestation, Soul is simultaneously present to all levels of the World-Idea.

The power of the Overself or Mind is modalized by the World-Idea into various powers of Reasoning, Memory, Perception, etc., and we can distinguish three broad grades of subjectivity which correspond to the levels of the World-Idea. If we include the Soul unity itself as the first stage, we can set out our understanding of the soul's presence in four stages. [*fig. 92*]

1. The I AM of the individual and the I AM of the cosmic soul are conjoined in the Pure Essence, Pure Consciousness. A description as either One or Many is not adequate. All souls are one Soul. We can call this our own Selfhood—what the Hindus call *atman.* ▽

2. The Sun soul, in trying to grasp the total presence of the Nous, manifests to itself the World-Idea within it through its own Intellectuality. At this level, the intelligence of the unit souls is inseparable from the cosmic soul, and they are pervasive of the entire cosmic thought, just as a mind is present entire with its imagination. When the Overself remains identified with the cosmic soul's thought, the power emanant in it is pervasive of the entire archetypal thought-form of the solar system held by the cosmic soul. The Intellect of the individual soul then has for its object the universal reason-principles (sidereal zodiac) which constitute the World-Idea that the soul has brought into manifestation. This formless, universal reasoning of the cosmic soul is participated by the highest aspect of our individual soul's understanding. This is our Intellection or *buddhi.* This is equivalent to spiritual subjectivity.

3. At the next stage we can speak of the soul as associating with the Planetary Reasoning —our rationality.

   ... we are not to think of it [soul] as going out from itself (losing its unity by any partition): the fact is simply that the element within it, which is apt to entrance into body, has the power of immediate projection any whither ... (Plotinus, IV.3.8)

   When it projects its power of understanding into the psychical, the planet Earth's sphere of reason-principles, then our soul's power is assimilated to the understanding of those reason-principles. Here the experience of the soul is contracted to the living activity of the planetary soul. This corresponds to the second level or psychical subjectivity. This we speak of as Undifferenced Self or Witness-I (*ahamkara*), which includes within itself Differenced Self. This is the level of the reasoning.

4. Further, because of its desire for embodiment and experience—complete differentiation of thought—Soul presides over particular embodiments on this planet, shining its light and power into the psychosomatic, and ego subjectivity emerges. The Body of

▽ Anthony Damiani, transcript, 1/6/84: We have to think of the soul (the Overself) as the substratum, the I AM, the unit consciousness, the infinite consciousness. Within that, there is a determinate content which we're calling the World-Idea, and it has a threefold division. That substratum, the universal consciousness, or the consciousness of the I AM, is simultaneously present throughout the World-Idea, so that the universal Ideas, the unembodied reason, and the reason-principles embodied as a psychosomatic organism represent three levels of knowledge that could exist within that World-Idea. Soul is substratum to all, so it can be simultaneously present to any one of those levels and at all three levels and can have simultaneous knowledge of universal Ideas and sense particulars. That's the difficult part—the simultaneity of knowledge in the soul.

Mentality is within the Planetary Aura. This is sensible subjectivity, or the power of Soul functioning through this mentality or organism, and it corresponds to the Ego or what Plotinus calls "the Animate." ▽

▽ PLOTINUS, I.1.7: From the organized body and something else, let us say a light, which the Soul gives forth from itself, it forms a distinct Principle, the Animate; and in this Principle are vested Sense-Perception and all the other experiences found to belong to the Animate.

Thus we find Plotinus speaks of the "We," not "I," for it is this union of Soul and Ideas within us that is properly a "reasoning soul"—which leads him to remark that each grade of life is also a grade of thought. (cf. III.8.8) The highest grade of life is Intellection, and all the grades are a conjoint of life-power and thought. The point we must bear in mind here is that the unity of the soul is present entire and undivided to all the various powers of reasoning, memory, and so on, and simultaneously the total mind or soul is present in its essence to all the grades. The presence of the World-Idea to the soul will evoke and differentiate its undifferenced infinite power without sundering the Soul essence, and the power actualizes the world-thought for itself.

Without the power of the soul we could not have the appearance of the various aspects of the World-Idea in perception, reason, and so on. Without the presence of the Idea, the soul's power remains infinite, but undifferentiated and ineffective. In an illustration we could think of the energy of the Sun (representing in our analogy, the power of Soul) being turned by the plant into chlorophyll, heating up the Earth, and giving the power of sight to our vision. Distinguished into these various powers—life, light, heat—the power of sunlight in itself, derived from the Sun, is unfailing and undifferenced. Through these various forms, its power is getting actualized.

In summary: a) The power which is projected by Soul into the activity of the cosmic circuit does not sunder the identity of the soul's essence, and b) the soul's power, conjoined with the activity of the circuit, is given determination as/by/through perception, memory, etc.

If we think that this essence, soul, remains aloof while sending its powers forth, then this teaching of the soul or growth of understanding would remain unavailable. But if they remain attached, we are again forced to the conclusion that the entire life essence is present with any of the soul's powers. Each of its unique powers can transmit its life without breaking apart the unity of soul.

Soul is omnipresent to the World-Idea because it *thinks* that Idea. As a consequence, manifestation is the unfolding of the Soul's life, and not an illusion. The presence of Soul is not like some quality plastered onto the world. Nor is it spatially distributed or a container of Ideas as in a box. As a principle of manifestness, Soul itself is not manifest. It is Real-Being, infinite power which sustains the cosmos, but it is important to understand the relation.

The cosmos is contiguous with Soul. There is no separation of Soul and cosmos in any way, as there is none between mind and its dreams. Only as embraced by the mind's presence does a thought have existence; only as present in the solar light is the cosmos manifested. To be existent means to be thought by the cosmic soul; this is, as Plotinus says, the "first condition of existence." (VI.4.2) But the existence of the World-Idea or cosmos, sustained by Soul, does not in any way part the Soul. Its power to think the universe into existence leaves it essentially undisturbed by the multiplicity of thought forms, as a dreamer's mind is unparted by a dream. Soul does not need to spread itself out to accommodate the entire universe. Because it is not limited by the cosmos or space, Soul can be present entirely to the whole of the cosmos and present entire through any of its powers.

The meaning of *authentic existent* here includes that the mind or Soul does not become other, for whatever can become other does not remain itself—and so is inauthentic. It is the Ideas within Soul which are *becoming* the cosmos. Soul does not become the cosmic appearance which yet is appearing to it and within it. The placelessness of Soul ensures that everything within it will have a place in its presence without fragmenting Soul. Right here we are concerned with preserving the unity of Soul. Bearing constantly in mind the thread theme of its sublime comprehensiveness, this will be the foundation for the theory of the soul's knowledge of multiple states of existence—which will become available at various levels of the World-Idea.

As mentioned earlier, the soul has the "power of immediate projection," which can be conjoined with the reason-principles of this Earth or anywhere in the cosmos. The assimilation of reason-principles by this Witness-I effectively differentiates it, and this is retained by the soul as its own understanding. The World-Idea is given to the soul but is not given with an explanation. To the extent that the soul understands, it assimilates and penetrates the World-Idea.

We have now to understand why all "beings" are not equally possessed of a true understanding of the World-Idea (what some schools of Vedanta call "the contingency of omniscience")—why there are grades of participants. One and the same World-Idea presents itself to various beings, but they will understand it more or less completely due to the "fitness" or ripeness of the soul. The understanding of the World-Idea by each soul will depend on the ability of the soul to penetrate the Idea, and in this process the soul's power gets effectively differentiated. (This does not cancel the fact that the essence of Soul is present entire.) Each soul is a distinct logos—an indeterminate life hermetically wedded to the World-Idea and evolving through it. The differences in their knowledge are due to the differences in the grade of thought in the soul or life of each—all of which are included as a multiplicity in the All-Soul as well. The World-Idea, in all its depth and complexity, is differentiating the understanding of the soul, and the soul's understanding then can better understand that Idea. This understanding itself is what will canalize further assimilation of meaning. The organ is itself determining the agent. ▼ The example of sunlight building an eye organ for the human race, which in turn enables the beings to distinguish more of the forms within that sunlight, illustrates how the fitness of the recipient is included in the World-Idea, develops out of the All, and determines the further participation in that World-Idea. Thus, the World-Idea elicits the soul's potential—differentiating and developing its faculty of understanding to the point of self-cognition. ■

As a consequence, the whole series of divisible lives which live by the power of Soul can never exhaust it. The ego draws its own finite life from Soul but never can lessen it. [43, p. 338] The development of the personality goes on almost indefinitely, differentiating itself, supported by the soul's omnipresent life. [44, p. 339]

▼ Editors: Anthony comments further: "An astrological example provides an important instance in our lives: A function manifests an idea, and the idea differentiates the function." (Notes and Papers, p. S18)

■ Editors: On the connection between the *psyche* (soul) and *soma* (body), Anthony comments: "When an individual I AM or unit of life manifests a world for itself, included within this manifestation is the link to a reason-principle or the logos, ultimately traceable to the soul in the Intellectual-Principle. And the individual soul unfolds this logos, its World-Idea. This manifestation also includes the cosmic soul's functioning, which is an elaboration of the Universal Soul's sketch of the universe to be. Both are included in the manifestation by this unit of life, and within manifestation it is difficult to distinguish these various components." (Notes and Papers, p. S70)

Editors' Extension V

Unit Soul as Absolute Individual

What could be more important than to know that at the core, you really *are?* Between the views of Self as absolute reality and self as illusory, is the profound and beautiful view of the unit soul as an intimate emanant of the divine.

Anthony uses a rich and varied language to articulate the nature of the unit soul. He bases his own view of the unit soul on that of Plotinus, as well as on the idea of "Overself" in the writings of Paul Brunton. In different contexts he calls unit soul an authentic existent, a unit of being-consciousness, I AM, absolute individual, unit mind, and Overself. Its nature is paradoxically transcendent and immanent, one and many, universal and individual, inviolable and also evolving. Sometimes the term Overself means a way of being, a shift of identity, or a power of intelligence. This variety helps us to grasp through feeling and intuition something about the nature of this non-objective no-thing. Anthony's ideas are suggestive, exploratory, extremely detailed avenues to reality.

The logic and language that apply to sensible objective experience can not convey the nature of spiritual reality. Paul Brunton writes with technical accuracy when he describes the Overself as "paradoxical." Paradox means beyond (Gk: *para*) the sense-based logic of opinion (Gk: *doxa*). Plotinus says that Soul's nature is such that it is both divisible and indivisible: impartible even in its partibility. (cf. IV.1 and IV.2) While such paradoxes can refer to the relation of Absolute Soul to the emanant, Soul's double nature can also be appreciated as applying equally to the emanant unit soul itself.

Three essays in *Astronoesis*—"Essence of Soul as Indivisible and Divisible" (p. 172), "Overview of Levels of Soul" (p. 175), and "Deliberate Repetition: Cosmic Epistemology" (p. 232)—present the culmination of Anthony's thinking on unit soul. Commentary on passages selected from those essays focuses on five ways to think about soul's double nature. Unless otherwise specified, all passages quoted are by Anthony Damiani.

Soul as Transcendent and Emanating

> Plotinus points out that there is a part of the soul that remains in the Nous, but it emanates a reflex of itself, a principle of intellectual Being that is similar to itself. It is this unit soul, the God within us, that may be referred to as our soul, our ultimate individuality, and the true knower within us. It is "man made in the image of God." (*Astronoesis,* p. 177)

Soul is both transcendent to the cosmos and immanent in the cosmos. But spatial metaphors may be misleading here. Anthony often remarked that "Soul is in the Nous and the Nous is in the soul." Indeed, it is because Soul is present in the Nous that soul as emanant includes the presence of the Ideas.

> Each and every soul, offspring of Absolute Soul, is similar to but not identical with its prior. (*Astronoesis,* p. 173)

To distinguish it from the unit souls, Anthony calls the principle and source of the individual unit souls (in the Divine Mind) Absolute Soul. "From this principle," Anthony explains, "emanate the many souls, each an intellectual reality, boundless, timeless, and eternal in its immutability." (Notes and Papers, p. S89) This is emanation as an intrinsic act of the soul, a natural outcome of Soul "lending itself to divisional existence." (Plotinus, IV.1.1)

Paul Brunton uses a similar metaphor to describe "Overself" as emanant:

> Overself is the inner or true self . . . reflecting the divine being and attributes. The Overself is an emanation from the ultimate reality but is neither a division nor a detached fragment of it. It is a ray shining forth but not the sun itself. (*Perspectives,* p. 296) [*fig. 93*]

To distinguish the transcendent and immanent phases of Soul is not to separate them. "On the one hand the soul as transcendent, and on the other hand the emanant from that transcendent soul [are together what] Plotinus refers to as the 'We.'" (transcript, 3/9/84)

Unit soul is thus simultaneously both inviolable and becoming:

> This unit of Being-Life-Consciousness goes forth, "formed from the undivided essence and the essence divided among bodies." (*Astronoesis,* p. 173, quoting Plotinus, IV.1.1)

Focusing on this double nature of the unit soul, Plotinus says that we must accept that soul is of such a nature. (cf. IV.2.2) Anthony gives a moving description:

> Each unit soul has an inviolable sanctity, in the sense that it is an image of God. But that it can be individuated and determined and fulfill a certain potential which is in the Mind of God, I think is a concept which is extraordinarily beautiful. (transcript, 1/6/84)

FIG. 93

Soul and World-Idea

> We may refer to this as the individual soul or mind; ... the combination of the mind or soul with the World-Idea, is our absolute individuality or Overself. (*Astronoesis*, p. 173)

Each unit soul as emanated forms a hermetic union with the World-Idea. "Your true being would be constituted of, on the one hand, unit soul or individual Overself or individual mind; and two, the World-Idea which is being manifested within that." (transcript, 12/7/83) [*fig. 94*] Anthony also concedes that the distinction between Soul and Ideas is a "theoretical abstraction," and that there can't be "an emanant, a unit soul, without the World-Idea radiating into it." (transcript, 12/7/83) He frequently addressed this key idea:

> We referred to the Soul in Plotinus as a One. The radiation of the World-Idea which occurs in the Soul, we referred to as Being or the Ideas. A combination of these two was necessary in order to understand the Overself in its universality. (transcript, 12/7/83)

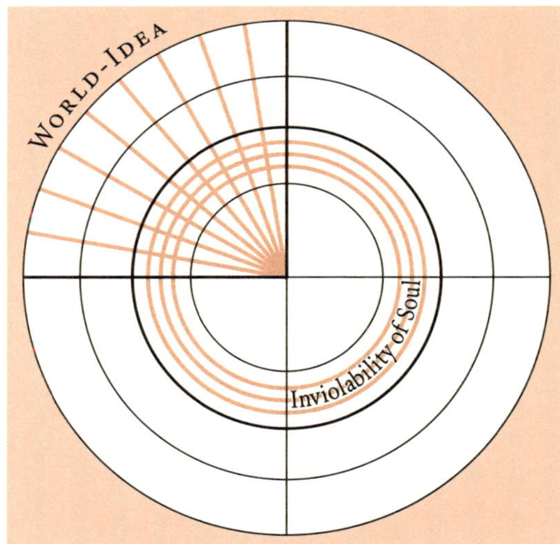

Fig. 94

> When you speak about Absolute Soul, it's not the soul *of* anything. But as soon as you start speaking about life in the sense that it's in manifestation, then you're speaking about either the soul of the cosmos or the soul of this plant, this animal, whatever. You're speaking about a unit soul. And if you have a unit soul, the World-Idea is inevitably associated with it. (transcript, 2/15/84)

> *Nirvikalpa samadhi* means precisely that you have become unit soul, Soul *per se,* void of the World-Idea as a content. And on the other hand, the development of the second knower brings about individuation. Or let's say, what God had in mind for this individual soul to become is being accomplished through the medium of the World-Idea, so that each unit or individual soul is being individualized according to the World-Idea. (transcript, 1/6/84)

> Think of wisdom in the sense that the mind is becoming the universe. . . . In that process of assimilating the World-Idea, the mind is assimilating the wisdom that is inherent in it. The mind, in becoming the world, is also unconsciously incarnating that wisdom. (*Living Wisdom,* p. 59)

> Soul incarnates or participates in the Ideas of the planetary mind by perceiving the world and experiencing, through a body, what the gods think. (*Astronoesis,* p. 177)

Anthony often uses the term "mind" for "soul" in the epistemological context. On the one hand, he comments that "in the process of mind becoming the world, it assimilates a certain wisdom which is inherent in the world." (*Living Wisdom,* p. 59) On the other hand, soul is actualizing its infinite potential: Insofar as the mind is becoming the World-Idea, "the infinite Consciousness that the unit soul is, is getting specified." (transcript, 1/6/84) In this context, Plotinus refers to each soul as a "Logos" of the Nous in IV.3.5. While being omnipresent to the entire World-Idea, each soul also identifies with a part of the World-Idea in order to experience the cosmos and fulfill its specific vision (of the Nous).

Soul as Omnipresent

> Furthermore, regardless of what it is occupied with—either the Idea of the universe or a body within it—Soul gives itself to either or both without abdicating its unity. . . . it is totally present in each of the items without its unity being sundered. In other words, the omnipresence of Soul does not fragment its identity, for it is at once both divisible and indivisible. (*Astronoesis,* p. 173-174)

Soul's omnipresence has three aspects. First, the essence of Soul is present with each unit soul, as what Anthony calls "its inviolable integrity." (transcript, 1/6/84) Second, Soul is omnipresent with the cosmic soul or World-Mind, as explained in "Omnipresence of Soul to the Cosmos" (p. 205). Third, Soul or Mind is present entire to each particle of manifestation, just as the mind is present to each of its thoughts.

> . . . the soul is omnipresent everywhere. When you put a universe in it, there's going to be no spot in that universe where the Overself is not. . . . Now the soul is omnipresent throughout the solar system: The highest phase is concerned with the interpretation of the universal Ideas, the degrees, that the Sun traverses. The next phase, the reasoning phase of this same soul, is that it could project forth from itself and identify with the reason-principles of a planet. And it could do that with all seven planets. Each and all, one at a time. Listen to what Plotinus says in IV.3.8 about this projecting power: "We are not to think of it [soul] as going out from itself (losing its unity by participation): the fact is simply that the element within it, which is apt to entrance into body, has the power of immediate projection any whither." (transcript, 1/27/84)

Unit Soul as a Double Knower

> For Plotinus, the unit soul or mind was a double knower. That is, it had self-cognition, undifferentiated being-consciousness, which included cognition of its source or origin. . . .The lower knower or secondary phase—that is, the faculty of understanding and reasoning—would evolve by its journey through the cosmic circuit, which provided the circumstances,

situations, and events that would educe its potential and make it actual through experience. (*Astronoesis*, p. 232)

Based on the double nature of Soul, Anthony sometimes speaks of the unit soul as a double knower. [*fig. 95*] He also often uses the terms "insight" and "understanding" (as used by Paul Brunton) as synonyms for Plotinus' two kinds of knowing.

> When there is that awareness that is knowing, and it knows its own Selfhood, then we say *insight*. . . . When we say that it is not aware of its own Selfhood but only knows the external—whether that external is the Ideas or the rational processes or perception of the sensible world—then we say that it is the faculty of *understanding* or the reasoning faculty or the reasoning soul. . . . One faculty can be referred to as the higher knower and the other as the lower knower. (transcript, 6/29/83)

> Insight gives you the actual recognition . . . the mystical perception . . . that the perceiver and what is perceived have for their substratum, consciousness. But intuition is reasoning upon objects . . . understanding of the reason-principles. With these two faculties you could find within yourself those attributes which are reflections from the Intellectual-Principle. (transcript, 7/6/83)

> One is Self-knowing and the other is other-knowing. Knowing-knowing and knowing-the-other. . . . They are not meant as a hierarchy. It's better to think of them as two aspects of the soul (I AM). (transcript, 7/6/83)

> When we speak of the *double knower*, [including the most exalted states of mystical philosophic insight]

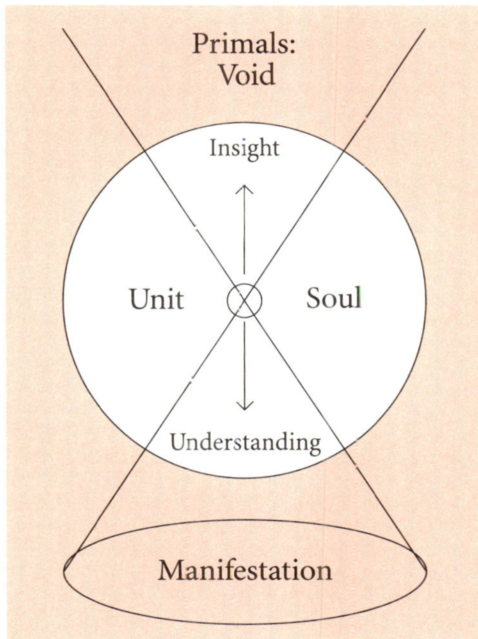

FIG. 95

> we're not dividing the knower: We're speaking about two different aspects of one and the same I AM. (transcript, 12/7/83)

Anthony equates the lower knower with the evolution of understanding in relation to the World-Idea. When the three functions of knowing, willing, and feeling are optimally developed and balanced, insight emerges.

> Intuition, reasoning, understanding are being used in the same way by different authors as the faculty of the reasoning soul by which it understands, interacts with the World-Idea (which soul has manifested). At the most basic level, intuition is modalized as absolute Knowing, Willing, and Feeling. They all operate silently. . . . Feeling developed to the utmost (or willing or knowing) will become intuitive. It won't then operate as we know it in the discursive intellect. . . . When you have a real intuition, knowing some idea, a real aesthetic perception or appreciation, it will be silent, calm. This is what Plotinus refers to as reasoning about the objects, or the reasoning faculty in the soul, which is an image of the Intellectual-Principle, having the ideas unrolled and separate. Intuition does not apply to mind itself, but arises in the soul's relation to the World-Idea. . . . That phase of the soul which seeks embodiment—also spoken of as a power of projection associating with the manifesting World-Idea—is the phase which must develop this understanding. This reembodying soul is a witness of experience, is the Overself as associated with the stages and processes of manifestation. (*Living Wisdom,* pp. 259–260)

> Eventually, the soul's reasoning and understanding will lead to the recognition of the presence of the Intellectual-Principle within itself. Repetitive and constant identification with these Ideas culminates in the appropriation and use of the Intellectual-Principle within itself. This is its true Selfhood, i.e., the I AM principle is an Intellectual reality or a logos. Thus self-knowledge involves a knowledge of its Source. It must be made clear that the similarity is not an identity. This is the mystery of man made in the image of God, with its own unity, intellectuality, and soul. (Notes and Papers, p. S93)

The skillful integration of these two kinds of knowing, insight and understanding, is an essential characteristic of the philosopher, the realized sage.

PB [Paul Brunton] speaks about willing, knowing, and feeling and how they have to be evolved and refined and brought to their highest pitch, before they can get fused and become insight. Now, reverse the process and think of the possibility that insight exists and that it could have a triple mode—from insight comes this absolute knowing, willing, and feeling. Now a person who has insight hasn't canceled out those three functions. They still function. The sage could still will, feel, and know. The insight has to remain. He is a sage and the ultimate reality is always available to him. (transcript, 2/3/84)

Now when a sage tries to find out something about the second law of thermodynamics, insight would be useless here. The sage has to use the faculty of understanding. So on the one hand, the sage has insight into the ultimate substratum of what reality is and, on the other hand, a developed critical faculty where he or she can understand the application of reason to all three levels: a) universal Ideas, b) the way they're embodied by the reason-principles in the middle domain, and c) the way they're manifested as sensible. Let's say the sage can reason all this out and hold both of them together. So you don't have two knowers, there's only one knower. That's the amazing thing about the so-called double knower. (transcript, 1/6/84)

. . . that's the philosophic experience: the experience of the I AM which includes that part or power that the I AM has to project forth. . . . Now the combination of these two—the experience of one's center of being and the experience of one as just being—is what Plotinus referred to as the "We." (transcript, 2/15/84)

Anthony left a work in progress elaborating the following passage from Plotinus.

But how are we related to the Intellect [Nous]? I mean by "Intellect" not that state of the soul which is one of the things which derive from Intellect, but Intellect itself. We possess this too, as something that transcends us. We have it either as common to all or particular to ourselves, or both common and particular; common because it is without parts and one and everywhere the same, particular to ourselves because each has the whole of it in the primary part of his soul. So we also possess the forms in two ways, in our soul, in a manner of speaking unfolded and separated, in Intellect all together. (Plotinus, I.1.8, A.H. Armstrong trans.)

Here is an excerpt from the paper as it appears in his notebooks:

> Plotinus' theory of knowledge is based on Self-realization—of who and what we are. Let us make clear to ourselves his initial premises regarding the indivisible whole of knowledge and the soul's relation to the Intellectual-Principle.
>
> A twofold order in knowledge is presented. The highest is the primary identity of Being and Knowledge in the Nous. This, he says, we possess as common to all insofar as the source of the second soul is the Absolute Soul in Intellectual-Principle itself—One Entire. In the Nous the Ideas are experienced as unified, the sheer presence of mighty powers indistinguishable one from another.
>
> The second order of knowledge has to now be considered distinguished from this. The Nous "in the Soul as it were unrolled and separate" (Plotinus, I.1.8) is represented by the inerratic sphere (Ideas) within each individual emanant soul, that is, the whole of the firmament representing the Ideas. More precisely, it has been explained elsewhere that the immaterial bodies which are given to the various souls as their seat in the firmament are each illuminated by an individual cosmic soul, each an Intellectual-Principle in its own right. So the entirety of the firmament is an unfolding of the Ideas, a revelation of the Nous in the heavens. This is all manifested within the infinity of the I AM principle and makes possible the twofold division of knowledge which is properly knowledge of the Nous. (Notes and Papers, pp. S137–138)

▼ PAUL BRUNTON, *PERSPECTIVES*, P. 342: The unit of mind is differentiated out and undergoes its long evolution through numerous changes of state, not to merge so utterly in its source again as to be virtually annihilated, but to be consciously harmonized with that source whilst yet retaining its individuality.

UNIT SOUL IN RELATION TO EVOLUTION

> We hope to explain later on how this formless and infinite consciousness gets contracted and starts on a journey of self-realization. . . . We may refer to this contracted consciousness as impersonal or witnessing consciousness, associated with the planetary mind from which it receives the image of the World-Idea. (*Astronoesis*, p. 177)

In the next excerpt, Anthony describes the relation between the unit soul's evolution and its eternal selfhood in the Nous. ▼

Soul has a paradoxical dual nature. On the one hand, we speak about the transcendent aspect of Soul as abiding in serene tranquillity, an undifferentiated consciousness. On the other hand, we say this consciousness has the power to project from itself a unit of life, or a stream of life which can be included in the world circuit and experience the World-Idea. . . .

On the one hand, the Nous gives birth to the soul, is the generator of the soul—it's an eternal generation. On the other hand, the Nous is *in* the soul. There is an Intellectual Cosmos in the soul, and in order for the soul to get true self-knowledge, it has to get to know the Nous which is within it. . . .

Now keep in mind that "within" and "without" are *metaphors.* We could say that the Intellectual-Principle illuminates the soul, or we could say that it radiates its presence into the ground of the soul, but no matter what language we use, we are not going to fathom this mystery. . . .

So, in order to know something about the Nous, the World-Idea, this transcendent being which is our I AM sends its offspring into the World-Idea, to evolve along with it. . . . We must think of this power in the sense of a continuous stream of life, a germinal intelligence or a germ of life from itself to that part of itself which is going to traverse all the kingdoms of nature. . . . Soul projects from itself a unit or a stream of life which reaches all the way down to Earth, starts out as an amoeba and goes through all the various kingdoms of nature. . . . When it reaches the human level, that means it's entered the reasoning phase. . . .

In order for this transcendent being that our soul is, to have experience of the world, it emanates part of itself into that world. We call that part the ego. And it goes through all the kingdoms of nature. But in doing so, it is discovering and finding out something about the World-Idea. . . . This unit of life is continuous and consequently can be referred to as a karmic continuity. . . .

That traversing through the realms of nature develops and unfolds that individual consciousness, brings it out more and more. In other words, the very experience of the world forces the consciousness of an individual to evolve, the faculties to get developed. . . . The richness, variety, and incomprehensibility of the World-Idea will satisfy individuals at every level. There's a deepening, an inwardization, a widening, a comprehension of the World-Idea. . . .

Isn't it paradoxical and almost uroboric when we conceive of this? On the one hand, the Nous is within the soul and we are speaking of the soul as this transcendent being. On the other hand, this transcendent being has to emanate a part of itself, get involved in the World-Idea, penetrate into the *depths* of the World-Idea, so that it can develop *prajna* consciousness, penetrate into the void intelligence of the Nous, and come to a self-recognition within itself. (transcript, 3/9/84)

Summary of Four Levels of the Soul or Overself [*fig. 96*]

Absolute Soul is the principle of Soul or Overself in the Divine Mind.

Unit soul as I AM is the term Anthony uses for the deepest phase of the individual as an emanant of the Divine Mind. I AM means the Overself in the context of being "a ray... but not the sun itself." (Paul Brunton, *Perspectives,* p. 296) It is pure consciousness and real Being, the spark of divinity, the source of our limited phases of consciousness and existence. Overself or I AM is sometimes used to refer only to the pure consciousness, sometimes to refer to both phases of pure-consciousness and Witness-I together. Since the soul must identify with the cosmos in order to assimilate it, Anthony uses the yoga term *asmita* (literally *I-am*) for the emanant soul. (Paul Brunton uses the term I AM at the very end of his chapter on "The Yoga of the Discerning Mind," in the Serpent's Path Exercise, in *The Wisdom of the Overself.*)

Unit soul as Witness-I is the function of the soul to contract itself to some part of a World-Idea. Witness means the hidden observer of experience, as well as the repository of the distilled essence of experience: the experiencer of that world, and the power in us that actually understands that world. This phase of the soul is the creative activity in us, receiving and transforming the energy of the World-Idea into experience. It is also the soul's power to contract its consciousness to an individual center of consciousness, identify with a body, manifest and experience a world (idea). It is felt in us as the life-current, or I-current, which Paul Brunton describes as a triune stream of life (life-current), intelligence, and individuality. (*The Quest of the Overself*, "The Mystery of the Heart," p. 180) The Plotinian equivalent term is "rational soul."

THE EGO-SELF is the personal identity, or I-ness. On the one hand, the ego is an evolving matrix of thoughts which are part of the cosmic process. As such, it includes the body, the personal memories, thoughts, images, and feelings, the unconscious tendencies and patterns of behavior. On the other hand, the ego is the reflection in or presence of or togetherness with the life-consciousness of the soul. Anthony liked to use Plotinus' term "Conjoint" or "living Animate," which is a conjunction of a ray of Soul and the organized body. (cf. *Ennead* I.1.7) Ultimately, the ego is the soul's individuation process; it unfolds the wisdom potential in the Overself, and assimilates wisdom from the World-Idea.

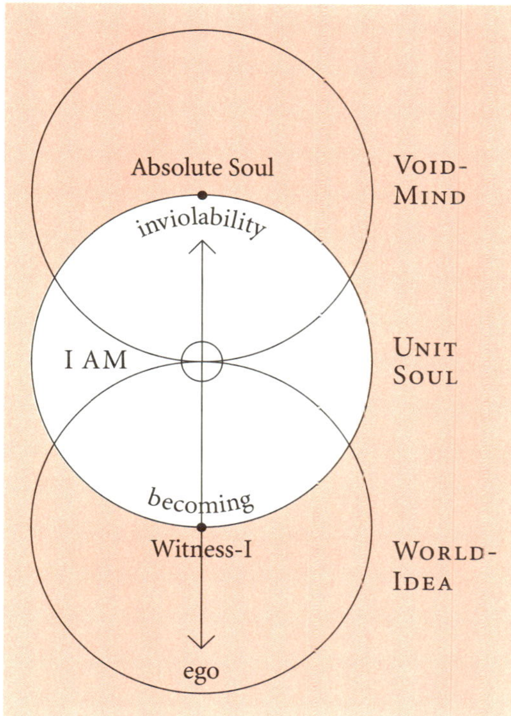

FIG. 96

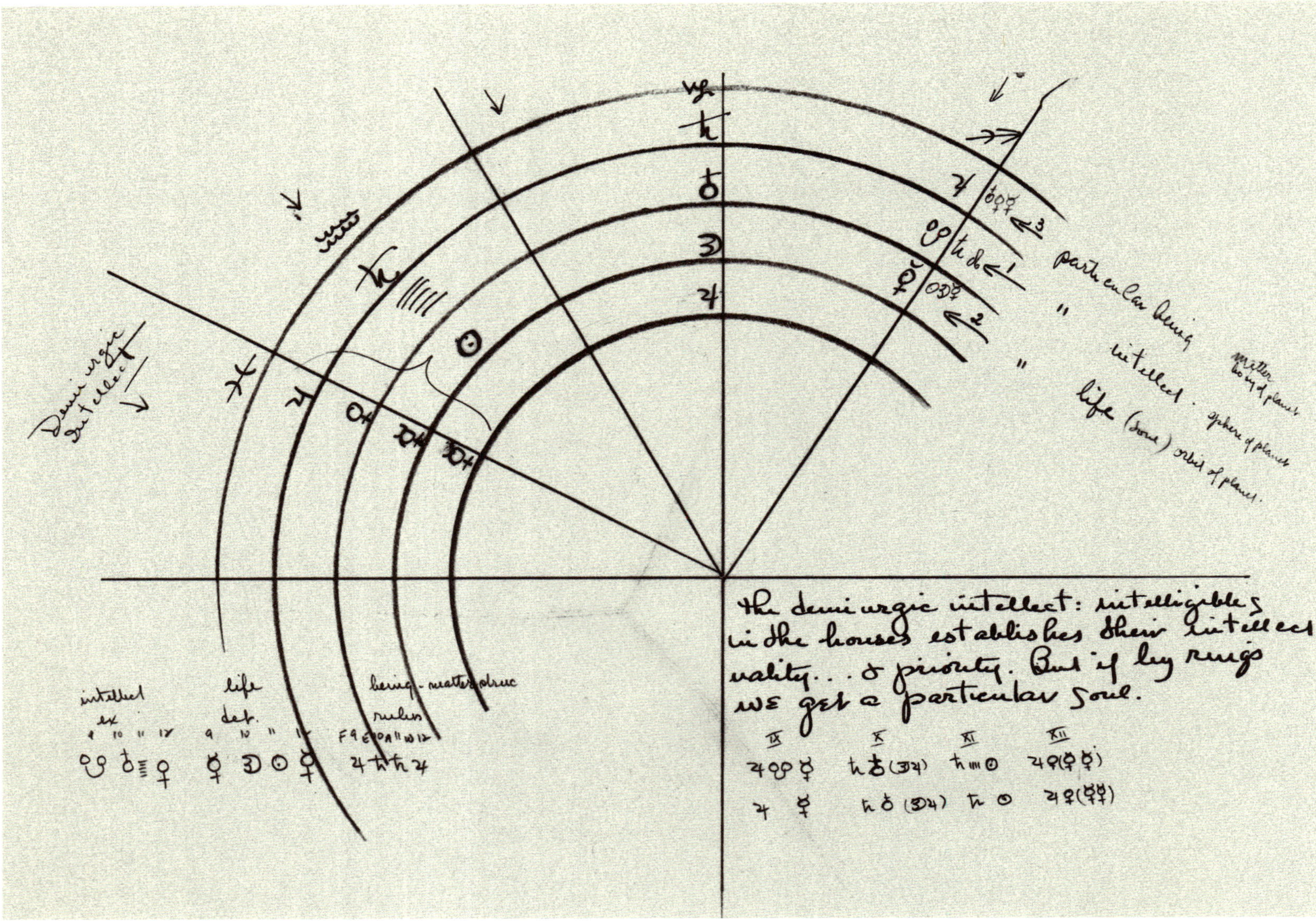
Demiurgic Intellect
particular being
" intellect · sphere of planet
" life (soul) orbit of planet
matter being of planet
intellect
ex
life
det.
being-matter-struc
rules
the demiurgic intellect: intelligibles
in the houses establishes their intellect
uality... & priority. But if by rings
we get a particular soul.

Chapter Four

THE SYSTEM OF NATURE

An Outline

*A*ND IF ANYONE *asks nature for what end she produces (*poiei*), if she consents to hear the questioner and to speak, she will say: "You must not question me; you must understand, and yourself be silent, just as I am silent and am not accustomed to speak. What, then, must you understand? That the thing produced is my product of contemplation (*theama*), silence, and my natural theorema, and I too am born from a similar contemplation, and have the nature of a lover of contemplation."*

PLOTINUS, III.8.4 (John Deck trans.)

Following the discussion of the three primal hypostases, we find that these also exist in and for the system of Nature. Plotinus is speaking here of the entirety of universal manifestation to which he applies the principles of Unity, Intellect or Being, Soul, and Body. (V.1.10) ▽

The Unity of universal manifestation is found in the Universal Soul or Demiurge. Universal Soul contemplates the Ideas in the Nous, and the assimilation of these Ideas by Soul is an abiding wisdom. The reflex of this abiding wisdom is the Logos, the reason-principle of the universe (World-Idea). (III.3.5, III.2.16)

The "in-forming of matter" by these reason-principles, or more precisely their instantiation in the mirror of non-Being, results in their stratification into the so-called various grades of matter. Plotinus refers to this as the "tentative illumining of matter," the tracing in matter of the Idea of the cosmos. (VI.7.7) It is the organization of these Ideas in the realm of non-Being and this is the Being of universal manifestation.

The Soul is the universal principle of animation. It presides over and functions in the seven levels or degrees of existence. This vital functioning reflects the immutable Ideas. The last of the powers of the Universal Soul, represented by Sun in Aquarius, illuminates the seven levels [seven *lokas*] or degrees of existence and is the means through which individual Sun souls superimpose on the universal primitive sketch of the universe and elaborate it into particular worlds. (II.3.13, III.3.1)

The fourth aspect or Body of universal manifestation is the manifest cosmos, in which we will also find a reflection of the three prior principles. Its Unity is the eternal life of the Demiurge. Its Intellect [or Being] is the inerratic sphere, its Soul is the planetary spheres, and its Body is the sublunary realm.

The one life of the Demiurgic soul is the unity aspect of the universal Body and sustains and pervades the entire corporeal manifestation. It is referred to as the paternal Jupiter [in Pisces]; Ether. (IV.4.10, VI.5.8)

The Five Great Elements when multiplied by the seventy-two powers of God (informed by reason-principles) become the 360 degrees of the zodiac or the Ideas, which are then sensibly represented by the intellectuality of each individual cosmic soul. The inerratic sphere represents our Sun soul's grasp of those Ideas.

The individual cosmic soul, or our Ishvara, is situated within this scheme of things. It is going to manifest its own world according to the universal reason plan, by projecting from within of its own store of reason-principles and articulating more precisely and completely its own world. (II.3.17–18)

The World Soul, with its graded powers (the cosmocrators) and the planetary spheres, is the Soul aspect of our world. Its powers extend into the vegetable and mineral kingdoms and up to the physical body of Man. (III.4.6) The functioning of the planetary spheres is Nature or the Soul of our world, which also can be referred to as the Infinite Mind of our world—the last of the reason-principles. (IV.3.7)

Nature is the medium through which our soul, or rather a light from it, is associated with various vehicles in the cosmos, principally the astral body. Our personalities are thus bound up with the cosmocrators. (II.3.9, IV.3.12)

▽ EDITORS: All references in parentheses within this outline are to Plotinus' *Enneads*. Anthony left this outline in a rough state. We have added the entries indicated by ❖ on pages 228 and 229, which include some language from Anthony's later text.

The cosmocrators in the system of Nature are vehicles of the higher soul's will. They are the agents in the fabrication of the phenomenal subject and object, and provide the necessary knowledge for their continuing functioning. (II.3.9, IV.3.9, IV.4.20)

Nature would refer to all possible combinations of reason-principles into various immaterial forms of species which organize the Earth's memory or imaginal essence. (IV. 4.13) This imaginal essence then becomes an entity constantly renewed or always in flux, but nonetheless, through which these reason-principles manifest themselves and become the experienced World. The bifurcation that takes place brings about self-reflection and awareness of an empirical subject and object or the ego and its world. (IV. 4.23)

❖ The stage is now set for the individual actor. Within the cosmos, the individual unit of life will evolve to full humanity. Deliberate repetition and further amplification is necessary to emphasize the intertwining of cosmology and epistemology which Plotinus has as a backdrop for the acquisition of knowledge by the human soul in the sensible world.

❖ The association of the individual with the cosmos unfolds in four stages:

- The unific life of the Universal Soul, which includes the interpenetration of the individual Solar Logoi;
- The sidereal zodiac—the form of the World-Idea adjusted to the intellect of each individual star soul (and its spiritual powers of feeling, willing, and knowing);
- The Chaldean (reasoning) powers of the solar system, each star taken as a solar system distinct from others;
- The tropical imagination within the Earth sphere, or the memorized impressions of previous functioning.

Soul gives from itself a light to the formation of living entities when conjoined to body, and thus produces living images of its powers. (I.1.7) These images are seated in the Animate (Dragon) principle, and are what we refer to as Body or appearance. (I.1.9)

❖ Within the context of the world and the individual ray of Soul, we can address the question, "What is the purpose of this evolution?" The development of understanding follows the sections of Plato's Divided Line, viewed as the relation of individual knowledge and cosmic wisdom.

❖ Using the symbolism of astrology, we can illustrate the principles of metaphysics as they apply right within our experience. The individual mind's development within the principles of cosmic epistemology can be illustrated in the astrological chart, both natal and transiting; and the structure of the mind can be correlated with the principal features of the astrological chart. Here we will see the final justification of Mentalism: that all our life is the unfoldment of Ideas.

Editors' Introduction to Nature

Cosmos and the Unit of Life

This final section of *Astronoesis* explores the nature of cosmos and its meaning for an individual. Cosmic manifestation as a whole is the entire body of the One, manifested within the Soul of the One. Here, within the system of Nature, we see the full import of Anthony's introductory remarks about the presence of prior principles in our experience.

Anthony interprets the cosmology of Plotinus in terms of cosmic epistemology. Each cosmos is a unique manifestation of Divine Ideas that Plotinus calls a universal reason-principle and Anthony calls "World-Idea." As an Idea, the universe manifests within and by means of the hierarchy of cosmic souls, which are both immanent in and transcend the cosmos they manifest. Taken as a whole, the cosmos is a living organism, instantiating Divine Ideas and moving according to the orderly orchestration of Number principles.

Ideas in the mode of Soul are called reason-principles (*logoi*). Each is a multiple unity in which imagination, relation, order, and form combine. At every level, creation (*poiesis*) is knowledge (*noesis*)—it is a contemplation. As Plotinus points out, Nature also contemplates (cf. *Ennead* III.8, especially John Deck's translations). Production results from contemplation and is its outward face: Knowledge is the result and aim of all production. This view of manifestation shows a complete circle: Thought transforms the formless into form, and returns all form (manifestation) to Mind.

As a manifestation of Mind, the world is neither eternal Being nor illusory. It is an "ever-becoming." Seen as thought-activity rather than as unintelligent matter, the universe becomes a place of divinity. Ultimately, the universe is a moving system of Intelligence.

Life is meaningful because it is the continual instantiation of Intelligence. Experience is a playing out of Ideas, of a more and more relational nature, linked from the lowest state of fantasy to the Primals. Within the cosmos each individual unit of life will develop to maturity, to what Buddhism honors as Buddhahood: Each human soul will fulfill and unfold its Idea. The living World-Idea is thus termed the "Womb of the Buddha." Anthony remarked:

> We think of the Overself as absolutely transcendent, having nothing to do with the World-Idea, abiding and resting in repose in itself. But then we must also keep in mind that the World-Mind is within the soul, just like the soul is within World-Mind. Now, theoretically, we can abstract these things and we can say that if the World-Mind with the World-Idea is present in the soul, then in order for this transcendent being that our soul is to have experience of that world, it emanates a part of itself into that world, which we call the ego. And it goes through all the kingdoms of nature. But in doing so, it is discovering and finding out something about the World-Idea. . . .
>
> Traversing through the realms of Nature is what's going to develop and unfold that consciousness, bring it out more and more. In other words, the very experience of the world forces the consciousness of an individual to evolve, so the faculties get developed.
>
> The wonder of all of this is that the cosmic circuit, or the purpose of manifestation, becomes much more overwhelming in importance. In other words, when they say that the purpose of manifestation is to guide and instruct the soul—and only the World-Mind can do that, nothing less than that can do it—then this manifestation is for that purpose, and manifestation becomes meaningful. (*Living Wisdom,* pp. 229–231)

The principles of cosmic mentalism, and the transformation of the individual unit of life within the cosmos, can be seen through an astrology rooted in the cosmic dimensions of epistemology. In this context, astrology provides direct evidence that all our experience is an unfolding of the Ideas.

Deliberate Repetition: Cosmic Epistemology

▼ Plotinus, v.3.4: Thus the self-knower is a double person: there is the one that takes cognizance of the principle in virtue of which understanding occurs in the Soul or mind; and there is the higher, knowing himself by the Intellectual-Principle with which he becomes identical: this latter knows the self as no longer man but as a being that has become something other through and through: he has thrown himself as one thing over into the superior order, taking with him only that better part of the Soul which alone is winged for the Intellectual Act and gives the man, once established There, the power to appropriate what he has seen.

Deliberate repetition and further amplification are necessary to emphasize the intertwining of cosmology and epistemology which Plotinus has as a backdrop for the acquisition of knowledge by the human soul in the sensible world. The manifestation of the World-Idea is the object of knowledge to the inhabitants of that world, and the interlocking of the various levels of reality within the structure of that manifestation is an infinitely complex situation that requires to be mentioned here.

The human soul inhabits and identifies with a body fabricated by the planetary mind. This planetary mind is manifesting its world according to the Solar Logos, which in turn is one principle of intellectuality among many others, and is situated within the groundplan or sketch of the universe provided by the Universal Soul. This is the context within which we must place the object of knowledge. Now at this point we place the human soul which seeks to gain experience of this sensible world or the sublunary sphere.

It was a tradition among the Platonists that the birth or generation of souls took place in a star (such as our own). (cf. Plotinus, III.4.6) The powers of a cosmic soul instigated the unit of being-consciousness to develop its faculty of understanding by providing the necessary environment within which its potential powers were evolved. For Plotinus, the unit soul or mind was a double knower. (cf. I.1.8, V.3.4, IV.3.10, IV.3.31) ▼ That is, it had self-cognition, undifferentiated being-consciousness, which included cognition of its source or origin. [This he called the higher knower.] The lower knower or secondary phase—that is, the faculty of understanding and reasoning—would evolve by its journey through the cosmic circuit, which provided the circumstances, situations, and events that would educe its potential and make it actual through experience.

In other words, the evolving World-Idea of our Solar Logos, our Sun, in its totality, is the stage upon which the soul's unfoldment occurs, the womb within which the gestation and birth of the sage are finalized. Now the individual soul, whose selfhood is similar to the soul of our Sun and conjoined to its "elder sister" (cf. Plotinus, II.9.18), is the principle of our being—an eternal authentic existence—which manifests the World-Idea and simultaneously includes the knowledge coordinate with the various levels of being contained in that World-Idea. To the contemporary mentality, this is a superstitious myth invented by the overactive and heated imagination of our frightened forefathers. For the achieved sage, these silent and detached intuitions are self-validating and authentic instances of a growing body of knowledge which has been deposited as an esoteric tradition or a "metaphysics of truth."

We must return to the specifics involved in order to complete our short synopsis of heavenly wisdom. Neither allusions or innuendoes can be substituted for definitive working principles. The sly remarks and one-liners by many of our contemporaries that imply a greater hidden knowledge on their part has been frustrating and found wanting. We, for our part, must be bold and make explicit these principles necessary for spiritual growth and deny the hostility and negation inherent in their attitude.

The object of knowledge for the soul incarnating *here* is the world-ideogram which is given to an individual mind. This original world-thought is transparently lucid and subtle, a constantly metamorphosing logical unity or whole of its own intellectuality or the Ideas which are the same as the 360 degrees. These Ideas or gods of the planetary mind are the very substance or stuff included in the organization of the psychosomatic organism and the vehicles used by souls to understand the World-Idea. Thus the soul is a receptacle of the gods.

Now the powers of the undivided mind of the Earth, which are symbolized by the Chaldean planets and represented by our instinctual functioning, organize those Ideas into specific and adventitious unities—i.e., psychosomatic organisms. This combination of function and substance is an ever-ongoing process that results in the creation of the World-Idea and includes entities that we may refer to as an organization of the Earth's memories into the imagination. Now, through this very medium, the organization of the Earth's memories into a functioning entity, is imparted a ray of light or mind by the individual soul; or we may say that this entity or intruder captures that light from a soul and appropriates it as its own. This partial consciousness, streaming out through the organism, gives rise to an appearance, the image of the world, within which it finds itself situated; these images in their totality will constitute the lowest level of knowledge as, for instance, in Plato's Divided Line. So we must bear in mind that these bodies are the medium through which a soul becomes sensitive to the sensible world, but these bodies are at the same time agents of the powers that are continuously being organized by the planetary powers.

Thus the entire cosmos provided by the cosmic soul, in cooperation with the Universal Soul, may be referred to as the Womb of the Buddha—that which will develop to perfection the many individual unit souls. The immense framework which we are presented with here is necessary as a background for our discussion of this our most intimate nature, which is what we are—an indivisible unit of Life or I AM.

The System of Nature: Houses 9–12

FOLLOWING THE DISCUSSION of the three primal hypostases, we find that these also exist in and for the system of Nature. ▼ This "fourth principle" is the Body of the One—which is equivalent to the entirety of universal manifestation. In v.1.10, Plotinus refers to the system of Nature as itself having a Unity, Intellect or Being, and Soul. Together with its own fourth aspect, the Sensible Cosmos or Nature itself, we have a fourfold division of universal manifestation which is contained entirely within the Universal Soul.

If we place the Unity of the universal manifestation in the logos principle within the Demiurge or Universal Soul, symbolized as residing in the 9th house of our chart, then the next three houses correspond to the Universal Intellect or Being, Soul, and sensible manifestation of the cosmos. [*fig. 97*]

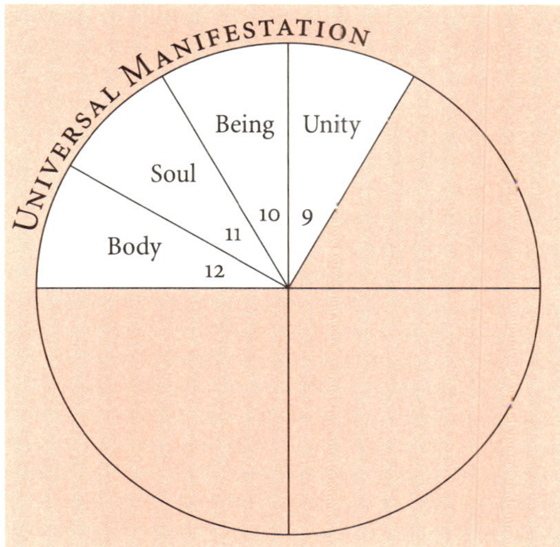

FIG. 97

▼ JOHN DECK, *NATURE, CONTEMPLATION, AND THE ONE*, PP. 110–111: Plotinus does not have two worlds, but only one. His world of true being is not, except metaphorically, a world above the everyday world. It *is* the everyday world, not as experienced by sense, by opinion, or by discursive reasoning, but as known by intellect, the Nous, the Knower.

Plotinus' world of true being is, therefore, the real world of everyday experience when the latter is known by the best knowing power. The more real turns out to be in the real, in fact, to be the reality of the real. Being, which seemed at one time to be for Plotinus a technical term less extensive than reality, seems able now to coincide with reality, as common sense would appear to demand.

We are making a distinction between two levels of cosmology. In the first view [houses 9–12], we are discussing the principles of universal manifestation—the fourfold organization of [every possible] cosmos. The second view [that of the particular 12th house manifestation, to which we can apply the principles of space, time, and causality ▼] will be nested within the first and prior, that of the eternity of the World-Idea in the Universal Soul. Only the fourth aspect of the system of Nature (i.e., 12th house) is the sensible universe or *Nature proper.* Within this sensible universe we find a reduplication of the prior—Unity, Intellect, Soul, and Body of the cosmic principles. [*fig. 98*] [45, p. 339]

▼ THOMAS TAYLOR, *THE HYMNS OF ORPHEUS*, P. 129: But since, from its material imperfection, it cannot receive the whole of divine infinity at once, but in a manner accommodated to its temporal nature: it can only derive it gradually and partially, as it were by drops, in a momentary succession. So that the corporeal world is in a continual state of flowing and formation, but never possesses real being; and is like the image of a lofty tree seen in a rapid torrent, which has the appearance of a tree without the reality; and which seems to endure perpetually the same, yet is continually renewed by the continual renovation of the stream. (footnote to the "Hymn to Nature")

■ EDITORS: Figure 167, page 347, gives an expanded view of the levels of the 12th house as presented in figure 98.

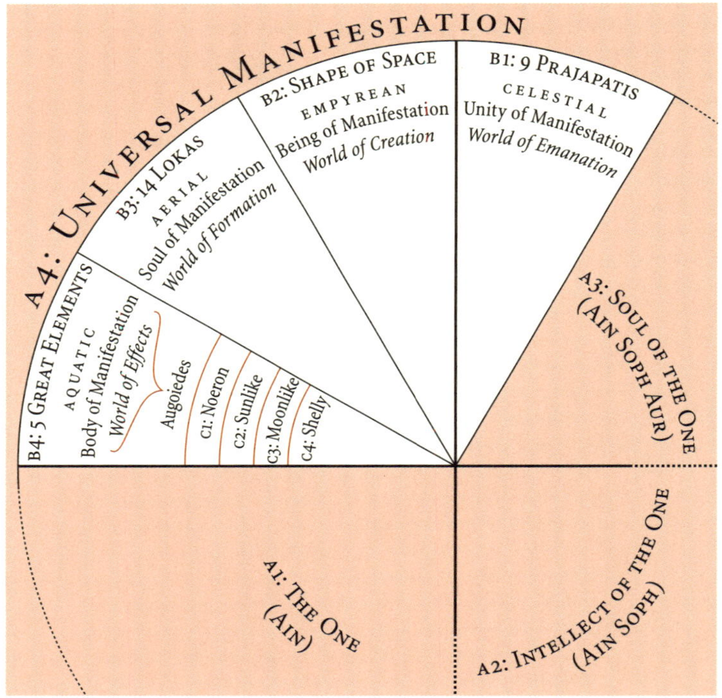

FIG. 98 ■

House 9: Unity of Universal Manifestation

Using the symbol of the third quadrant as the Soul of the One, we can see the dependence of the Body of the One as eternally illumined by and rooted in the Soul of the One. This entire system of Nature is contained in the Demiurge of wholes [represented by Jupiter in Sagittarius ♃♐]. [*fig. 99*]

> ... what must be stands shaped before the Cosmos, and is ordered without any setting in order ... the principle that brings them into being is Order itself; this production is an act of a soul linked with an unchangeably established wisdom whose reflection in that soul is Order. (Plotinus, IV.4.10)

The Unity of the universal manifestation is found within the Universal Soul. The assimilation of the Ideas which the Universal Soul has contemplated in the Nous is an abiding wisdom, which gives rise to the logos or unified reason-principle of the Universe.

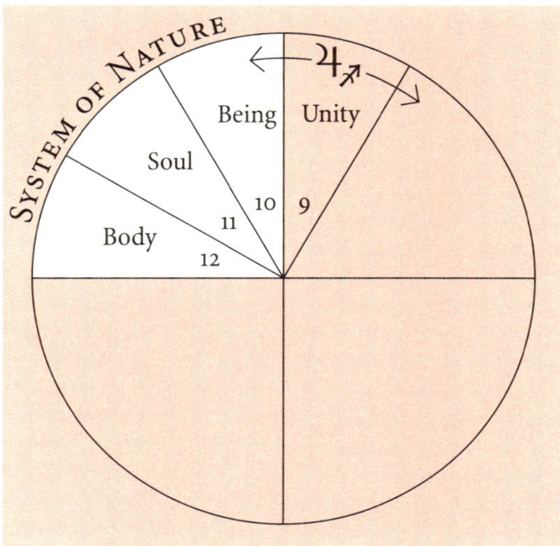

Fig. 99

Thomas Taylor's introduction to Plato's *Parmenides* speaks of the "absorption of Phanes" or the intelligible plan of the Universe, by the Demiurge of wholes. (p. 19) This plan is said to be fourfold, corresponding to the divine plan of Unity, Intellect, Soul, and Body. This is the inner vision of the Demiurge, Zeus, whose vision is the basis of universal manifestation—which we have translated as the last four houses. Thus universal manifestation is within the soul's contemplation: "All that is past, and all that e'er shall be, / Occultly, and in fair connection lies / In Jove's vast belly, ruler of the skies." (Orphic verses cited by Proclus, in T. Taylor, *Hymns and Initiations,* p. 57) [38, p. 333]

House 10: The Being of Manifestation is the Being of Non-Being

Within the 10th house we represent the Being [or Intellect] of universal manifestation. Here is the projective expansion of this logos, the informing of matter by these reason-principles, or more precisely their growth in the "mirror of non-Being." (cf. III.6.7 ff.) This results in the stratification into various grades, which Plotinus, following Plato, refers to as the "tentative illumining of Matter." ▼ ⊞

> The power of the All-Soul, as Reason-Principle of the universe, may be considered as laying down a pattern before the effective separate powers go forth from it: this plan would be something like a tentative illumining of Matter; the elaborating soul would follow the tracing and give them minute articulation . . . (VI.7.7)

T. Subba Row calls this illumination the Shape of the Material Universe in the Mind of the Demiurge. This must be a Thought, and it is this Thought which has shape. This is the substratum of universal manifestation, the structure of the reason-principles in the realm of non-Being. The strata of illumined matter are the image of the four states of Being found in the Intelligible world. [46, p. 341; 47, p. 341]

▼ Anthony Damiani, transcript, 10/14/83: The organization of the *samskaras* by the reason-principles represented the primitive "stuff" or primitive matter that is going to be guided by the Universal Soul into the production of the universe. It was like the ground-plan of the universe. Plotinus called this a *tentative* illumination of matter because these *samskaras* of God or so-called matter would not [necessarily] retain those reason-principles—it was possible to shed them.

We could think of this notion of the matter of God like this: When we refer to the Transcendent as having a fourfold aspect—the Unity, a Being or Intellect, a Nature, and a Body—we're thinking of the body of the transcendent One as the matter which is coeval with the One. Matter at this level is no kind of *stuff,* obviously. But if we were to ask where universal manifestation could take place, we'd have to say that in some way or another it takes place in God. Where else could it take place? And if it takes place in God, then when the entirety of universal manifestation dissolves, or when parts of it dissolve, they leave behind the traces of their presence. . . . And if this has been going on forever, and it has no beginning, then these traces which have been left behind in the body

House 11: Planes of Existence

The Soul of the system of Nature is the 11th house Idea. Here are the seven levels of objective existence and their corresponding subjective states that we refer to as degrees of existence with their corresponding epistemologies. These *lokas* and *talas* [objective levels and subjective states] symbolize the prefiguration of what can be manifested, all the planes of existence. They contain the vital and separable forms within which any of seven levels of any manifestation can appear. [*fig. 100*] ⊕

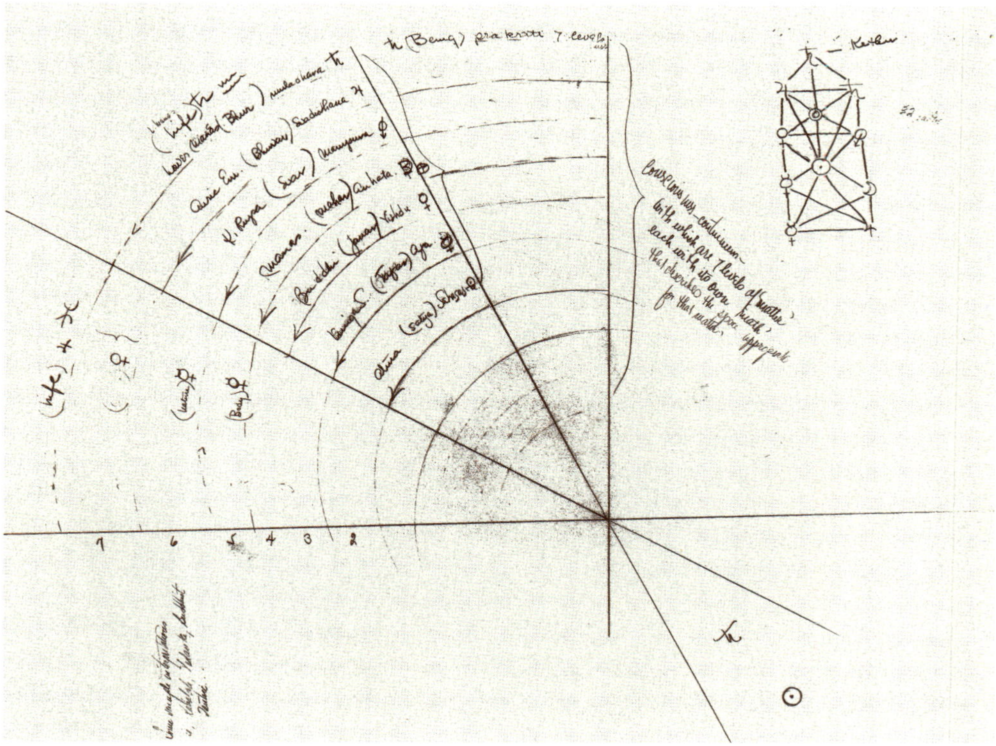

Fig. 100: Detail of a drawing by Anthony Damiani

of God—which would be subtler than space, the primal space . . . would remain there. This is what we're calling the matter, the disordered motion, that the Demiurge has to organize into the primitive matter.

⊞ Anthony Damiani, Notes and Papers, p. S159: This illumined "matter," or streams of living energy, is conceived of as the Absolute Mind of the World. In Samkhya this is hypostatized into a self-subsisting principle (*Prakriti*). According to the Ancients, if Mind is the ultimate principle, any expression of Mind would ultimately have to be reducible to energy, an attribute of Mind and not "matter."

⊕ Editors: The laws of Space, Time, and Causality, the orderly unfoldment and interrelation of Ideas, are determined in the 11th house. In Kabbalah, this is called the World of Formation, meaning that the actual form of the manifest world is determined here. At this level, the World-Idea is potentially infinite. Each reason-principle can be conceived of as an infinite sequence, susceptible to an indefinite refinement and reasoned investigation.

House 12: Appearance of a Cosmos

The 12th house represents the living body of the universal manifestation, the appearance of the combined reason-principles. Within this imaginal manifestation, completely contained within the light of Universal Soul, we will find the prior principles, reflected as the three material worlds: The appearance of the inerratic, planetary, and sublunary spheres corresponds to the Intellect, Soul, and Body of sensible manifestation.

> ... [Soul] ... generate[s] a place for itself. ... The Cosmos is like a net which takes all its life, as far as ever it stretches, from being wet in the water ... the Soul is of so far-reaching a nature—a thing unbounded—as to embrace the entire body of the All in the one extension. ... The universe spreads as broad as the presence of soul ... a shadow as broad as the Reason-Principle proceeding from soul ... (Plotinus, IV.3.9)

Within the 12th house is the entirety of the illumined or intelligized space, the ordered motion, which the Demiurge provides for all the individual cosmic souls—and within which they will manifest and elaborate their worlds. Whatever the Universal Soul, or Demiurge, produces is eternal; but the manifesting activity of the cosmic souls and their powers completes the fabrication of all the beings within their sensible world, in conjunction with this principle of Nature. The combined view gives us a picture of this last realm of cosmic manifestation as an efflux of the Sun in Aquarius ☉♒ or the Intellectual Imagination of Universal Soul, whose light radiates our cosmos from beyond the starry sphere.

> And it is that loftier light—falling variously upon the stars, to each in a certain proportion—that gives them their characteristic differences, as well in magnitude as in colour; just such light constitutes also the still higher heavenly bodies which, however, like clear air, are invisible because of the subtle texture and unresisting transparency of their material substance and also by their very distance. (Plotinus, II.1.7)

If we use the Ring diagram of the Universal Soul, its active or outer production is the activity of the Intellectual Imagination—the Sun in Aquarius ☉♒ illumining the last Idea, the Five Mahabhutas or Force Fields. The manifesting activity [☉♒] of the Universal Soul is organizing and ordering the disordered motion [♓] into intelligized space—the field within which all the celestial Ideas are inserted.▼ This field of reason-principles, or illumined products of Universal Soul, is Nature. Conceived of as the locus for the elaborating activity of any cosmic or individual soul, this field [the self-reproductive motion or life of the Universal Soul] is prior even to the starry heavens. It is [included in] the perpetual illumination by the Soul of the One of the "matter" of the One. [*fig. 101*] ⊞

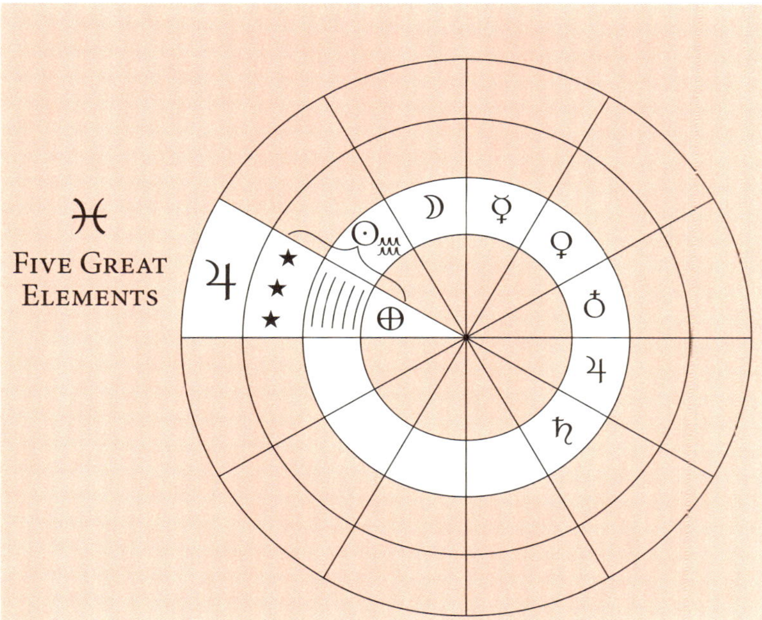

FIG. 101

▼ EDITORS: Anthony often referred to these 12th house Mahabhutas as "directionalized space":

> We can define Jupiter in Pisces as . . . the Life of the universe. For Life must have some vehicle, some ultimate matter. Directionalized space . . . is inherent in the space itself. The form of the Jupiter in Pisces is the particular configuration of the matter. . . . Space takes the form of perception of the object, and has a quality of direction and motion. —transcript, 9/24/81

⊞ EDITORS: One representation of the Universal Soul—as the third quadrant with Jupiter in Sagittarius ♃♐—emphasizes Soul as hypostasis.

In another representation, Universal Soul encompasses the dignities called "soul" in the Soul ring as in fig. 101. The last of these is the Sun in Aquarius ☉♒, which has also been called the central Sun and the Intellectual Imagination of Universal Soul. This Sun has an inner and an outer aspect: Inwardly it is an abiding Wisdom (deploying to the prior principles in the ring and drawing upon the opposite 5th house Idea). Outwardly, as Intellectual Imagination, this central Sun provides the ground-plan and intelligence of the whole cosmos, to be subsequently elaborated by the star souls. The 12th house is Nature *per se*, the last of the reason-principles, the Idea-image of the universe.

Nature is referred to in *Ennead* III.8.3 as the last of the reason-principles or lower logos. The outer radiation of the Universal Soul's lowest power, Nature is distinguished from the higher logos or Universal Soul.

> Taking the Reason-Principle, the Logos, in all its phases, the lowest and last springs from a mental act (in the higher Logos) and is itself a contemplation, though only in the sense of being contemplated . . . but above it stands the total Logos with its two distinguishable phases, first, that identified not as Nature but as All-Soul and, next, that operating in Nature and being itself the Nature-Principle. (Plotinus, III.8.3)

The intelligence of Universal Soul, operative as the seventy-two powers of God (the Demiurge), informs the Five Mahabhutas which are the 12th house Idea. This results in the illumined space, the underlying force field pervading the cosmos. This is the Universal Alchehist and Ether of the alchemists, the primal life substance. This is the Ether, the disorganized motion that has been provided with reason-principles by the Demiurge. We think of the vast expanse as the Universal Aether [*sic*]—an in-depth emptiness. What has to be noted is that the forms (numbers) within the Universal Soul combine the Five Mahabhutas and organize the elemental life into the immutable Ideas. These bodies of the stars are the sensible representation of the Ideas. ▼

▼ EDITORS: This is a tentative description of the way that the Idea-principles are instantiated in the Sun.

The central Sun, the Sun in Aquarius, which has projected forth the blueprint for the universe, has for its substratum the Universal Life of the Five Great Elements. Plotinus (cf. V.8.12, IV.3.12) makes explicit what is slyly hinted at in Plato, where this Universal Life is called Pater Jupiter [here represented by Jupiter in Pisces, the universal motion]. The central Sun has the power to know and organize that elemental (mental) substance into the Idea of Man, the inerratic spheres. Within the unific principle of the Five Mahabhutas are the forms necessary for the Universal Soul to draw upon in order to bring about the Idea of Man. Insofar as the central Sun contains within itself the Five Wisdoms of the Dhyani Buddhas, with that paradigm it can extrapolate from the Five Great Tattvas the Idea of the Zodiacal Man. The seventy-two powers of the central Sun, combined with the wisdom of the Five Dhyani Buddhas, is the form of the Universal Zodiacal Man within the soul; but when the seventy-two powers are combined with the Five Tattvas, we have a sensible representation of the Zodiacal Man in the starry spheres. [48, p. 343]

To understand more fully the complexity of this manifest cosmos, we should plumb to the depths of the wisdom that has generated the Seats of the Gods, and we must keep in mind that there are seven levels or degrees of existence and these are provided by the Demiurgic Intellect or Universal Soul. Prior to the manifesting activity of the cosmic souls, the Demiurgic Intellect organized the disordered motion, the *samskaras* of God, as an immaterial or bodily essence, with its attendant functioning. This illumined matter is received by cosmic souls as their seat in the firmament, and is manifested and qualified according to each cosmic soul's peculiar individuality and characteristics. The totality of these manifestations is equivalent to the starry host, representing in Image the Intellectual Cosmos. Each star is the body of a unique cosmic soul, and manifests within its light-emanation its own cosmos. This gives us only some small feeling for the immensity of the inerratic sphere, which symbolizes the Ideas that in their totality constitute universal manifestation.

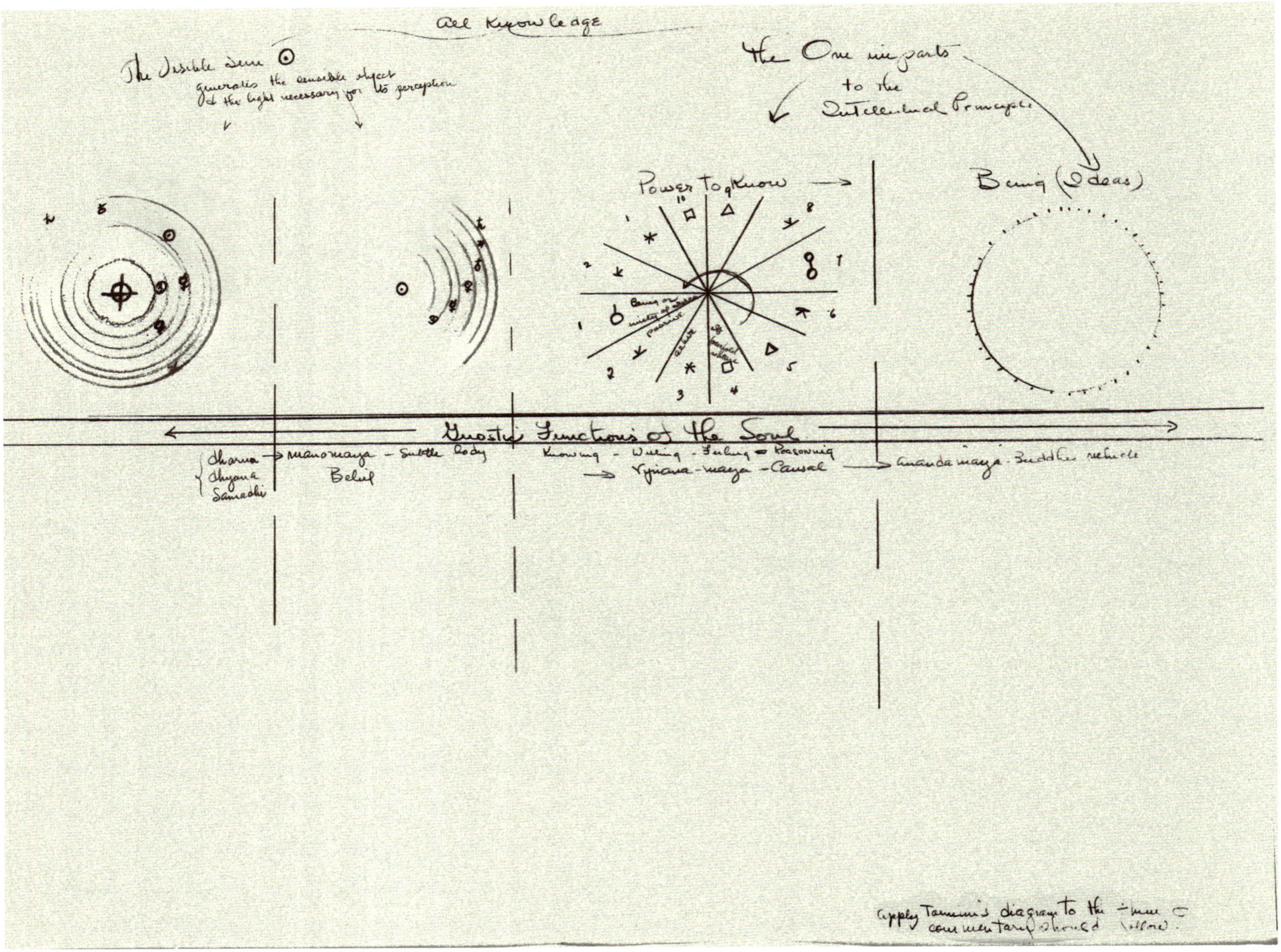

FIG. 102: A DRAWING BY ANTHONY DAMIANI

Nature: House 12

THE VISION PROCEEDS TO DESCRIBE this cluster of stars we call the Milky Way. Situated within that galaxy is our Sun. Surrounding our Sun are the planetary spheres which are representative of its rational understanding. The Sun's intellectuality is indicated by its relationship to the universe as a whole, whereas the rational powers are concerned with the relationship of all the parts within the solar system to one another and as a whole.

Let us take our solar system, our world, and place it or situate it in a prefigured environment provided by the Universal Soul—a directionalized space which already contains the primary reason-principles that are a preliminary sketch of the world to be, within which an individual soul or Sun is located. Now, an individual Sun or soul is determined by the Intellectual-Principle; as a recipient of the Ideas, it contains its own store of forms which it will use to elaborate or articulate its own cosmos. The sensible Sun, "our sister soul," has therefore the necessary forms by which it can manifest a world following the sketch laid down by the Universal Soul. ▽ [49, p. 344]

> In sum, then, the Intellectual-Principle gives from itself to the Soul of the All which follows immediately upon it: this again gives forth from itself to its next, illuminated and imprinted by it; and that secondary Soul at once begins to create, as under order . . . (Plotinus, II.3.17)

Thus, the principle of universal manifestation will be the basis for cosmic soul's activity, and the three aspects immanent in it will be reduplicated in the levels of the cosmos. In the symbolism, Jupiter in Pisces ♃♓ refers to the universal light and life of the cosmos as a whole. This contains the seats of the souls, their perpetual vehicle (or Augoiedes), which each star soul interprets according to its own intelligence. Each of the other three dignities

▽ ANTHONY DAMIANI, transcript, 10/14/83: The Demiurge, by organizing and imposing reason-principles on the matter, laid the ground-plan or the plan of the universe. From that the next step would be to think of the Sun souls (cosmic souls or star souls), as inhabiting the realm within the plan which was laid out by the Universal Soul. . . .

If we think of each one of these star souls as an Idea, or as having its own intellectuality, then each of these star souls, a Sun in its own right . . . is evolving its own world from within. The cosmic soul or the Sun soul has the powers by which it evolves the potentiality or the logos principle [latent Ideas] within it. . . .

Each one of the powers of the Sun, let's say each planet, is in its turn a soul. So the Earth of our solar system is a soul. And this soul also has the same kind of fourfold division that the Sun soul [*fig. 102*] has. . . . We're speaking just of our Sun; in another Sun the arrangement may be different—binary systems, etc.

in the 12th house represents a level of manifestation, a level of the universal reason-principle elaborated or interpreted by one of the faculties of a cosmic soul's own I AM or manifesting phase. [*fig. 103*]

The formless light-intelligence of the Sun contains a subordination of its own Ideas and of the reason powers it contains. The Sun's intellectuality is represented in the way *it* interprets the [entire] starry system. The Sun's intellectual vision [of the Intellectual-Principle] is represented sensibly by its singular motion about its axis—the Pole of Vesta,▼ and the Augoiedes is the [portion of the] sphere resulting from the interaction of the universal reason-principles [the entire sphere of stars] with the Soul's own intellectual vision. Each Sun's vision of the Intellect is peculiar to it—includes all the gods in a unique way.

▼ EDITORS: Vesta is the goddess who is said to rule the subsistence of things about their own essence. See *The Hymns of Orpheus,* and Thomas Taylor's first note to the "Hymn to Vesta" (p. 220). Anthony made a connection between Vesta and the famous description of the "spindle of Necessity" in Plato's *Republic* (616–617, "Myth of Er") and the "Spindle-destiny" in Plotinus II.3.9.

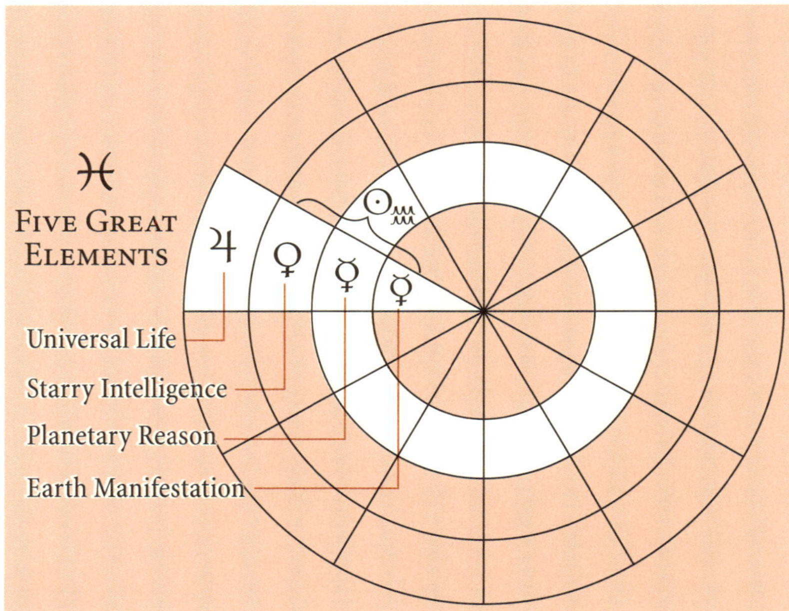

FIG. 103: THE CENTRAL SUN ☉♒ AND THE LEVELS OF PISCES ♓

Thus its participation in the intelligence of the Universe [and the other logoi] is a unique zodiac [termed the sidereal zodiac]. [*fig. 104*] ⊞

> The Soul exists in revolution around God to whom it clings in love, holding itself to the utmost of its power near to Him as the Being on which all depends; and since it cannot coincide with God, it circles about Him. . . .
>
> Similarly Plato attributes to the stars not only the spheric movement belonging to the universe as a whole but also to each a revolution around their common centre; each—not by way of thought but by links of natural necessity—has in its own place taken hold of God and exults. (Plotinus, II.2.2)

The rational powers—the activity of the planets—are both projecting the inner vision in an orderly sequence and fabricating the vehicles for the individual souls to occupy. All the individual souls in the solar system will participate in the intellectuality and rationality of the cosmic soul, just as it is participating in the Universal Soul's illumination.

⊞ Editors: The primal motion of the universe is the Life activity (♃♓) of the Universal Soul. Each of the star souls participates in the Universal, and at the same time is in revolution about the Universal as a common center. The inerratic sphere is an ancient term for the intellectual space (locus) of the star souls and the intelligible motion of the universal Idea of the cosmos. Motion in Soul is of Living Minds, and place is its place in Thought. The motion and location of the heavens are a sensible image of the motion and locus of the soul.

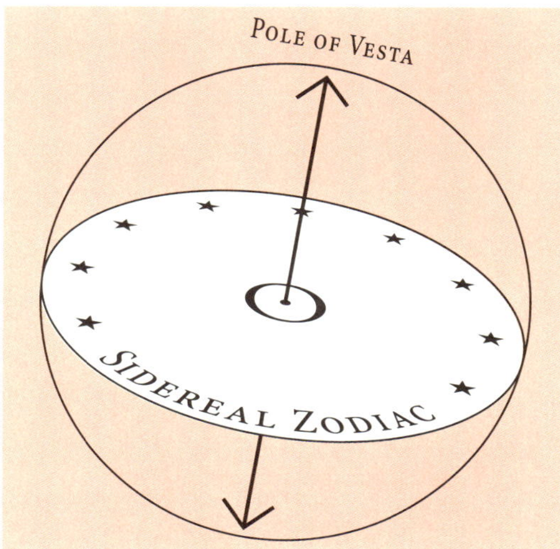

Fig. 104

The [physical] motion of the planets around the Sun is an image of the rational activity of the Sun soul. The motion is circular—but not fixed. From the standpoint of the Sun, there is no retrograde motion [of the planets]. [This motion is a representation of the soul's life motion, as it grasps the intelligence of its prior.] ▼

> . . . the Soul is ceaselessly leading the Cosmos towards itself: the continuous attraction communicates a continuous movement—not to some outside space but towards the Soul and in the one sphere with it . . . (Plotinus, II.2.1) [*fig. 105*]

▼ Editors: Note the parallel with the 11th house universal framework and this solar framework, in which the principles of combination, sequence, and relation are translated into a living unfoldment of the cosmos.

Following the fixed sphere or Intellect of the cosmos, and the planetary (spheres) or rational soul, we have the third material realm—the sublunar, the realm of being and assimilation. Only from the standpoint of [any] one of the planetary bodies [e.g., Earth, or the

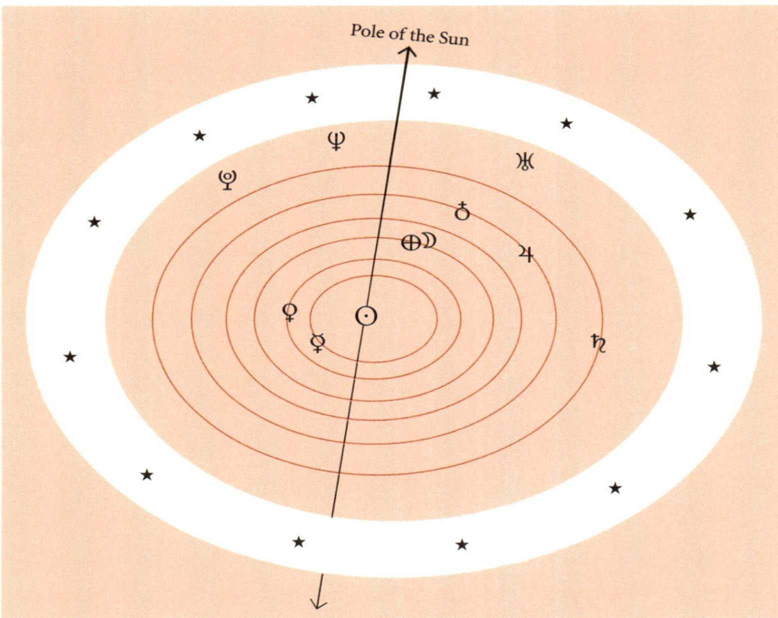

Fig. 105: Heliocentric view

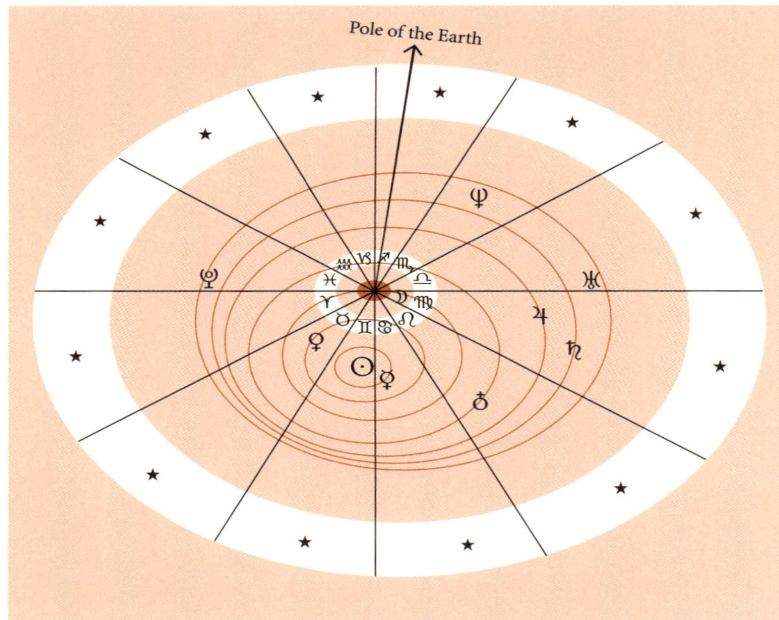

Fig. 106: The geocentric view from Body Mercury in Pisces ☿♓B

geocentric viewpoint]—the third level of cosmic soul's activity—is there the irrational activity, the lower or "animal soul," and retrograde motion of the relative position of the planets. [*fig. 106*] ⊞

This last realm of the cosmos comprises the elemental bodies of the particular worlds—our Earth, for example—and therefore all the creatures on them as well. Within the elements are contained in a subtle way the traces of all the previous living activity of the creatures on the Earth, and it is these which the cosmocrators will mold and remold into the psychosomatic bodies for the individual soul's purposes of Earth-experience. In the Earth's aura we find the last of the reason-principles—the realm of Tartarus and ultimate images. "Thus the body will be Matter associated with an indwelling Reason-Principle which will be in itself immaterial, pure Form, even though irremovably attached to the body." (Plotinus, II.7.3) These traces or tropes are also called *vasanas*—the ultimate material habit-energies which we inherit with our bodies, and which have a compulsive nature ordered by the cosmic circuit. [*fig. 107*]

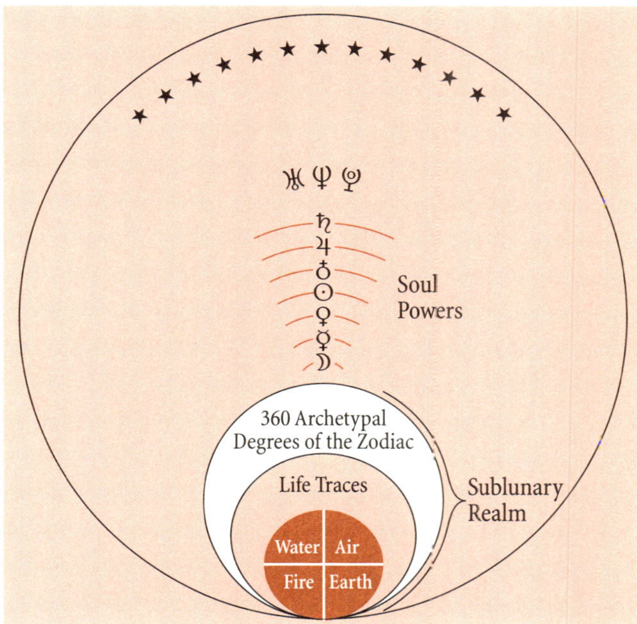

FIG. 107

⊞ EDITORS: The pole of the Earth determines a plane (which contains the equator). This plane intersects the plane of the solar system—the ecliptic—and symbolically represents the Earth's participation in the Sun's vision. This intersection moves with a slight wobble, as the Earth's pole makes a complete revolution in about 25,000 years. This causes what is called the "precession of equinoxes." The Earth's World-Idea (including the multiple reason-principles available in the Earth mentality) is based on its interpretation of the Sun's Idea. We can symbolize the relation of the Sun's reasoning and the Earth's reasoning as two overlapping systems in a plane, as shown in figure 106. [50, p. 344]

Each of the planets is a god in its own right, manifesting or unfolding the master plan inherent in the Sun's intellectuality. Each provides in its respective sphere habitations for the next degree or level of Soul. ("Habitations" means the vehicles or sheaths that are provided by the planetary gods to individual souls which will then experience that particular world by the bodies thus supplied.)

Our Earth is a god, is also included within our Sun's manifestation, and has a Body, Soul, Intellect, and Unity. The grossest aspect corresponds to the elements, and this is equivalent to the Body of the Earth. The traces of life that have been left behind in the ele-

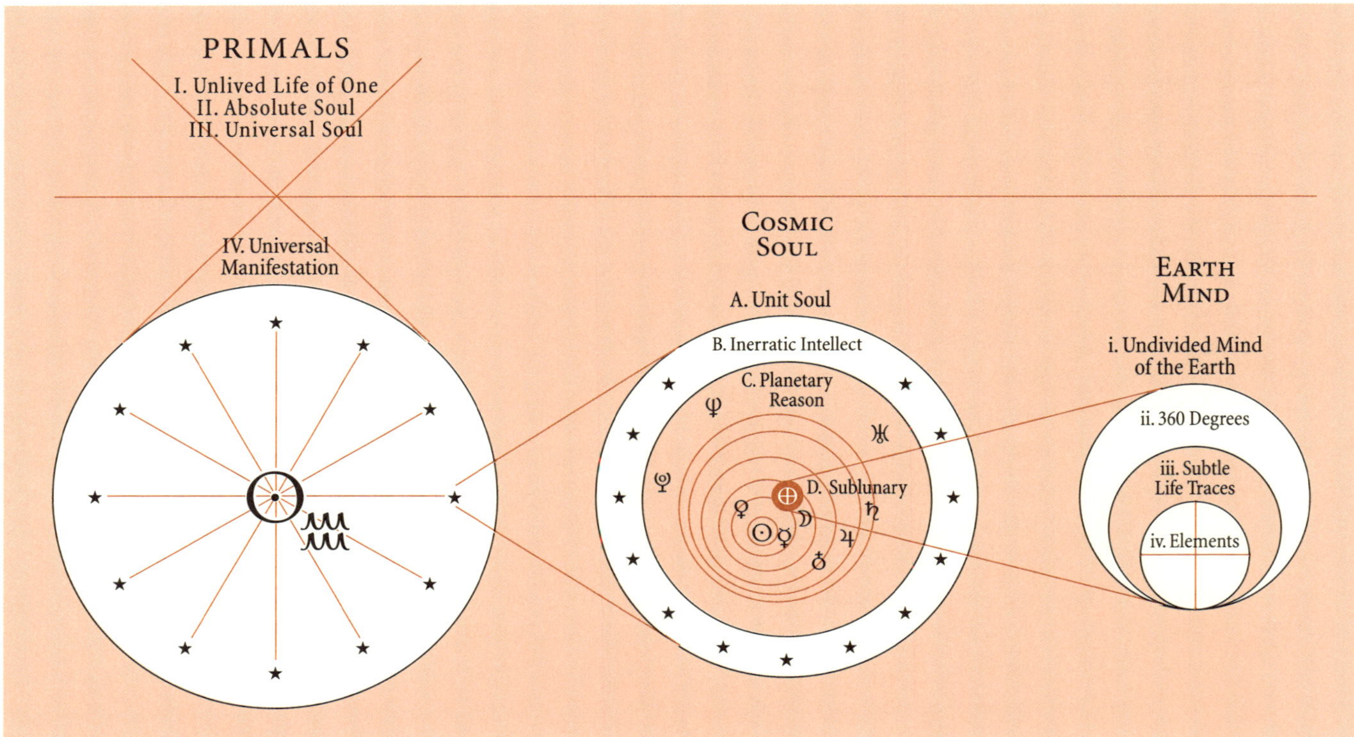

Fig. 108a

ments may be referred to as its soul. These memories retained by the planetary mind constitute our subconscious and correspond to the ancient notion of the Dragon. At a third level are the degrees of the zodiac around the Earth, which constitute its intellectuality and are often referred to in astrological literature as the differentiation of the Earth's aura. Lastly, the unity of each of the Mundane gods is a participation in the Sun soul's unitary functioning. There is no contradiction in referring to the unity aspect of each of the Mundane gods to be the same as the wisdom-knowledge functioning of the Sun soul's powers. [*figs. 108A and 108B*]

NESTED LEVELS		
Primals	**Cosmic Soul**	**Earth Mind**
I. Unlived Life in the One		
II. Absolute Soul in Intellectual-Principle		
III. Universal Soul	A. Unit Soul of Sun	
IV. System of Nature (Universal Manifestation)	B. Inerratic Intellect (causal)	
	C. Rational Powers (subtle)	i. Undivided Mind of Earth
	D. Sublunary Sphere (gross)	ii. Intellect of Earth = 360 Degrees
		iii. Soul of Earth = Life of Elements
		iv. Body of Earth = Elements

FIG. 108B

The Third Phase of Soul

▼ EDITORS: The World-Idea is given to and included in the soul, and the soul contracts itself to experience the World-Idea it manifests. We may think of the way the mind creates a dream then enters into identity with the dream person. Body is not a thing, but the process of embodiment, a lived experience.

⊞ PLOTINUS, I.6.9: Withdraw into yourself and look. And if you do not find yourself beautiful yet, act as does the creator of a statue that is to be made beautiful: he cuts away here, he smoothes there, he makes this line lighter, this other purer, until a lovely face has grown upon his work. So do you also: cut away all that is excessive, straighten all that is crooked, bring light to all that is overcast, labour to make all one glow of beauty and never cease chiselling your statue, until there shall shine out on you from it the godlike splendour of virtue, until you shall see the perfect goodness surely established in the stainless shrine.

THE THIRD PHASE OF THE SOUL, which we may call Overself or celestial Mercurius, needs to be thoroughly understood and explained, and its functioning amplified by illustrating its relation to a body which is part of the cosmos and, therefore, part of the World-Idea. ▼

It is a spark from that soul in the Nous, an individual unit of life and consciousness and a perpetually evolving offspring, which receives or manifests the World-Idea and projects it forth as its pure Idea. This evolving World-Idea that it unfolds leads it through an indefinite development and ultimately forces the vehicles it operates through to qualitatively resemble its highest phase through the practice of the "imitation of God" or the philosophic disciplines, including yoga. (cf. Plotinus I.6.9, V.3.9) ⊞

Without understanding the fundamental nature of this eternal twin, the theories of reincarnation and karma become hopelessly muddled. Some contemporary writers' precise understanding of the philosophic doctrines falters here. Their criticism of a misunderstood theory of reincarnation is not of service to philosophic students, and may even have serious consequences for themselves and their followers. Their intense and overwhelming aspiration to explain everything by reducing it to the nondual is an admirable display of devotion; but it is not the balanced view required by philosophy, for it leaves certain very important and practical matters unexplained and unexplainable.

The fact remains—and no words can abolish the situation—that our identification with our ego includes a meaning that cannot be whisked away at the stroke of a pen. Yet, many writers have not properly inquired into the value and significance of the ego. Obviously, the interval between the Self or *Atman* (Soul) and its representative, the ego, is a no-man's land, and it is very noticeably left unexplained in many texts. Why contemplative "Nature" put so much travail into producing the necessary vehicles by which the Mercurial phase may evolve is one of those questions which, if left unanswered, will forever rankle in our mind. It must be squarely faced. [51, p. 345]

The ancients conceived the Soul as tripartite, but it can just as easily be conceived of as dual. The higher Cognitive Isness or principle of Awareness and its intelligence (*cit-shakti*) cannot directly experience the sensible world; it therefore acts through the lower phase, or Mercurius, within which the manifestation of the World-Idea will take place. This lower phase may be referred to as the function of manifestness, or *asmita* (I-ness) in Yoga terminology, which "descends" completely into the realms of manifestation, into the womb of the world, called by the Buddhists the *Tathagatagarbha,* where Buddhas are born.

This germinal life from the higher soul will evolve or gestate through indeterminable time concomitantly with the Idea of the World. The World-Idea is fabricated and organized by the Universal Soul and cosmic soul, and is projected by and through the individual soul as its thought. The perpetual unfoldment and development of the World-Idea is irreversible in its direction and therefore guarantees the evolution of that individual soul. The evolving "double" of the soul, which is included as part of the World-Idea, is often spoken of as the "second soul." ⊕

⊕ Editors: Unit of life, second soul, Mercurial phase of soul, continuum of mind—all are equivalent terms. The ego is the manifestation of the accumulated experience of this unit of life.

Through its function of manifestness, this soul projects its image of the Intellectual Cosmos, which is given to it from within. Its experience of that Cosmos is the modification of its own functioning, referred to as the World-Idea, or the world as thought (in Hinduism, *citta-vrittis*). The archetypal or intellectual Idea of the universe is thus manifested through the secondary phase of the soul. This projection encompasses the entire manifest universe for an individual soul, which will project it within itself by continuously transforming itself. This projected universe will include the necessary sheaths with which the soul will identify and through which it will operate. [52, p. 346]

The function of manifestness or *asmita* is also the means by which the individual soul takes itself, at any level, to be part of the cosmos. These mental modes are

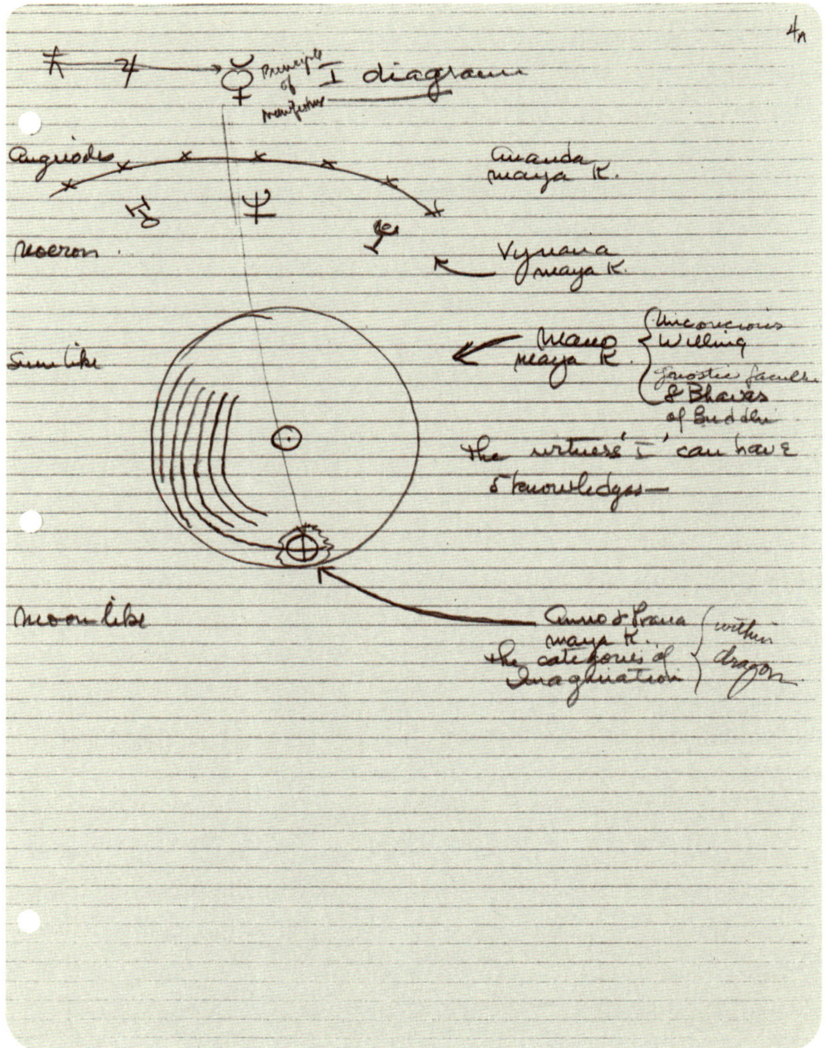

FIG. 109: A DRAWING BY ANTHONY DAMIANI OF THE LEVELS OF THE COSMOS AND THE SHEATHS

instantiating or imaging the Ideas as various grades of thought, and all these are manifest to, and appear within, the Mercurial phase of soul. In the Vedantic classification these grades are called the *Annomaya, Pranomaya, Manomaya, Vijnanamaya,* and *Anandamaya koshas.* The Platonic equivalents are the Shelly, Moonlike, Sunlike, Noeron, and Augoiedes bodies. The imaged cosmos itself provides a framework within which we can locate these vehicles. [*fig. 109*] We have tabulated and correlated these classifications within the chart in an attempt to pictorialize somewhat the ancients' concept of Man.

This conception of I AM (Soul or Mind) may be likened to a transparent crystalline sphere of life-radiating consciousness, which is a reflex or emanation from the Absolute Soul. The vision of the pure Soul in the Nous is held within the boundless consciousness, and this is represented by the inerratic sphere. The planetary spheres represent our rational soul, the discursive form of the abiding wisdom of the Soul in the Nous. And finally, the individual mind contains within it the sublunary realm, the medium in which the two priors are expressed; this realm provides the means by which the soul may come to know the reason-principles through its experience of the sensible world.

Thus, the entire manifested world can be conceived of as a huge thought projected forth from each individual soul. Within this whole we may conceive of a hierarchy of levels, each level distinguished from the whole and from one another. The highest level is Life Intellectual, or the soul as embodying reason-principles. Among the lowest is the vegetative life—a life dim and obscure, that is, the soul manifesting through the kingdom of Nature. The laws that apply to one level are different from those of another. All the various levels can be considered as compressed and operatively represented within a single entity, the microcosm, which is a miniature cosmos. (cf. Plotinus IV.3.17)

Thus, we are to remember that at all of these levels, ourselves included, we are situated within thought—Mind's functioning. The Mercurial phase of individual soul may be thought of as [first] associated or identified with the lowest of Nature's kingdoms and gradually rising through the hierarchy, consequently assimilating a knowledge of the reason-principles from each level. ▽

▽ EDITORS: There is a distinction between Soul's association with the system of Nature as a whole, and the embodiment taking place in any particular life:

> The entry of soul into body takes place under two forms.
>
> Firstly, there is the entry—metensomatosis—of a soul present in body by change from one (wholly material) frame to another, or the entry—not known as metensomatosis, since the nature of the earlier habitacle is not certainly definable—of a soul leaving an aerial or fiery body for one of earth.
>
> Secondly, there is the entry from the wholly bodiless into any kind of body; this is the earliest form of any dealing between body and soul, and this entry especially demands investigation. —Plotinus, IV.3.9

The first kind of birth is the entry into the Earth from a previous birth on Earth, or the entry from some other subtle realm of experience. The second kind we are calling the association of the soul with the Augoiedes.

Evolution of the Individual Ego: Embodiment in the Earth Sphere

We can also start at the other pole, [an approach] much more in line with our contemporary modes of thinking. When this I-ness is inserted into an animal body which has gone through a phylogenetic and ontogenetic development, the inherent functioning and vitality of that organism is initially superimposed upon this "I." This state corresponds to the "man of flesh and blood," the separative ego that regards the world as completely other and as that which must be the source of satisfaction to its own desires, a whirlwind of never-ending activity. This egoistic position gives rise to a realistic epistemology and a geocentric world-view.

However, constantly intruding into this "natural man" is the functioning of the cosmic and higher individual mind. This functioning is indicated by the motion of the Chaldean planets—altering, modifying, transforming that matrix of tendencies (despite the claims of inertia due to the psychosomatic organism). Ultimately, a sufficient deposit of the higher mind's functioning orientates that individual soul towards a more rational or heliocentric position. Thus, we are retelling an ancient view about the mystery of Man, which started with Orpheus and continued through the philosophic mythology that culminated in the *Enneads.*

Dragon: Earth's Memory as 360 Degrees or Reason-Principles

If we conceive of the Mother Earth as a goddess within whose domain has transpired the manifestation of innumerable species of life, their immediate origin would be her body itself. This body will contain the necessary ingredients for the building up of the two lowest sheaths, what are referred to in Neoplatonism as the Shelly and Moonlike bodies (i.e., the psychosomatic). This "material" has been recycled innumerable times; it therefore has certain tendencies or proclivities which are indications of past functioning and which may be referred to as "tropes." ⊞ The combination of the Great Elements into organized living bodies includes these selected responses or tendencies that manifest themselves in the body's functioning. In other words, within this mysterious imaginal essence are retained

⊞ Editors: Tropes are traces of life activity. Trope means to turn, or return, and indicates the torque put into the energy complexes we know as memory. They are the "living reason-principles" (*logoi*) Plotinus speaks of in III.8.2.

traces of the entity's functioning after its dissolution, and the totality of these traces constitutes the memory of the Earth. ⊕ This beginningless ontogenetic and phylogenetic development is the source from which and within which the bodies of all living creatures are produced and live.

This indefinite and indeterminable storehouse of images, this seething cauldron of innumerable disembodied functionings, is the material of the soul's fabricative activity and is suggestive of the nature of the Dragon. [*fig. 110*] The Moonlike and Shelly bodies of living creatures are an embodiment of it, and their activity is refunded into this "subconscious." The Dragon lives and breathes through the individual organisms; not only is it constitutive of the physiological functioning of the organisms, but also of the significance attributed to that functioning which is understood later on when the reasoning of the soul judges it. This is the lowest vehicle which is supplied to a soul with which it will acquire a rudimentary understanding of the sensible world. The soul as an inhabitant of such a body is becoming acquainted with the traces of the intelligence of the Ideas which have been left behind as part of the Earth's memory.

⊕ Editors: The totality of these traces of life constitutes the 360 degrees or reason-principles of the tropical zodiac.

ROSEMARIE QUEBRAL

Fig. 110: The dragon around the Earth

These traces (of the Ideas) give rise to images; each (non-optical) image constitutes a matrix of disembodied tendencies that have been misnamed "archetypes." Imperceptible in themselves, these traces can nevertheless be inferred through reflection on their activity by the soul's gnostic powers when it [the soul] reaches the level of reflective conscious judgement. These contents have become the environment or the Idea of the world for that individual soul inhabiting that individual body, and within it the presence of the soul within the World-Idea is indicated as well.

A preliminary inquiry into the nature of the all-encircling zodiacal degrees will require extended investigation and research. For now, we are simply suggesting that they represent an intrusion of the Intellectual, of the Ideas, flowing into our world in the peculiar way described previously. (Each of these degrees is often symbolized by an image of multivalent significances, an image which includes the formative sources of archetypal Ideas and the presuppositions of the genera and species of all life.) ▼ We have the intellectual mind of the cosmic soul (its power to know the Ideas in the archetypal realm). Through its functioning, which is both gnostic and fabricative, it organizes the Earth's memories into a world picture, which is an alogical whole that includes adventitious unities whose very organization brings about the experience of a phenomenal subject and object. ⊞ ⊕

A selection or specific combination of the totality of the zodiacal degrees [the tropes or reason-principles in Nature] and their portrayal by the functioning of

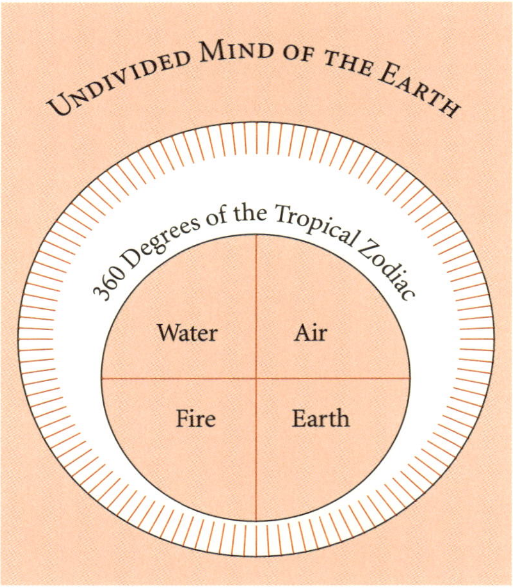

Fig. 111: The Undivided Mind of the Earth and the 360 degrees

▼ Editors: These degrees (reason-principles) are represented by various sets of images. Anthony most often used the Marc Edmund Jones "Sabian" symbols. Our Epilogue (p. 285) explores the use of these images.

⊞ Editors: The paper "Deliberate Repetition: Cosmic Epistemology" (p. 232) describes how these "adventitious unities" come to be and experience their world.

⊕ Editors: In Notes and Papers, pages S44, S167, Anthony writes:

> Let us consider the Earth's unity as identical with or coinciding with the unitary functioning of the Sun's powers. This will be equivalent to referring to the unitary aspect of the Earth as its undivided or infinite mind. . . . We may say that the undivided mind of the Earth, or its [Earth's] unity, coincides with the functioning of the powers of the Sun soul, its [Earth's] intellectuality [with] the archetypal degrees or Ideas, and [it has] a life principle [soul] which extends and permeates its gross body, the elements.

Figure 111 pictures this situation as it is hierarchically prior to the subject-object experience. The "powers of the Sun soul" in the quote above refers to the planets as they operate for the whole Earth, prior to being phenomenalized by an individual embodying at a specific location within the Earth's body.

the categories of the imagination [represented by the planets] will constitute a specific psychosomatic individuality, which will provide the soul with a body to which its karma entitles it. Only after this acquisition by the soul, which can then operate through this body, can we speak of the phenomenalization of the primary image—that is, the operation of reflective judgement on and through the functioning of this entity.

This entity is mortal; there is no reincarnation for it. This point must be understood in order to prevent an animistic conception of reincarnation. The ongoing continuity of the individual soul's karma—that is, its next psychosomatic organism—is provided by the cosmic circuit. However the continuity of karma, which a succession of bodies provides, refers to the self-identical "I AM" or individual consciousness.

It is important at this junction to point out that the moment-by-moment manifestation of the World-Idea includes these bodies which individual souls will inhabit to experience a world appearance. These bodies are being produced and manifested moment-by-moment and have a sort of continuity which is provided by the World-Idea. The individual soul in turn imparts to its associated body a subjectivity or consciousness which is characteristically self-identical. The combination of these two distinct levels gives rise to the personal ego.

So, within this same entity there are both an image or traces of the cosmic soul powers that fabricated it and also traces of an individual soul consciousness (and its evolutionary status). It should be noticed that the basis of the ego—its body or psychosomatic organism—has no self-identity from one moment to the next for it is part of the vibrating World-Idea, but the feeling of there being a self-identity comes from a completely different level of reality; that is, the ego appropriates this consciousness of the lower phase of the soul as its very own. △

△ Editors: Two kinds of consciousness or intelligence characterize (contribute to) the individual's experience of self and world. One is that of the Soul, or "self-identity." This consciousness is represented by the Soul ring, Soul's living consciousness as a *continuum* from Venus in Aries ♀♈ to Mercury in Pisces ☿♓.

The other operates from the side of the World-Idea, or system of Nature. It is *discontinuous,* "vibrating," and "moment-to-moment," instantiating the Demiurge's Ideas in the Five Mahabhutas.

Anthony insists that either of these positions is one-sided if taken exclusively; both are needed to explain the experience of soul-in-manifestation and the personal ego. This issue of the two "kinds" of consciousness is related to the discussion of the double nature of soul in the Editors' Extension "Unit Soul as Absolute Individual," p. 211. (See also figure 115, p. 267.)

▼ EDITORS: A natal chart represents an individual center of experience. Plotinus describes such a center as the "Animate": a conjoint of an "organized body" and a "ray" of soul.

> This, however, is not to say that the Soul gives itself as it is in itself to form either the Couplement or the body.
>
> No; from the organized body and something else, let us say a light, which the Soul gives forth from itself, it forms a distinct Principle, the Animate; and in that Principle are vested Sense-Perception and all the other experiences found to belong to the Animate. —I.1.7

■ EDITORS: The "ray" of soul includes life force, character, power, and intelligence. The "organized body" is determined at the moment of birth: Mental, psychic, and physical potentialities are built into the organism. At the same time, the chart expresses the "ray" of the soul and is infused with the soul's intelligence, life-current, and character. Soul expresses its vision and experiences the world through this psychosomatic organism. Thus in order to fully appreciate a chart's meaning, we must also consider it as an expression of the individual soul's inner vision.

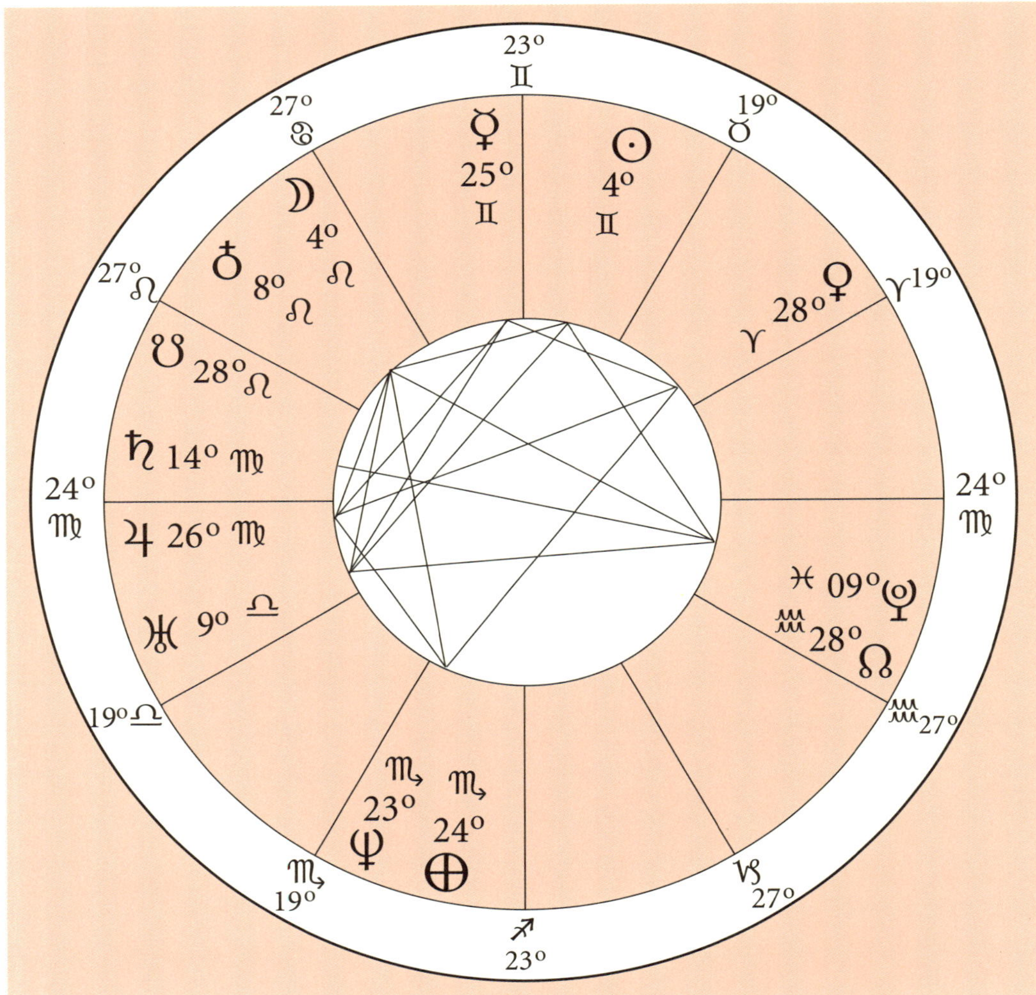

FIG. 112: RALPH WALDO EMERSON ◆ MAY 25, 1803, 1:01 PM EST ◆ CONCORD, MA, USA
LATITUDE 42:28 N, LONGITUDE 71:21 W ◆ PLACIDUS HOUSES

We are gradually approaching an understanding of how the ego is a thought made of memories, the stuff of dreams, yet nonetheless reflects macrocosmic principles. The totality of the memories of the Earth may be referred to as our subconscious, which is a vast intelligence. The specificity of the individual ego refers to the significances attached to the functioning of its tropes. This ego will be used as a vehicle by an individual soul, whose knowledge and consciousness—or, in other words, its "I AM"—will be unfolded through the resulting interaction.

If we use picture language, mind, which is formless, can be conceived of as coextensive with the planetary spheres. Its entrance into a body is indicated at the moment of birth [by the natal chart. Figure 112 is a chart of R.W. Emerson's birth moment. —eds.]

The first breath imprints the quality of the rational soul on the organism. At this moment that aspect of the soul which is divisible about bodies is identified with the organized body, forming the Animate or ego. ▽ The rational remains [as coextensive with the cosmos].

A moment of spontaneous activity of the undivided mind contains the peculiar characteristics or traces that coincide with the peculiar evolutionary status of the individual incarnating soul. The individual soul at that moment brings with it into the body those peculiar compulsions, fixed ideas, etc., and it will be the source of life and consciousness for that individual entity. This *ray of individual soul* will be the source of the subjectivity, the individual I-ness, which tethers together all the instinctual functioning of that entity. ⊞

The cosmocrators, powers of the cosmic soul to which our mind is conjoined, have organized the tropes into a body of imagination. This is the psychosomatic ⊕ organism that includes within itself traces of the Ideas. Within the Dragon, each category of the imagination △ bears the impress of the corresponding cosmocrator which results in the different ways this entity of imagination can function and know.

The functioning of the cosmic soul, to which is conjoined the individual soul, will create the necessary organs in the imaginal essence to manifest that function. Thus, something of the intellectual nature of the soul's powers is transmitted or left behind

⊕ EDITORS: The "psychosomatic" organism is a combination of two levels of reality: *psyche* (soul) and *soma* (body). *Psyche* (soul) refers to the planets in the astrological *signs,* a reality that is common to all creatures on the Earth. *Soma* (body) refers to the more restricted viewpoint of the planets in the system of astrological *houses,* which is unique to the bodily perspective of a given individual. "Psychosomatic organism" thus refers to the temporal/temporary compound of cosmic and personal perspectives.

△ EDITORS: From the geocentric viewpoint, each of the planets (cosmocrators) represents a certain soul power operating within the conditions of the Earth sphere. At the moment of a person's birth, each planet occupies a specific degree in the geocentric (tropical) zodiac. These positions constitute the individual's natal chart, his or her unique "matrix of thoughts."

While the planets themselves move onward immediately, the tendencies symbolized in the natal positions persist as dominant influences on the way that person will create and experience his or her image of the world. As unfolding that image in terms of their natal aspects, the planets are symbols of the soul or psychological functions. Anthony uses the term *categories of the imagination* for these functions.

as a trace in that entity or adventitious unity. The reduplication of these soul powers within the organized imaginal essence serves as the ground of our psychological experience. [*fig. 113*] ▼

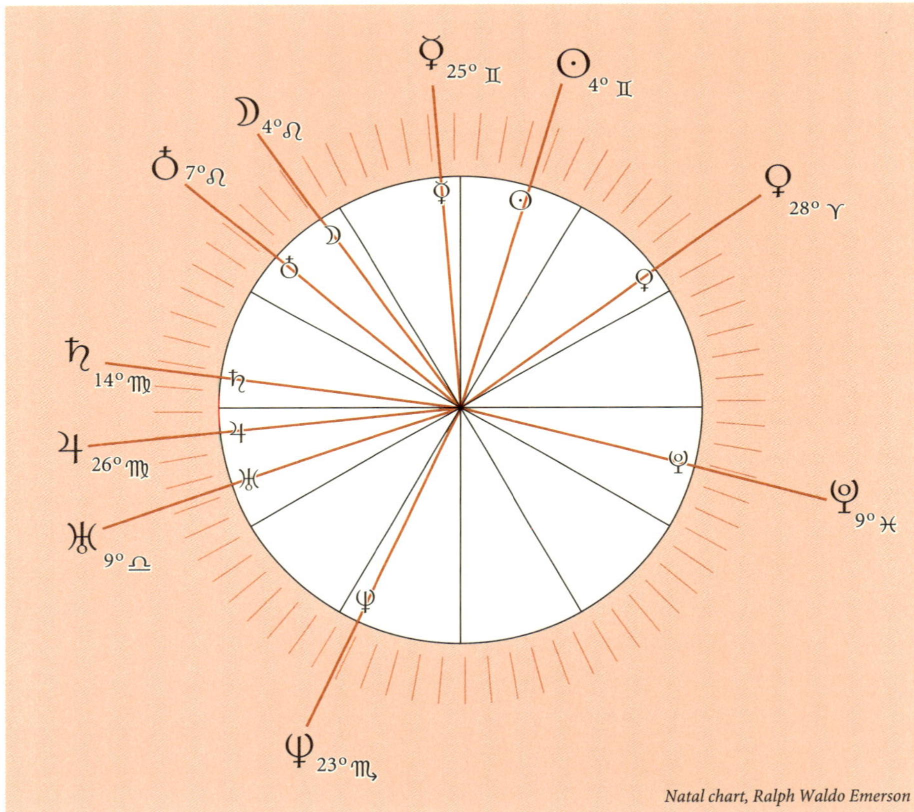

FIG. 113: THE PLANETS IN A NATAL CHART AS AN IMAGE OF THE COSMIC POWERS

▼ EDITORS: Each natal chart is simultaneously a moment of the continuous cosmic activity and a moment of enduring significance for individuals born at that moment. In figure 113, the planets outside the circle occupy the zodiacal degrees in the *signs* at the time of Emerson's birth. These positions are valid for the whole Earth at that moment. The planets inside the circle show "the reduplication of these soul powers within the organized imaginal essence," represented by the system of *houses.* The positions in the houses are determined by the specific place in Concord, MA, where Emerson was born. As so organized, the totality of these "living essences" constitutes Emerson's natal chart.

> [The undivided mind] organizes the categories of the imagination: These organized categories of the imagination are what we call our natal chart, and of course the aspects they [the planets] have among themselves. —Anthony Damiani, transcript, 4/4/83

Astrology: The Topography of the Subtle World

IT IS OUR BELIEF THAT THE ASTROLOGICAL MANDALA is a fragment of an archaic wisdom, and was devised by the profoundly wise sages of the past. This mandala of the signs and planets is a rare instance of—and product of—the archetypal imagination; it presents the topography of the subtle world through the forms available in sensible becoming. [53, p. 351] *This root symbol of astrology represents the formulation by reason-principles of a symbolic method through which those very reason-principles are being manifested. Reason-principles are those living meanings or categories of thought that are the substantial aspect [forms] of the soul.*

ANTHONY DAMIANI, NOTES AND PAPERS, P. J-1

The Natal Chart

A thoughtful consideration of the natal horoscope—our cosmic identity—suggests the particularity of the individual soul's status, which is a description rather than an explanation. [*fig. 114*]

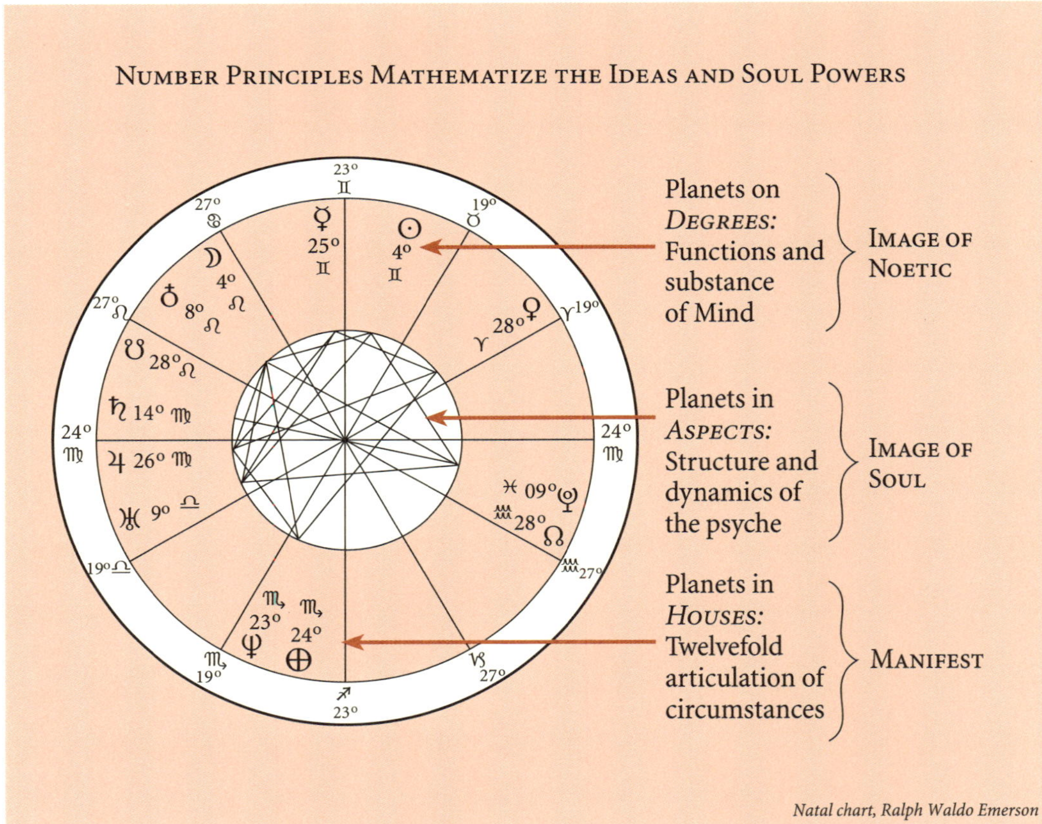

Fig. 114: A suggestive schema of the components in a natal chart—Number, Planets, Degrees, Aspects, Houses

At the moment of birth, the planets are conjunct with certain degrees and an aspectual relationship arises between all the planets and the degrees they conjunct. All the geometrical and arithmetical relationships are precisely indicated and announce the nature of this imitative being which participates in the functioning of the soul powers figured in the chart.

Thus, at the very origin at birth, the entity is already endowed with meanings that will be manifest in the course of the person's life.

Meaningful experience always contains both functional and substantial aspects. In the chart we will represent the functional part by the planets and the substantial part by the degrees. That is, in a given life there is a development both of the powers of the soul as specific abilities and of reason-principles as specific ideas.

Let us call the seven Chaldean planets representative of the powers of the soul, and the degrees of the zodiac representative of the reason-principles (or units of life-wisdom). These, taken together, are the basic meanings to be established in the native's life; these reason-principles and soul powers are invincible potencies at birth, and will become actual perception drawn from the life itself.

In the chart, a *planet conjunct a degree* indicates, on the one hand, how that function which is represented by the organ or category of imagination is influenced by the trope [degree], and on the other hand, how that planet will manifest the significances within the trope [degree]. ▽

To recapitulate: In the *degree symbolism* is inextricably buried a hierarchy of meanings—significances that are attached to the functioning of the tropes. It can be shown that the understanding of a given entity, in all cases except the person liberated from the ego, can be traced back to these ideas which the categories of the imagination have manifested. It can also be shown that a deeper understanding accrues when the natal chart makes certain aspects with the planets by transit. ⊞ It can further be demonstrated that a vastly superior and universal consciousness is transmitted through an individual soul that has particularized itself within the system of tropes that represents an individual entity.

▽ Editors: A planet and its degree are intrinsically tied together. The planets represent the activity that will manifest the Ideas as a person's life. The life will be an Idea working its way out on various levels: emotional, physical, mental, etc.

⊞ Editors: "...certain aspects with the planets by transit" refers to the continual and changing formation of significant angular relationships by the ongoing motions of the planets. The process of "transiting" is briefly elaborated in our Epilogue, pp. 294 ff.

We naturally assume that there is a significance and meaning to life; that there is some sense to the sensible. Beyond that, there are principles ordering that sense in the sensible, and these principles are the native's unconscious presuppositions. ▼ It is difficult to emphasize the magnitude of this point. Even with a sufficient maturity of philosophic understanding, we cannot stand aside from our presuppositions, or even recognize them. These presuppositions are behind both the perception of the world and the understanding of that perception; they underlie and order both the psychic and the somatic elements of experience. The structure of these presuppositions is indicated by the natal chart.

The origin of these presuppositions is in the higher soul's vision of the intelligible world, or what classical Greek philosophers call the Divine Mind. The chart is a reflection of that, and is part of the working out of that vision. So the chart is like a miniature of the metaphysical mandala.

Each soul's development is unique, as is its vision of universal Truth. Hence all it expresses in the symbol of the chart is from within itself. So, as the mentalist points out, the entire unfolding of the chart (including the chart itself) is a projection from within the individual soul, and as such reflects the soul's unique vision of the Truth. This is quite different from thinking of the individual as some structureless innocent enshrouded by a clockwork cosmos. Rather there is an organized interiority to the subject, one which develops the powers of the individual soul.

To be precise, one must add that this organization of the soul is coordinated with the evolution of the world or the World-Idea—and it is through this same World-Idea that the soul manifests its vision of the intelligible world as sensible. The sensible image of the World-Idea is the cosmos itself—the stars and the planets. [*fig. 115*]

Here we are stating that these reason-principles (the totality of the 360 degrees) are being arranged and combined in an indefinite variety of ways by the intermediary powers of the soul according to the Ideas in the zodiac. ⊞

▼ Editors: An individual's "unconscious presuppositions" are indicated by the planets on certain degrees, the aspect pattern, house positions, and so on. These presuppositions determine our experience, and also—as expressions of the soul's functioning—bear the meaning of that experience. The structure of the natal chart indicates the form of our individual mind, which receives and manifests the World-Idea, and also understands that world.

⊞ Editors: Anthony is implicitly referring to three levels of the cosmos: "reason-principles" are degrees in the tropical zodiac; the "intermediary powers of the soul" are the planetary cosmic powers, primarily as transiting; and the "Ideas in the zodiac" are the sidereal zodiac, referred to in his Invocatory Prelude (p. 14) as the "Nous in the heavens." The arrangement "according to the Ideas in the zodiac" means according to the vision Soul has of the Nous, which is represented by the paradigm of the twelve signs of the zodiac. This description parallels the one in the Intellectual-Principle chapter (pp. 105–107, in particular figure 41) of Nous looking to the One and receiving "successive impressions." These levels (with unity or void as the first level) are discussed on pp. 206–208 of the paper "Omnipresence of Soul to the Cosmos."

It is these reason-principles projected into physical circumstances that are the very understanding one must assimilate in order to become the knower, the higher soul. The astrological chart therefore attempts to join the processes going on in the subtle world with the experiences going on in the sensible world. ⊕ These experiences have a

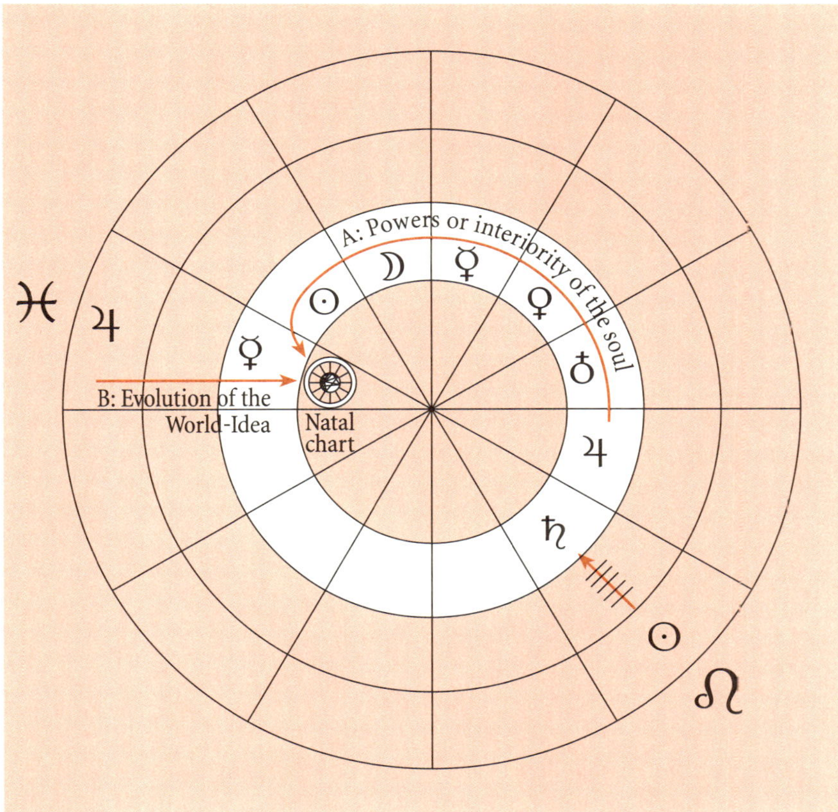

Fig. 115: The two streams of soul, a) from Saturn in Leo through the Soul ring, and b) World-Idea (Pisces/house 12). These combine to constitute the natal chart.

⊕ Editors: The individual mind, represented as the natal chart, is the link and continuity between inner meaning and outer manifestation. Each moment, the individual mind transforms the consciousness of the cosmos into an individual subject and object—like the dream mind manifesting the dream. As the mind transforms the inner cosmic process (World-Idea forces) into experience, the transformer is itself transformed in the process (cf. C.G. Jung in part 7 of "On the Nature of the Psyche": "consciousness is not only itself a transformation of the original instinctual image, but also its transformer."). As Anthony puts it in *Living Wisdom:* "Think of wisdom in the sense that the mind. . . . Your mind is becoming the universe. . . . In that process of assimilating the World-Idea, the mind is assimilating the wisdom that is inherent in it." (p. 59) By the very act of being manifest as experience, the reason-principles in the World-Idea permeate the experience. Something of this cosmic meaning is indicated by the transiting soul powers. Our Epilogue, p. 285, gives an overview of this process.

compulsive nature that cannot be avoided and that forces the individual's attention towards understanding them. [*fig. 116*]

Through this process, the individual soul powers are developed and refined until a full understanding is attained. ▽ [54, p. 351]

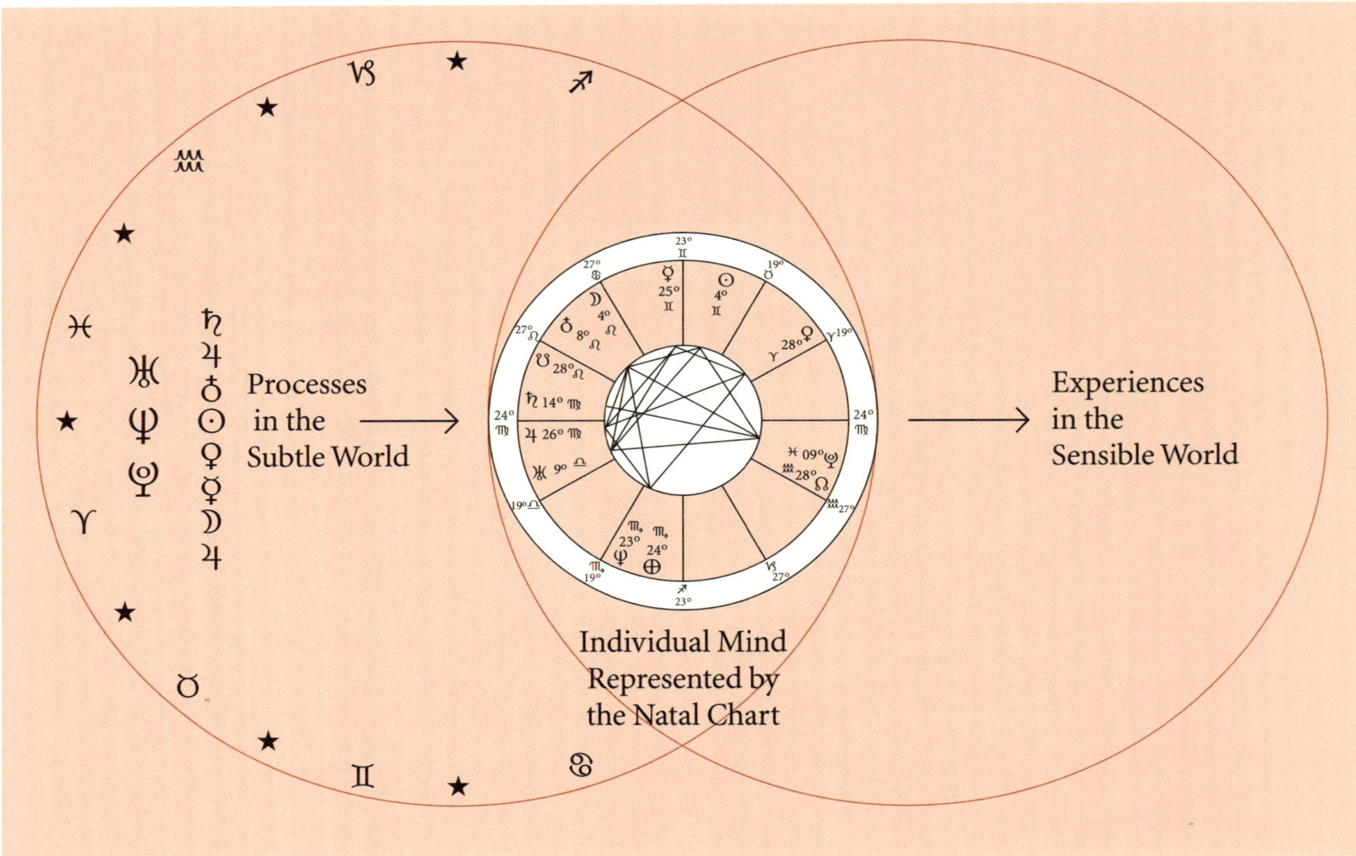

Fig. 116: Individual mind transforms subtle forces of the World-Mind into our experience

This implies that along with the manifestation of the individual, the World-Idea as represented by these reason-principles is being manifested and known through this progressive and individual unfoldment. (Otherwise the commonality of experiences cannot be explained.) Here at this level is the instance of identity of knowing and being; that is, of those reason-principles which are becoming the sensible being, which in turn is assimilating those reason-principles.

So at birth the chart is given to the soul, but is yet unavailable to the psyche; it indicates the meanings that the soul is going to try to assimilate through the life experiences. Ultimately there is the possibility of the universalization of the soul's outlook in the maturation of this development. This is tantamount to the soul's recognizing its original vision in the Divine Mind. So the refinement of the soul powers and the actualization of the reason-principles are directed toward the goal of becoming a true knower.

▽ Editors: The functioning of each of the soul powers ranges from instinctual to rational to spiritual. "Planetary Powers as Rational and Instinctual" (Note 54, p. 351) notes a correspondence to the virtues of the rational soul as expressed in Plotinus I.2.

Our central concern is with the rational soul's *intentions,* not with the personality's interpretation of those intentions. What the classical Greek philosophers called the rational soul might be explained as a permanent subjectivity, timeless but containing time, alternatively aware of the eternal and of the transient; in fact, it is this subject which is itself the *Ratio* between eternity and time, and so takes its name as "rational" from this function.

The fundamental meanings of the horoscope will always become understood by the rational soul of the native. On the other hand, whether the empirical personality will or can understand these meanings is a very difficult question.

A more urgent difficulty, and one which emphasizes the importance of astrology, is that the rational soul ⊞ itself is not free from its own presuppositions—symbolized by the chart. These presuppositions of the rational soul will always affect its inquiry into Truth.

⊞ Editors: Anthony here means the rational soul as individualized, as conjoined with and conditioned by the degrees of an individual's natal planets. He does not mean rational soul *per se.*

As we continue retracing our steps on the upward path of that part of the soul which is said to be divisible among bodies, we can further our understanding by pointing out the relationship that exists between the transiting Chaldean planets and the natal chart. The transiting planets, carrying with them their natal modifications, represent those powers of the soul which will rationalize these fixed attitudes of the ego—that is, the transits may make these psychic functions responsive to the unfoldment of the higher soul's vision.

This relationship indicates how the individual's greater mind is attempting to become aware of itself in the reflecting medium that the psychosomatic organism supplies. The soul must come to an understanding of itself through this organism with its phylogenetic and ontogenetic background, but nonetheless to which it is heir. The soul's task in each incarnation is to erase the interferences and distortions that intervene between the subtle image produced by the greater mind's functioning and its reflection in that empirical subject which is its vehicle, until finally the ego is subservient to the God within.

The rich storehouse of information that the horoscope contains can be a source of spiritual and metaphysical knowledge that could inspire and enrich our lives, for in this mysterious way the transcendental is included or is immanent in the empirical (and not embodied in footnotes). Perhaps in delineating the charts of well-known people whose lives are well-documented, we could, through a *postmortem* astrological analysis, illustrate something of what is meant by the famous dictums that, "We do not have ideas, rather they have us"—or—"What superior beings think, we must perceive."

This then is a bare outline of the meaning of soul-making and its expression in the symbolic language of astrology.

Three Levels of Subject

We hope to establish a basis for three levels of subjectivity in the chart and gain some insight into their interaction.▼ The most immediate is the empirical [bodily] subject; then there is the rational soul [or psychical subject]; and the highest is the intellectual soul [or spiritual subject]. These are all levels or grades of the same individual, and they cannot be entirely separated or distinguished from one another. Hence there is always some fluidity in their definition, as well as in our ability to distinguish them within our-

▼ Editors: Through the individual mind's participation in the cosmic mind, there is a presence of hierarchically prior principles right in our life. The levels of sensible (ego), psychical, and spiritual subjectivity were discussed earlier in the essay "Omnipresence of Soul to the Cosmos" on pages 206–208. For a further explanation and basis of three levels (bodily, psychic, and spiritual) see "The Subject as Freedom" by K.C. Bhattacharyya in *Studies in Philosophy.*

■ Editors: For the individual, Uranus, Neptune, and Pluto represent spiritual or intuitive Knowing, Feeling, and Willing, respectively. Each function operates through a spectrum of levels of meaning and activity. Sometimes Anthony described these powers as belonging to the cosmos, and sometimes as being expressions of the unit of life. In both respects they are facets of Understanding (*dianoia*) through which the soul assimilates the World-Idea.

> That phase of the soul which seeks embodiment—also spoken of as a power of projection associating with the manifesting World-Idea—is the phase which must develop this understanding. This reembodying soul is a witness of experience, is the Overself as associated with the stages and processes of manifestation. Intuition, reasoning, understanding are being used in the same way by different authors as the faculty of the reasoning soul by which it understands, interacts with the World-Idea (which soul has manifested). At the

selves. Nevertheless, we shall try to understand how and where these three levels of the soul may be seen in a chart.

Let us begin by equating the empirical with the natal Chaldean planet-pattern, the rational soul powers with the transiting Chaldeans, and the Intellectual soul with the [transiting] trans-Saturnians [Uranus ♅, Neptune ♆, Pluto ♇]. [*fig. 117*]

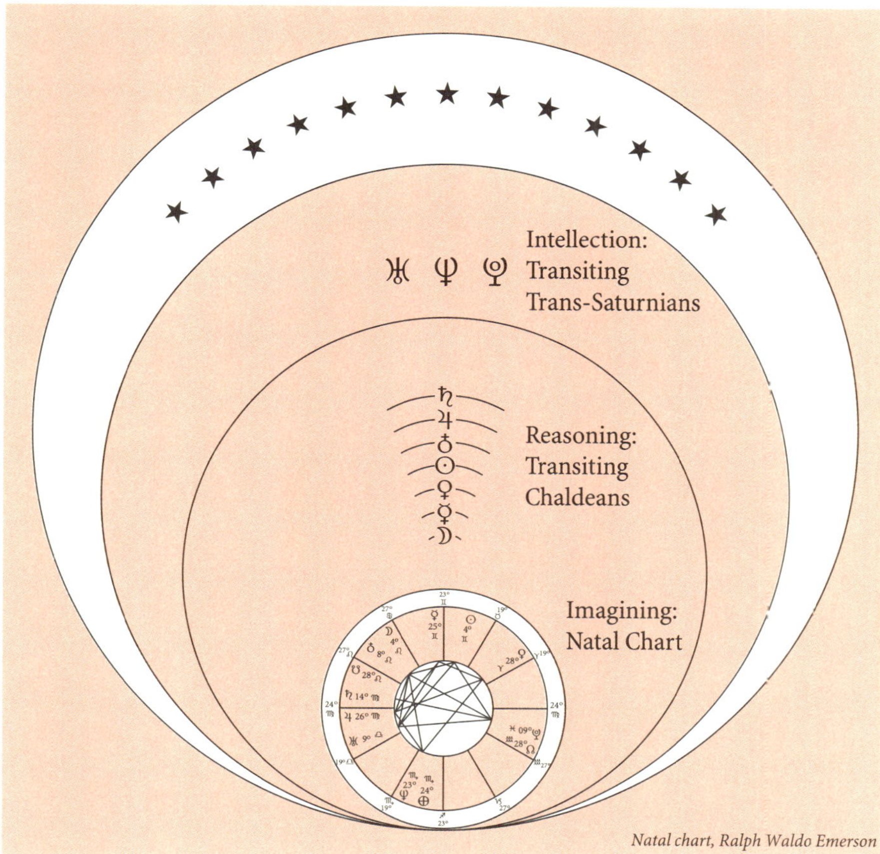

FIG. 117: THE THREE LEVELS OF THE COSMOS AND ASTROLOGY

most basic level, intuition is modalized as absolute Knowing, Willing, and Feeling....

Feeling developed to the utmost (or willing or knowing) will become intuitive. It won't then operate as we know it in the discursive intellect....

When you have a real intuition, knowing some idea, a real aesthetic perception or appreciation, it will be silent, calm. This is what Plotinus refers to as reasoning about the objects, or the reasoning faculty in the soul, which is an image of the Intellectual-Principle, having the ideas unrolled and separate. Intuition does not apply to Mind itself, but arises in the soul's relation to the World-Idea....

For the sage, there can be direct perception of the substratum reality as well as the operation of the understanding simultaneously, at any of the levels of ideation. In order to know anything about the contents of the World-Idea, or in order to articulate and formulate what insight delivers, the sage too will need to use understanding. —Anthony Damiani, *Living Wisdom*, pp. 259–260

Paul Brunton describes how these same three functions are developed and harmonized and eventually evoke the synthetic, all-encompassing "faculty" of Insight. Insight, the immediate and direct cognition of Mind into Mind, is the "higher knower" which integrates and transcends these three. See the Epilogue, pp. 296–297, for more on Uranus, Neptune, and Pluto.

A. The psychological subject has a given pattern of fixed functions; these functions are symbolized by the natal Chaldean planets.

B. The relationship between the transiting Chaldean planets and the natal chart indicates [as stated earlier] how the individual's greater mind is attempting to become aware of itself in the reflecting medium supplied by the psychosomatic organism.

Through the various transits to the natal chart—of the Chaldeans and the trans-Saturnians to the natal planets—we will begin to see something of the process of soul-making involving the complex interaction of these three levels of subject.

The transits by the Chaldean planets indicate the functioning of the rational soul:

1. adjusting, purifying, and transforming the psychosomatic person;
2. instantiating fresh essences into it;
3. experiencing its own content of essences either directly through its own powers, or indirectly through the medium of the empirical person and circumstances; and
4. penetrating, at times—perhaps indicated by the transits to the [natal] trans-Saturnians, into the pure thought held continually in soul's highest phase.

C. These [Chaldean] powers of the soul themselves have varying degrees of maturity and strength, since they are always evolving. For example, one's understanding of the meaning of Truth changes and becomes increasingly open as one's spiritual awareness deepens. So, there must be something which organizes and adjusts these rational processes, communicating to them something of the higher soul's vision of Truth. This higher order of functions is represented by the trans-Saturnians (Uranus ♅, Neptune ♆, Pluto ♇). This recognition of the original structure of the rational soul, together with its mode of alignment with the higher soul's contemplative act, is what completes our understanding of the chart as it is the motive for spiritual discipline. [*fig. 118*]

Thus spiritual astrology is a tool through which one may become aware of what is required. That is, it may provide an understanding of, and ability to work with, the Ideas being brought in by the trans-Saturnians. The trans-Saturnians not only inform the rational soul, they also work directly on the irrational soul—using its structure as the material and medium for the expression of the Divine Mind's knowledge, and in so doing the trans-Saturnians actualize the World-Idea through the empirical subject.▼

The transits by the trans-Saturnians show us something of the influx, even inuncation, of the empirical person by the light of the higher knower—the trans-rational knowing, willing, and feeling which have access to the Nous. These are regulating and infusing the lower understanding, and the wisdom stored in the lower.

There are those whose goal is to live a thoroughly symbolic life, who use symbolism—dreams, astrology, synchronistic phenomena—as the ground upon which to base

▼ EDITORS: This discussion is primarily about the transiting trans-Saturnians.

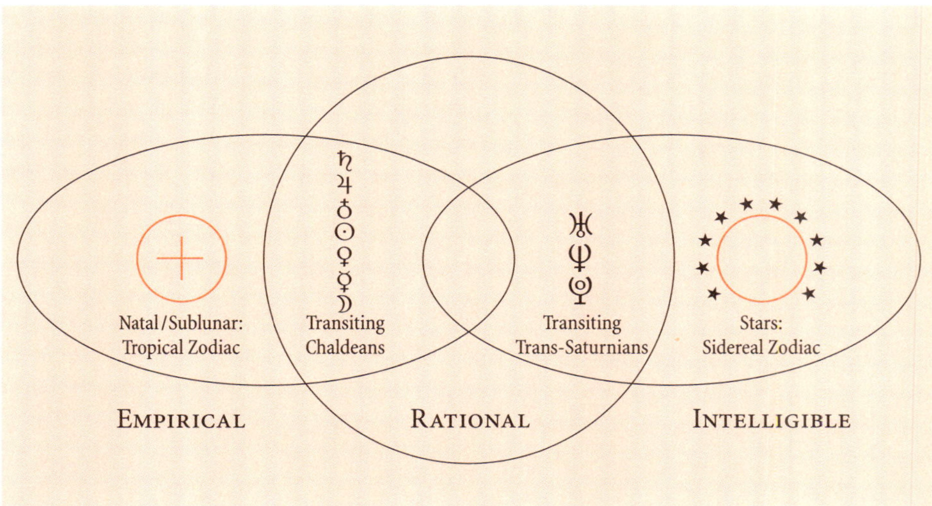

FIG. 118: LEVELS OF THE COSMOS—THE NATAL CHART, TRANSITING CHALDEAN PLANETS, TRANSITING TRANS-SATURNIAN PLANETS, AND STARS (SIDEREAL)

the meanings of their life. Although this is undoubtedly fruitful, we are adding to this a view at once more profound and more immediate. That is, there is a deeper significance to life, and in astrology we have an opportunity to see the horoscope as an embodiment of a deeper, *imageless* grade of meaning and purpose by which we may live our life.

Indeed, the perceived world is enough of a symbol without making a symbol of a symbol. Beginning from common experience, when one becomes aware of the rationality within, one begins to be aware of the rational soul's presence in one's life. If we are alert to the possibility of recognizing this presence where there are transiting aspects to the natal positions, we are capable of assimilating its operation, and so succeed in becoming rational ourselves.

Soul Making

Several quotes from Plotinus pinpoint distinctions which we wish to find in the workings of astrology. The first and most lucid of these is IV.3.30.

> Perhaps reason, whose actualization always accompanies thought, has the function of receiving it and transmitting it to imagination. (IV.3.30, K.S. Guthrie trans.)

Here reason is the activity [represented by the transiting movement] of the seven Chaldeans, the power of the rational soul. As we have pointed out, these contain all the necessary essences, the living forms needed to transmit the thought they receive.

"Thought" is the content of the intellectual soul, the higher soul's vision of the Ideas. Thought is accessible by the intelligence of the trans-Saturnians, but can only be evoked by the living essences which the soul powers use to portray the indivisible thought. "Imagination" here is the realm of the primal imagination in the Dragon. So Plotinus continues:

Indeed, thought is indivisible, and so long as it is not evoked from the depths of intelligence, it remains as it were hidden within it. Reason develops it, and making it pass from the state of thought to that of image spreads it out as it were in a mirror, for our imagination. (IV.3.30, K.S. Guthrie trans.) [*fig. 119*]

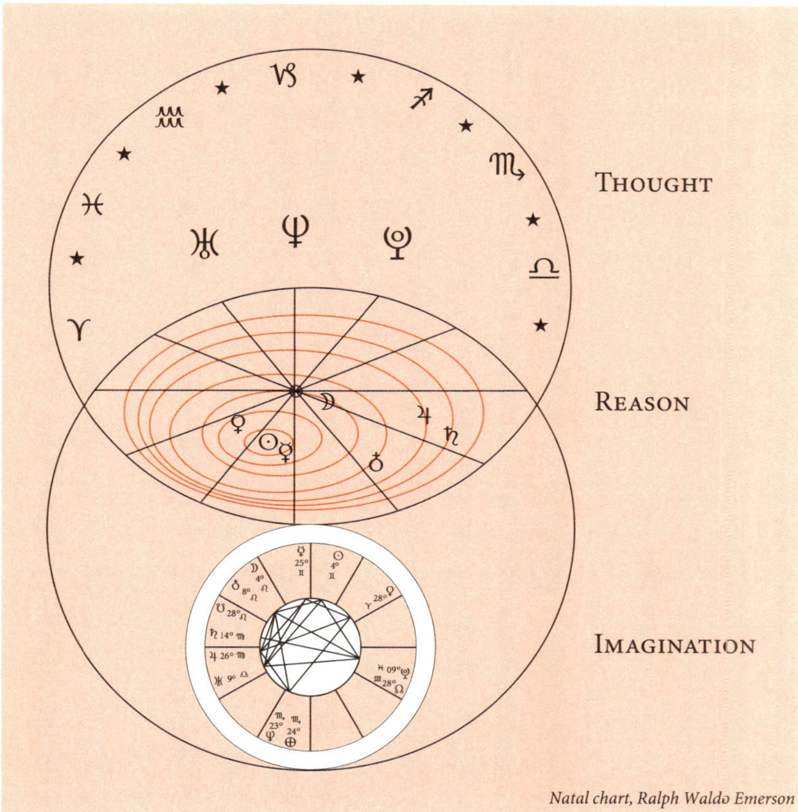

FIG. 119: THOUGHT, REASON, AND IMAGINATION IN COSMOLOGICAL AND ASTROLOGICAL CONTEXTS

Reason, the seven soul powers and the essences they contain, ▼ develops the thought. Various essences need to be organized by the soul powers in such a way as to then mold the imagination to portray the [inerratic] Ideas, to "spread them out as in a mirror," so that they may be experienced by our imagination—that is, in our empirical experience. Plotinus continues:

▼ Editors: "... essences they contain" refers to the intelligent forms in the degrees.

> That is why we grasp (the thought) only when the soul, which always desires rational thought, has achieved a thought. (IV.3.30, K.S. Guthrie trans.)

The achieved thought is the actual psychosomatic entity. This is the portrayal and result of the essences in the rational soul, which have taken the pure thought and transformed it into the imaginal realm. The "we," as including the lowest or sensible counterpart, only grasps the thought when it has been manifested in our experience. Plotinus concludes:

> There is a difference between thought and the perception of thought. We are always thinking, but we do not always perceive our thought. (IV.3.30, K.S. Guthrie trans.)

Again, the thought, or thinking that is continual is of the intellectual soul; but the *perception* of the thought referred to here must be the result of the rational operation just referred to. And since the rational powers are attending to the sensible as well, when they do this there is generally not a perception of the thought. The principle that perceives the thoughts, the rational, must occupy itself with the pure thought, and portray it, transmit it to the empirical. The empirical person's perception of meaning is dependent on the rational soul portraying it to him or her.

In cosmological terms, the soul's vision of the Nous (inerratic) is a kind of pure thought = *prajna* (the [transiting] Uranus, Neptune, Pluto), which is transformed by the reasoning = *taijasa* (seven [transiting] Chaldeans), into the state of a primordial Image which is based on the elements and life traces of our Earth planet (and so our particular organism [represented by the natal chart] which is part of it as well). Thus spread out in Image, the reproductive personality (which is part of the image) can have sensations, thoughts, feelings—in short, empirical experience (*vaishvanara*). [*fig. 120*]

In VI.7.6 Plotinus says:

> Man as sense percipient becomes aware of these correspondences and accommodates the sense-realm to the lowest extremity of its counterpart There [in the intellectual], proceeding from the fire here to the fire Intellectual which was perceptible to the higher soul in a manner corresponding to its own nature as Intellectual fire.

In V.3.2 Plotinus further elaborates the function of reason:

> The reasoning-principle in the Soul acts upon the representations standing before it as the result of sense-perception; these it judges, combining, distinguishing . . .

This activity of the reason is to make some sense of the representations, the presented circumstances of the empirical experience. Or, it may also "observe the impressions, so to speak, rising from the Intellectual-Principle, and has the same power of handling these." (V.3.2) The impressions rising from the Intellectual-Principle are the thought which is the

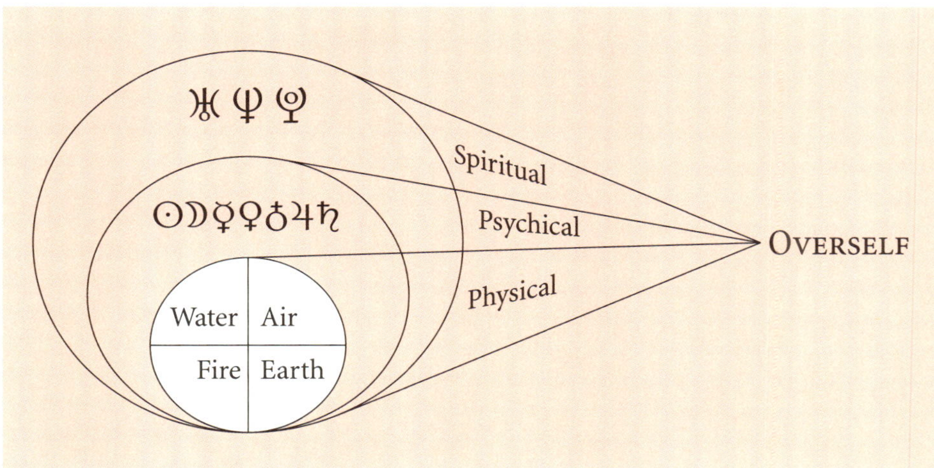

FIG. 120: THE OVERSELF AND SPIRITUAL, PSYCHICAL, AND PHYSICAL SUBJECTIVITY

content of the Uranus, Neptune, Pluto, and which the Chaldeans can handle and distinguish, combine and express.

Further,

> . . . reasoning will develop to wisdom where it recognizes the new and late-coming impressions (those of sense) and adapts them, so to speak, to those it holds from long before—the act which may be described as the Soul's Reminiscence. (v.3.2)

Reason, the soul powers, may take the very sense experience and find in it that which can be developed to wisdom. Reason finds the very essences that have gone into the organization and form of the experience, and links the understanding of it back to the essences it holds within itself—the Reminiscence. In other words, we could perhaps say that the reason has the power to extract from the experience the essence [imaged in degrees], or quintessence of it, and thereby crystallize something of its own inherent store of essences.

There can be many causes or reasons which bring the ego to a temporary standstill. When the mind ceases to be preoccupied with living out the animal compulsions—even temporarily—it can become a medium or apparatus wherein the rational process can be reflected. Even more precisely, instances can be shown within a natal chart when such a crisis is reached and the normal functioning is inhibited so that reflection on that life can proceed. We can speak of this as an intrusion of a higher level of understanding or the presence of the rational mind mingling with and illuminating the instinctual life and evaluating it. Thus it is inevitable that a moral conflict ensues, demanding a growth of ethics and understanding that is spread over an indeterminable amount of time.

> Reflection is at all stages a freeing process, the mind freeing itself from the body or from itself as a subtle body of the self. The process is spiritual in the sense that it is as much conative as intellectual, a detaching activity which may be indifferently called the will to deny will (will to nivritti) or the knowing function. It is in reflection that the self-assertive will (will to pravritti)

> is known to have been one with the body that is sought to be freed from; and it is in reflection that the given object is negatively willed to be detachedly known. In Samkhya there is really no will to deny will as in Yoga: it is just the knowing process, knowledge as mental being a process. In Samkhya the consummation of this reversion of the will, which is the knowing process, is conceived as viveka which is not eternal quiescent knowledge which would be the self but the mind distinguishing, defining or detaching itself from the self and thereby losing itself in unknown or unmanifest Prakriti, the self consequently standing in its eternally manifest solitariness. (K.C. Bhattacharyya, "Studies in Samkhya Philosophy," para. 17)

In the midst of our embodiment, God has given us reason, "virtue the unconquerable." (Plotinus, II.3.9) Through a process of purification, or through a series of lives detaching us from the ego, this becomes more and more available. The virtues of the soul fully developed are the soul come into likeness to its divine model—which itself has no need of the virtues. At the lowest level we find the civic virtues, the psychological good-doing of a man or woman. But the virtues of the rational soul are the contemplative, aesthetic, and conative powers of the purified soul. Parallel to the actualization of the Idea of Man in it, is this development of the particular soul powers and their virtues, which begins right at the level of the imaginal experience. The continuous manifestation of Pure Thought [intelligence] as our experienced world, through the mediation of the reason, constitutes the basis of the true evolutionary process which this life demands of us. The freer our imagination is of the compulsions and discrepancies of the irrational animal-based life, the more clearly our soul faculties can manifest the Ideas or reason-principles to us without distortion, and the more fully the Ideas in us, governing our life, are revealed, the more the soul can be brought into alignment with that intelligence. And this is primarily Plotinus' meaning of virtue—the soul brought into conformity with the Nous.[54, p. 351] All this seems to imply the presence of the Nous transmitted in us by the medium of the light of Soul, and molding that very transmitter in its operation to portray that intelligible cosmos as this world.

We have to keep in mind here an important point: The entire series of thought is contained already in the Idea. Even the particular modes of its sensible expression are included in the Intellectual vision. It is important to try to see the continuity and the illuminating of the lower by the higher:

> The Man of the realm of birth has sense-perception: the higher soul enters to bestow a brighter life, or rather does not so much enter as simply impart itself; for soul does not leave the Intellectual but maintaining that contact holds the lower life as pendant from it, blending with it by the natural link of Reason-Principle to Reason-Principle: and man, the dimmer, brightens under that illumination. (Plotinus, VI.7.5)

> This is how the secondary Man, copy of Man in the Intellectual, contains the Reason-Principles in copy; and Man in the Intellectual-Principle contained the Man that existed before any man. The diviner shines out upon the secondary and the secondary upon the tertiary; and even the latest possesses them all—not in the sense of identifying itself with them all but as standing in under-parallel to them. (Plotinus, VI.7.6)

The illumination of the lower, the sensible man of the natal chart, by the various and continual transits, is made possible by the continuous link of the higher reason-principles and essences to the lower—the images. The diviner, the trans-Saturnians, shine out; and if some link through a transit or aspect can be made then a transmission of some light and understanding can come all the way to the empirical life. The chart "possesses [contains] them all [i.e., higher] . . . in under-parallel."

These three phases of the Soul's Act are required in order for knowledge to arise. The "natural man" has knowledge only of the sense world with an intermittent mild infusion of rationality. The level of knowledge available to him depends upon the dominant phase of Soul operative, though each individual has all three in potential. Any of these three phases can be the dominant principle in the life. What makes a sage unique is that *the sage operates with all three simultaneously if he or she so desires.* There is no hiatus between the intellectual and the sensible in the sage's experience. Because the sage has a complete and direct understanding of the nature of human being, the sage can delineate the necessary factors involved in knowledge and has in actuality all three [phases] at his or her disposal. This constitutes the sage's happiness, which cannot be taken away by the "blows of fortune" and the "nursery terrors." (cf. Plotinus, I.4.8)

Plotinus describes the continuity of image and Idea in IV.3.11:

> Every particular thing is the image within matter of a Reason-Principle which itself images a pre-material Reason-Principle: thus every particular entity is linked to that Divine Being in whose likeness it is made, the divine principle which the Soul contemplated and contained in the act of each creation. Such mediation and representation there must have been since it was equally impossible for the created to be without share in the Supreme, and for the Supreme to descend into the created.

The natural link of the lower, the imaginal categories to the pure powers, and of the tropical images to the living essences in the rational soul and the Ideas in the intellectual, is that the lower are the very expression of the higher, and the higher have the power to illumine and to pervade the lower. The Providence of the higher, and the unfoldment of the Ideas in the highest phase of the soul through the operation of the reasoning soul, ensure to us that astrology "works." For ultimately it is not chance which establishes this sensible cosmos and its creatures, but an all-pervasive divine Wisdom and Soul.

> For every human Being is of twofold character; there is that compromise-total and there is the Authentic Man: and it is so with the Cosmos as a whole; it is in the one phase a conjunction of body with a certain form of the Soul bound up in body; in the other phase it is the Universal Soul, that which is not itself embodied but flashes down its rays into the embodied Soul: and the same twofold quality belongs to the Sun and the other members of the heavenly system.
>
> To the remoter Soul, the pure, sun and stars communicate no baseness. In their efficacy upon the (material) All, they act . . . as ensouled bodies within it . . . body is the agent while, at the same time, it becomes the vehicle through which is transmitted something of the star's will and of that authentic Soul in it which is steadfastly in contemplation of the Highest. (Plotinus, II.3.9)

Levels of Understanding Traversed by the I AM

The purpose of life, according to Plotinus, is the acquisition of true knowledge and the individual's life in the sensible world is for that purpose. The astrological chart can and does reveal the specific ideas being manifested by that life and assimilated by the individual mind towards that goal. This is the fundamental spiritual nature of astrology.

There are many levels of meaning from the ideas hidden in the symbolism of the degrees, which we have referred to as the basic presuppositions and attitudes that are the person. These basic ideas orientate a person towards certain goals in his or her life, and this activity furthers the contemplation carried on by the mind. This contemplation or "in-seeing" is the silent assimilation of those reasons by the second soul from the experiences in the sensible; it is garnering them here because it could not achieve that assimilation directly in the Intellectual realm.

Plotinus goes on to say that the purpose of the journey through the cosmic circuit is to acquire some good. What possible good could this be? It is the acquisition of those ideas by the mind—the second soul—to complete the vision resident in the logos. So it is rather a straightforward affair to discern that the purpose of all this activity is to ultimately possess the object of contemplation, the Ideas or reason-principles that are being manifested by life. This understanding will eventuate in the fulfillment, the peaceful smile of a Buddha.

But now we must consider the more immediate consequences involved in acquiring this knowledge. The knowledge must be applied to the facts of experience and to the yet unresolved picture of the world. The mind must make its own that which it knows. Knowledge must be lived or it becomes a source of infection; therefore these ideas must be applied in their place in the lower world; that is, by being translated into ethics, morality, and the virtues; otherwise it is not yet knowledge actualized.

The briefness of this explanation should not delude us into conceiving that this is a process that can be accomplished within the mortality of the ego. It is a process spread over an indeterminable amount of time and involves the assumption of many bodies to secure the necessary observations. To grasp the meaning of this process, assume the existence of a secondary soul that has set out to fulfill its vision. Through its own activity it descends into the lower world and procures the body necessary in order to understand certain reason-principles which sensible experience can provide. The accompanying natal chart would indicate those reason-principles that will sensibly manifest as this entity and its experiences, from which the second soul extracts the essence. The repetition of an indefinite number of such bodies will provide the second soul with all the reason-principles necessary to achieve its vision. Many lives in the body are needed by the evolving understanding to achieve maturity in the philosophic life. Only in this very wide context can the last paragraph of this tractate of Plotinus be understood.

> The Sage, then, has gone through a process of reasoning when he expounds his act to others; but in relation to himself he is Vision: such a man is already set, not merely in regard to exterior things but also within himself, towards what is one and at rest: all his faculty and life are inward-bent. (III.8.6)

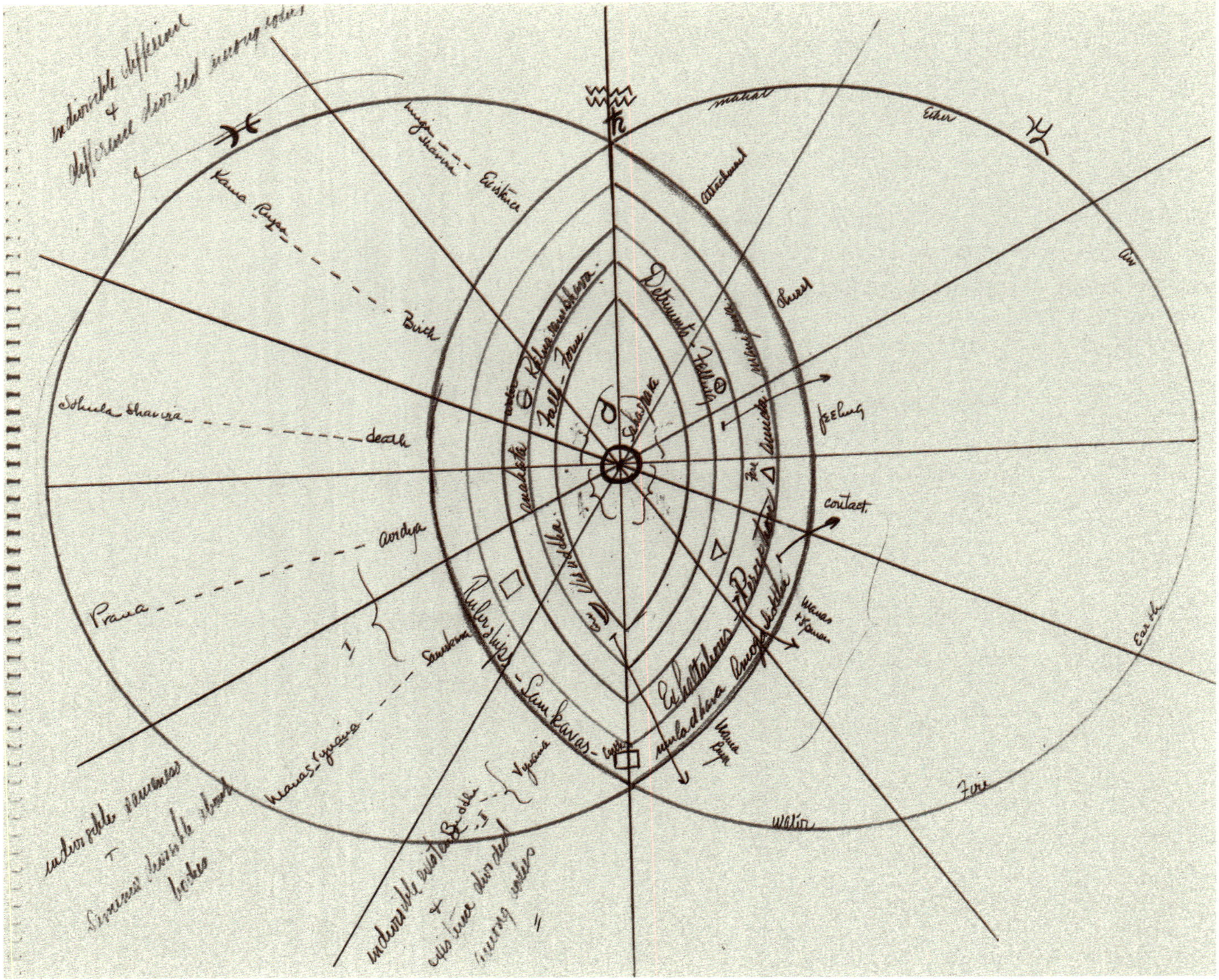
Existence
Birth
death
Prana
Ether
Air
feeling
contact
Earth
Fire
Water

Editors' Epilogue

Throughout history, many people have held that "We do not have ideas, rather they have us." Nearly every culture has its versions of this notion, sometimes also expressed as "What superior beings think, we must perceive."

Astronoesis makes major contributions to our understanding of where such ideas might originate, what they might be, how they might combine in (or as) a specific individual, and what purpose or significance they might give to a life such as yours or mine. Its implications for meaningful living are immense.

This point of view embraces the universe as meaningful and orderly. It insists that life matters, that we human beings are integrally involved in creating our experience, and that we do not create it alone. It celebrates the ongoing discovery that our greatest joy is in vividly experiencing and expressing all that goes into making us unique, intelligent, and alive.

The potential for such self-knowledge, Anthony held, is woven into the fabric of each soul's very being—along with a certain audacity that assures its self-actualization. Under the pressure of an innate longing, each soul journeys into cosmos for deeper communion with what it most desires. To this end, the intelligence in the cosmic order educes—in the sense of helping to draw out—each soul's power and vision, ultimately aligning it with the wisdom that the universe itself expresses.

This positive view of life's purpose does not fly in the face of well-known pessimistic views so much as it balances them. It doesn't pretend the universe is more perfect or that human life is less painful or joyful than either is. Rather it discovers meaning in the imperfection and the suffering and the joy, and uses that meaning as a stimulus to more effective action and more abundant life.

Anthony's vision of this process is so vast that it is literally mind-boggling. It allows for billions of people to have myriad experiences in their own uniquely meaningful ways, yet roots each one within a unified framework sufficiently universal for each to see itself and be seen as a part of an evolving whole.

What might it be that this soul most desires? What powers might this soul be developing through its involvement in the circumstances of this particular life? What might the cosmic mind itself be trying to accomplish in this person? What can this person's astrological chart show about his/her present stage of spiritual evolution, and which spiritual practices will be most useful for him/her right now? Questions such as these drove most of Anthony's research into astrology, and he pursued them in the fertile context of the model Plotinus provides: three primal hypostases (One, Intellectual-Principle, Soul) immanent in the system of Nature.

Oneness is present in the bare fact of individuality, of any*one* to speak of. For the ideas or meanings operating in a life, he looked to what he called the "substance" pole of experience; this he correlated with the 360 degrees of the zodiac as the presence or immanence of Intellectual-Principle (Nous) in individual mind. ▼ For the powers that sequentialize our actual experience of these ideas/meanings, he explored what he called the "function" pole of experience; this he correlated astrologically with the planets as the presence or immanence of Soul to individual mind.

▼ Anthony Damiani, Notes and Papers, p. 175: A look at the horoscope would reveal 360 degrees—Ideas in the Nous—that in their totality can be conceived as a single entity or monad. These Ideas which are beyond—and the reference is not spatial—are associated with certain portions of the stars and referred to as the zodiac: universal force fields or substances that have a boundless and radiant wisdom that a sensitive reverence recognizes as a source.

The continual motion of the planets through the zodiac at varying speeds relates planets to degrees, planets to other planets, and degrees to other degrees in ways that are constantly changing and unique to each individual. Anthony correlated this complex matrix with the "thinking" of a mind that Paul Brunton described as "cosmic in its spatial sweep and permanent in its embrace of time." (*Wisdom of the Overself*, p. 51) This "thinking" makes available or emphasizes or enlivens certain ideas / meanings for specific individuals at specific times and in specific sequence, as temporary combinations of substances and functions form and dissolve to provide the flow of meanings at the core of a life's events. Unique patterns of number (aspects) link the infinity of Soul's oneness with specific intelligible forms that reveal each person as *a* one among other ones, a distinct one of the many that are one.

This process of world-creating thought has a definite direction, though the timing and the routes through which it completes that direction in an individual can vary in many ways. It corresponds with what Paul Brunton speaks of as World-Mind manifesting as much of its World-Idea as is possible through the Overself of an individual. World-Mind provides *what* the individual will experience, and the individual determines *how* it will experience.

In the symbolism of classical astrology, Anthony found a framework that helped him enter more deeply into the profound mystery of this process—which he spoke of as the "arithmetization" or "mathematization" of the ideas. Principles of number and symmetry immanent in the visible motions of the universe helped him explore how corresponding contents of the subtle world of the psyche are evoked. He saw the planetary movement through the heavens as a visible image of the subtle life of the soul or mind.

Within the context of our experience here on Earth, for example, the degrees of the tropical zodiac indicate how specific features of the intelligence, or wisdom, of the Earth soul become available to individual human beings embodied here. The positions of the planets relative to the Earth indicate what powers the individual must exercise to assimilate that wisdom. The significant angular relationships indicate how the individual will experience the process of integrating that wisdom. These angular relationships are called "aspects" or "transits," depending on context.

The fertility of this schema becomes even more apparent when we see how Anthony viewed the zodiacal degrees. He did not consider them 360 fixed "things" that each individual would or could experience in the same way, or that would mean the same thing in every individual's constitution. Rather, he likened them (analogously only) to musical notes. C-sharp, for example, retains a certain identifiable quality regardless of in what musical piece it appears; but the particular meaning-quality of C-sharp revealed in a particular piece varies enormously according to the key of the piece, the tempo and touch with which it's played, its context in the melody and/or movement, the instrument being used, and—perhaps most important—the unique unifying theme that drives that musical piece as a whole.

ANTHONY DAMIANI, *ASTRONOESIS*, P. 263: It is our belief that the astrological mandala is a fragment of an archaic wisdom, and was devised by the profoundly wise sages of the past. This mandala of the signs and planets is a rare instance of—and product of—the archetypal imagination; it presents the topography of the subtle world through the forms available in sensible becoming.

ANTHONY DAMIANI, *ASTRONOESIS*, P. 270: The rich storehouse of information that the horoscope contains can be a source of spiritual and metaphysical knowledge that could inspire and enrich our lives, for in this mysterious way the transcendental is included or is immanent in the empirical . . .

To develop a deeper appreciation of how these degrees work to manifest ideas, Anthony studied and gave classes on the charts of many great individuals—Gandhi, Emerson, Napoleon, Vivekananda, Eleanor Roosevelt, Einstein, Oppenheimer, Fermi, Puccini, Verdi, Wagner, Freud, Jung, and many more. He suspected that the more highly developed soul powers (planets, on the function side) of such individuals would reveal the intelligence in the degrees more transparently than would be the case with less evolved people. Empirical study soon supported that intuition. It also became apparent that the essential or core idea of a particular life is easier to see after that life has completed itself.

This brief epilogue cannot illustrate the details of how Anthony compiled and analyzed these "spiritual dossiers." And giving just one or a few examples is much too limited to show how so-called coincidences proliferate beyond what a reasonable person could attribute simply to Chance. That work must be left to a series of books that will follow *Astronoesis* and use it as a root text. For now, however, a few more broad strokes may be helpful in the interim.

Degrees—Living Meaning

We use the word "degree" three ways.

1. As 1/360th of a circle, a degree is a geometrical division of the zodiac into unit intervals; all significant number relationships between the degrees is intrinsic to their relationships in the whole circle.
2. The word degree primarily means a unit of living meaning, a logos or reason-principle, in the mind or soul of the Earth (tropical zodiac) or Sun (sidereal zodiac).
3. "Degree" also refers to a symbolic image, or "degree symbol," that can be used as a seed to contemplate that meaning.

In the sense of degree as idea in the soul, each degree is a whole mandala or matrix of possibilities. Each is an expression of wisdom, a divine image, or archetype. These are divine seeds to be unfolded through the process of life itself, and simultaneously or subsequently understood. ▽ ⊞

Anthony used symbolic images or "degree symbols" as a main tool to help gain imaginative entry to the "intelligized space" of these reason-principles. Such images, skillfully used, help us feel the presence of the idea. They can help saturate our reflective consciousness with the intelligence-quality of the idea until we open to it at a deeper level and are lit up by it.

Each degree lends itself to many images, and each symbol-image is but one perspective on the multivalent immanent idea. Several collections of such images are available. Of them, Anthony most often used the Sabian symbols developed by Marc Edmund Jones. In particular, he preferred the *image* from the "long form" Sabian symbols from "Symbolical Astrology" in Jones' *Lecture Notes* and the *interpretation* that accompanies the "short form" in *The Sabian Symbols in Astrology.* He also used some of the other resources for comparison, especially the symbols published by Isidore Kosminsky.

Each degree also manifests differently through different soul powers. This difference becomes significant as different planets conjoin with any particular degree—spoken of as being "on" the degree. Each soul power reveals the intelligence of the degree in a way unique to its own functioning. When the Sun, for example, is "on" a degree in a natal chart, that degree operates as a core idea driving and giving its unique shape to the whole living of the life. If Venus, instead, were on that same degree, the intelligence of the degree would shape how one's aesthetic sensibility operates, but not necessarily dominate the whole purpose and direction of the life.

"In order to understand a symbol that your planet is on," Anthony said in a class (3/24/84),

> what is required is that you look at the symbol very intently but don't *think* about it, don't bring up any associations that the symbol may have. So if we took an example . . . my Uranus is on 13 Pisces: "An ancient sword in a glass case in a museum." ⊕ . . . I can read the psychological definition in Jones, but that doesn't satisfy me. . . . He's explaining the psychological behavior which is the manifestation of the idea. So I'm not to the idea yet.

▽ ANTHONY DAMIANI, *LOOKING INTO MIND*, PP. 255, 72: Whatever meanings you find in your life are indicated by these degrees, these ideas. . . . You're living these ideas, you are these ideas—why not try to consciously understand what's going on?

⊞ ANTHONY DAMIANI, TRANSCRIPT, 9/14/83: The intelligence of the Earth—what we spoke about as the forms in the soul of the Earth—that's awesome. . . . the 360 degrees or gods or archetypes are the Earth's interpretation of the Celestial Ideas. . . . The Ideas which are outside [of the dragon] in other words, the Celestial Ideas, are very well reflected—if we use that term—by the Earth's soul. They are objects, universal objects, universal substances worthy of contemplation. . . . They create everything here. Everything in our world is the result of, let's say, their ideation or thinking. So when you reach the level of contemplation, any one of these Ideas might reveal itself and it would be, you would say, the presence of a god.

⊕ EDITORS: The symbol in Jones' long form reads: "In a still corner of the museum are weapons of long ago and in a glass case by itself is a single venerable sword." His short form is: "A sword in a museum."

So I contemplate the degree. I keep thinking about it, but I think by not thinking. No association. No thought. . . . At first I memorize it, visualize it, retain it in my mind: first, the words, naturally. The words stand for some kind of meaning, and once I memorize the words then I'm preoccupied with the meaning and I forget the words. I try to gaze at that meaning without bringing in any associations, because any associations I bring in will be from my past and that's dead. I'll get no fresh knowledge that way. . . .

It is a kind of unfocused attention, diffused attention. . . . With an unfocused attention there is no determination, no judgement. ▼ What's there is there. . . . If an intuition comes, I'll be there. Whereas focused attention is a predetermination as to how you're going to see something. . . . If attention is focused and an intuition comes, I might block it, even deny it. . . . When I read a chart, I try to keep that kind of attitude.

▼ ANTHONY DAMIANI, TRANSCRIPT, 3/29/82: When you read the symbol, do not bring in your concepts. . . . An image has a totality and absolute presence. You cannot dissect it when you look at it. You must feel it. It won't tolerate conceptualization.

The meanings Anthony found for symbols, like the C-sharp in many compositions, were unique to specific individuals. He often criticized what he called "cookbook astrology" that gives standardized interpretations of symbols as if the meaning of images does not vary enormously by context. He insisted that there is a presence of the person in each chart and symbol and that meaning arises through a contemplation of the whole chart. He told a group of students in Ohio on 3/24/84, for example:

A symbol . . . is a whole cluster of meanings. And one of these may strike you immediately, come to you like a kind of intuition. You try to bring it into the reflective consciousness. . . . . [In the case of the Uranus at 13 Pisces in his own chart] ultimately what the sword really means is discrimination. . . . One of the ideas that's included in the way I function is that I must always discriminate between two things and try to get to the essence by understanding the difference between two things. . . . Discrimination in this sense is a kind of idea. I do that with each symbol until I begin to get to the idea that each symbol is trying to portray. If I do another chart, that very same symbol will alter. The symbol is a matrix of possible tendencies, and which tendency is going to become available to you depends upon the configuration of the whole chart.

Planets—Soul Powers

The kind of work that remains to be done in developing key applications of *Astronoesis* includes not only much more on understanding the degrees, but also getting a deeper appreciation of how we develop the soul powers that allow us to do so. To this end, Anthony's correlation of planetary powers in astrology with the traditional idea of three levels of Soul offers an enticing place to begin.

To reiterate from *Astronoesis,* Anthony first seized upon two key distinctions in traditional astrology's treatment of the planetary powers. One distinction is that between the "natal" planets and the "transiting" planets. The other notes an important difference between the activity of the Chaldean planets (from Saturn inward) and that of the trans-Saturnian planets (Uranus, Neptune, and Pluto).

When the planets are considered as "fixed" or as "staying put" on the degrees they occupied at the moment of an individual's birth, we speak of them as the "natal" planets for that individual. This configuration, when related to the specific place of the person's birth, is what astrology calls the natal horoscope. The degrees (and their opposites) occupied by the planets show a great deal about which ideas dominate the individual's presuppositions, habits, and compulsions (both inner and outer). Anthony correlated this configuration with what Paul Brunton calls the "ego."

In the flow of "real time," however, the planets continue to move through the zodiac. Their relationships to one another and to the positions of the natal planets change continually. The significant numerical relationships ("aspects") that emerge between each of these moving planets and any of the natal positions are called the individual's "transits." This ongoing process of the Earth-mind thinking its evolving world-idea into and through a given individual is what Anthony correlated with the level of the rational soul. This is the level at which, as Paul Brunton describes it, each individual receives his or her image of the world from his or her own Overself.

A third distinction naturally emerges, and that is how the activity of the trans-Saturnian planets differs from that of the Chaldean planets at the natal and transiting levels. In *Standing in Your Own Way,* Anthony addresses this question from Paul Brunton: "How can a man fully express himself unless he fully develops himself? The spiritual evolution which requires him to abandon the ego runs parallel to the mental evolution which requires him to perfect it." (*Notebooks,* vol. 6, pt. 1, p. 22)

Anthony comments:

> Now the rational soul is always going around that [the fixed natal positions] and it's trying to make necessary additions, subtractions—whatever is necessary to perfect the ego. That's a part of the mental perfection that he's speaking about. . . . So it would be the rational soul that would be [building up] the mental, trying to correct the categories of the imagination. . . . The development and refinement of the ego is taking place because you're constantly having [Chaldean] transits going on. . . . Whereas the spiritual evolution would be implemented through the outer three planets [Uranus, Neptune, Pluto]—which require abandoning the ego—the rational souls from Saturn inward are demanding that the ego get fulfilled. (*Standing in Your Own Way,* pp. 173–176)

Three primary distinctions, then, emerge within the functions represented by planets.

1. Natal positions = ego, within which are a reflection of the rational soul (the natal Chaldeans) and a reflection of the purely spiritual functions (Uranus, Neptune, Pluto).
2. Transiting Chaldeans = rational soul, developing the ego and its powers.
3. Transiting trans-Saturnians: Uranus, Neptune, Pluto. Anthony correlated these with the soul's higher functions of knowing, feeling, and willing, respectively, that release the individual from his/her exclusive identification with the personal ego.

Natal Planets and Degrees

The positions of the planets in relation to one another, to the degrees they occupy at the time of the individual's first breath, and to the specific place of the birth constitute, in their totality, the natal chart of that individual. This structure represents an individual center of experience and transformation. It serves as the locus or "container" of a rich complex of potential dynamics—both within this structure itself and in its relationships to and within the ongoing life of the cosmos. Anthony correlated these natal positions with the "organized body" part of the famous 1.1.7 passage from Plotinus. ▽

This structure reveals mental, psychic, and bodily presuppositions built in or "hard-wired," as it were, into the individual's constitution. If we think of the World-Mind's thinking as a vast field, then this individual structure may be thought of as one small cell immersed in that field. Each of us filters this intelligence through the structure of our own natal presuppositions. In the process, we reduce, or step down, its vastness to the fragment or fraction that we can handle (as Arjuna realizes in chapter eleven of the Bhagavad Gita). The relative size of that fraction is determined by the kind of body and intelligence we have, the selectivity of our attention, and the soul-force and character the individual soul itself brings into the mix at the moment of birth.

Anthony considered each of the planetary intelligences a deific power in its own right. Nonetheless, the natal relationship between the planet and its degree is fundamental to explore in order to see how that power is operating in the life of an individual. Planet and degree at this level are two sides of a coin, as it were: The degree gives a specific structure or form to how the planetary power operates, and the planetary power manifests that structure or form through the specific quality of a certain soul activity.

Of these pairings, the most vital in an individual is the relation between the Sun and its natal degree. The Sun and its degree indicate the central meaning of the life. Anthony wrote that "Soul is free when it moves toward the Good," (*Astronoesis,* p. 199) and for him the Sun is what lights and guides our path to the Good. The Sun is a manifest symbol of the Good, and its degree indicates the particular idea that essentializes life according to that Good. ⊞

The Sun provides the life energy of the chart, the center or feeling of I, the life-current itself. All the planets articulate something of this light, as a prism unfolds white light into colors. In an individual, this light or life has a very specific structure—the idea that the individual is compelled to work out in this life. This idea is indicated in the Sun's natal degree. Contemplating this degree gives us a view into the quality of the living consciousness

▽ PLOTINUS, 1.1.7: From the organized body and something else, let us say a light, which the Soul gives forth from itself, it forms a distinct principle, the Animate; and in this Principle are vested Sense-Perception and all the other experiences found to belong to the Animate.

⊞ ANTHONY DAMIANI, TRANSCRIPT, 1/30/84: We have to look at the Sun as life. . . . No one lives just life, he lives *a* life. As soon as you put in the word "a" . . . you're speaking of an idea that dominates an individual life. In other words, it is this idea which is dominating that life and which must be expressed in and through that life.

communicated continuously to that individual, and a way of appreciating how that individual finds or creates meaning in the world. Anthony correlated this relation of the Sun and its degree with what Jung called the principle of individuation. ▼ ⊞

Anthony's Sun degree, for example, is the first degree of Virgo. Jones' long form of this symbol is: "A man's head is revealed; it is a portrait in which the artist has sought to idealize the best of human character." This degree symbol helps us understand Anthony's lifelong preoccupation with understanding and exemplifying the Idea of Man: what it is to be a fully evolved human being. It also helps those who knew him well understand where the force in him came from that inspired them to work toward expressing their own higher potentials: It was his very life, and that life supported and catalyzed the individuation process in others as well as in himself.

Anthony usually also included the opposite degree symbol in his consideration of natal degree symbols. He explored the possibility that the degree symbol and its opposite might offer two faces of a complex meaning, representing the subjective and objective poles of a primal image that includes and sustains both. "The organized categories of the imagination . . . have an outer function and an inner functioning. . . . The inner function is the planet sitting on the degree and the outer function is represented by the opposite degree." (Anthony Damiani, transcript, 4/4/83)

In addition, he also correlated planetary dignities and non-dignities with the soul's gnostic and fabricative powers. This correlation may prove to be one of his most significant contributions to astrological research.

▼ Anthony Damiani, Notes and Papers, p. S58: The Sun in the solar system is the central force and individuality of the cosmos. Likewise the Sun as the central unity is present in the individuality as the central point of a person's ego.

⊞ Anthony Damiani, Notes and Papers, p. S128: The Sun degree in the chart may be used to symbolize the individuating quality of that life, brought to self-reflective awareness through the function of the other six powers within the organism. The resulting functioning of this psychosomatic organism brings about the appearance which is the field upon which the reasoning soul passes judgement.

Transiting Chaldeans

Anthony spoke often in classes about the importance of learning to shift one's sense of identity from the mental processes at the level of the natal planets (ego) to those that emerge from a deeper level through the activity of the transiting Chaldean planets (rational soul). For example:

> Read the natal planets from the transiting, not the other way around. . . . we as individuals are the transiting Chaldeans. That's what we really are. That is your undivided, spontaneously active mind. . . . Identify yourself with the transiting Chaldeans: That gives you the opportunity to identify with the illuminating idea which is coming in. . . .
>
> To identify with the transiting Chaldeans means that you live totally in the present, not dependent on the past. Every moment would be novel, a constant influx of knowledge. . . . when I have aspects from the transiting Chaldeans to my natal, I should put myself in a silence so I can see what is to be done. If I have a concept, I know I am coming from down below. If I have no concept, then an activity will take place that will be appropriate for the circumstances that are presented.
>
> I'm almost positive that if you operate without conceptually trying to grasp but you flow with the idea produced by the undivided mind, you'll have an experience of flowing with the Tao. There will be no need for any kind of thinking, yet you will do everything with perfect correctness. (transcript, 4/18/83)

The individual functions represented by the natal positions are both continuous with and distinct from the soul powers and ideas operating through transits. The transiting activity corresponds with the "light" of the Plotinus I.1.7 quote, and together the natal and transiting positions (with all the dynamics potential within them) form what Plotinus calls the "We." This "we" includes a spectrum of levels of function for each of the powers, ranging from the instinctual at one pole to the rational at the other.

Anthony also maintained that "the rational soul itself is not free from its own presuppositions." (*Astronoesis,* p. 269) By this he meant that the activity of the transiting planets (and this includes the transiting trans-Saturnians as well) is conditioned by the ideas (degrees) that infuse them from the natal chart. As the Sun transits, for example, it is not pure life but life infused with the idea in the degree of the natal Sun.

One possible exception to this continuing influence of natal presuppositions is the case of an individual whose cognitive powers have surpassed or become free of the ego's conditioning. Even in such a case, however, the extent to which the natal degrees continue to influence that individual's self-expression in this world must be acknowledged in what he or she is able to communicate to others. Anthony generally advised that if we cannot entirely free ourselves from such presuppositions, then at least we can become conscious of how they affect our perception and limit the possibility of our seeing things without bias.

Uranus, Neptune, Pluto

Anthony also emphasized the importance of learning to recognize moments of higher knowing, feeling, and willing as made available through the activity of Uranus, Neptune, and Pluto. He looked to them and their degrees for ways an individual can abandon the egocentric personal viewpoint and "see as God sees." In general he spoke of them as powers of the Overself and explored their activity through the motions of transiting trans-Saturnians; but he also acknowledged them as somehow present in the ego as represented by the natal positions of Uranus, Neptune, and Pluto.

Paul Brunton writes of these three:

> That element in his consciousness which enables him to understand that he exists, which causes him to pronounce the words "I Am," is the spiritual element, here called Overself. It is really his basic self for the three activities of thinking feeling and willing are derived from it, are ripples spreading out of it, are attributes and functions which belong to it. But as we ordinarily think feel and act, these activities do not express the Overself because they are under the control of a different entity, the personal ego. (*Notebooks*, v. 6, pt. 1, p. 3)

To be more explicit: Within the individual chart, the trans-Saturnians present images of the higher knowing, feeling, and willing, as well as three ways to abandon the personal viewpoint in favor of a cosmic or universal perspective. Uranus represents the possibility of awakening by shifting one's viewpoint from the warped perspective of egocentric self-interest to the higher knower's disinterested appreciation of genuine truth. Neptune acquaints each individual with how his or her life is permeated by and woven into the life and evolutionary needs of the entire cosmos. This mystical presence of universal reason pierces our self-centered feelings and opens us to a higher dimension of participation in the cosmic process. Pluto awakens us, often roughly, to the immediate presence of a world-idea that draws its power from an infinitely deeper necessity than that of our personal desires. Its presence in us makes us aware of what we might loosely call God's will, or with more sophistication, the inevitable direction of the universal idea underlying all our experience. This awareness often rends our imaginative projections with glimpses into the actual "state of the union" in God's universe.

Anthony describes one way that these higher functions intrude on our normal mental processes and interrupt their momentum:

> Suppose you have a trans-Saturnian aspect going onto one of your natals. . . . The way we generally experience a response to that is not a definitive knowing or willing or feeling, but an awareness that I don't know something. . . . The response to that transit will be in terms of an intuition, or a momentary silence. (transcript, 4/4/83)

Often these experiences bear intuitions whose details we work out over time, gradually altering our whole way of thinking or feeling or acting about something as we assimilate what came into us at that moment. Sometimes they act more immediately, turning us briefly into laughing Buddhas.

This area of research may prove the most interesting in the long run for what it can tell about which spiritual practices are most appropriate for a given individual at a given time.

Aspects

Aspects, whether natal or transiting, are occasions in which significant number relationships emerge through the movements of the planets. They are neither substance nor function, but they coordinate substance and function. Their arisal, abiding, and dissolution from thought-moment to thought-moment corresponds with the rational soul activity (or World-Mind's "thinking") that transforms innate intelligence into actual experience and/or understanding. One dimension of the astrologer's skill comes in seeing how this activity indicates which corresponding tendencies in the individual come into play.

Anthony explored three different aspect groups, based on dividing the zodiac into equal divisions of twelve, five, or seven.

The first aspect group, based on the division of the circle into twelve equal sections, he called "regular" aspects. This group includes the conjunction (0 degrees), the semi-sextile (30 degrees), the sextile (60 degrees), the square (90 degrees), the trine (120 degrees), the quincunx (150 degrees), and the opposition (180 degrees). Anthony normally looked to these aspects for information about complexes within the circumstances of the physical organism.

The second aspect group he explored in depth is the "quintile series," based on a division of the circle into five. This includes the conjunction (0 degrees), the quintile (72 degrees), and the biquintile (144 degrees). Anthony correlated aspects in this series with activity at the psychic level, where thought is formed and produces creative vision. He explored these as referring both to the rational activity itself and to the reflection of that activity in the psychic life of the individual.

Anthony also explored what he called the "septile series," based on dividing the circle by seven. This series includes the septile (51 3/7 degrees), biseptile (102 6/7 degrees), and the triseptile (154 2/7 degrees). He looked to these aspects as indicating activity from or at the noetic level, where fixed ideas are built into destiny:

> I think the septile indicates how we understand the functioning of our psyche—the way one part of the psyche understands another part of the psyche. . . . A septile is like a trace of former wisdom. These are your fixed ideas which carry over from life to life. . . . Within the natal chart, and only within the natal chart, the septile seems to be the standard of judgement you are using all the time. . . . It is a process and an innate idea, and you are always using that ruler to understand what is going on within you. (transcript, 7/11/83)

The three aspect groups introduce an even greater refinement and sophistication to researching how three levels of the soul coordinate their activity in producing and understanding experience.

Looking Ahead

"Is there a totally comprehensive schema or symbol," Anthony wrote in an early stage of *Astronoesis*,

> that will permit us to combine the varied and many facets of truth that are to be found in each tradition, to sort out those basic and truthful presuppositions into a working hypothesis which will also aid the philosophic student in the investigation into the truth of his or her life? Such a schema does exist, and it can be used in helping us to resolve some of the most abstract problems of metaphysics as well as those emerging crises of the everyday life. Philosophy has conceived the experiment: Astrology in its most inclusive context goes on trial." (Notes and Papers, p. 596)

Astronoesis shows how Anthony conceived the experiment and how far he got with it. The empirical work he initiated on the "spiritual dossiers" of great individuals encouraged him to continue the experiment, in light of astrology's promising response. The research was always a work in progress, an arena for enquiry. Deepening his own and others' appreciation of the key questions was in the long run of much greater value than any of the particular answers. People working with him in this context often marveled at the aptness of the Sabian symbol for his Mercury degree (17 Virgo) as an image of how he perceived and communicated: "A vast display of cosmic force is seen in the eruption of a volcano with dust clouds, flowing lava, earth rumblings."

For all the excitement and stimulation this study brought him and many students around him, Anthony remained humble throughout. His sense of perspective never left him. He reiterated often that ultimately we have only glimpses of the tremendous process through which our minds grow in fields of wisdom sown and tended by intelligences greater than we can ever fully fathom. "What is the relationship," he asked in a class (10/14/83) "between the spontaneous wisdom and functioning of the planetary spheres and the individual ego? What is going on here? I don't think we can ever finally say what is going on."

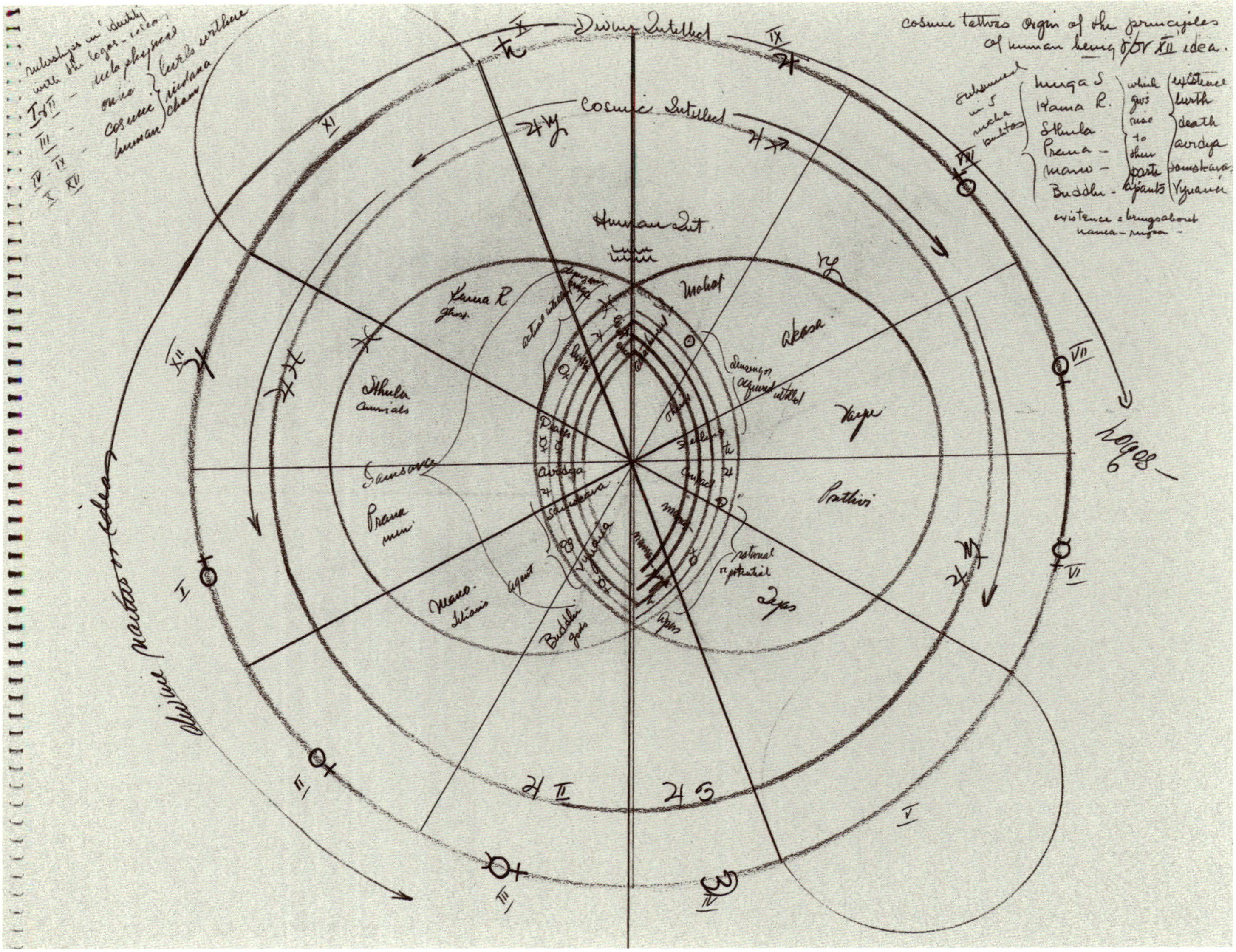
Divine Intellect
Cosmic Intellect
Human Int
cosmic tattvas origin of the principles
of human being for XII idea.
Linga S
Kama R.
Sthula
Prana
Manas
Buddhi
existence
birth
death
avidya
Logos
Mahat
Akasa
Vayu
Prithivi
Apas
Kama R.
Sthula
Samsara
Prana
Buddhi
divine principles of ideas

Notes

1

EDITORS

Sacred Revelations References

See Plato, *Republic*, Book VI, lines 508c–509c, for the "Idea of the Good"; John 1:14 in the New Testament, for the "Word made flesh"; Chaldean Oracle fragments in Hans Levy's *Chaldean Oracles and Theurgy* and in the *Shrine of Wisdom* (1925–1926) for "conceptions of the Father"; and the Mandukya Upanishad, verses 1–2, for "Pranava Aum," respectively. Anthony sometimes uses the term "conceptions of the Father" to describe the twelve divine Ideas. The following is translated by the editors of the *Shrine of Wisdom*, VI, no. 24, p. 105:

> The Mind of the Father burst thunderously forth, conceiving with His transcendent Will Omniform Ideas. Winging from one source they leapt forth. For from the Father was both the Will and the End. But they were differentiated, being made participable by intellectual fire to other intellectual natures.

2

EDITORS

On Glimpses

For Anthony, a "glimpse" indicates a direct, immediate experience of reality, beyond image. His insistence on the importance of this profound state of introversion can be best appreciated by comparing the stance of archetypal psychology with the divine visitations experienced temporarily in a mystical glimpse and with the permanent enlightenment of the philosophic sage. Plotinus provides a paradigm of such experience in VI.5.7: "To know without image is to be." See also *The Notebooks of Paul Brunton*, volume 14, chapters 4–8 (which contain detailed references to glimpses) and volume 16, part 1, ch. 2, which elaborates "the enlightenment which stays." In contrast, contemporary psychologists and philosophers commonly claim that there is no experience beyond the archetypal imagination. Roberts Avens, for example, examines the views of four such philosophers of the imagination in his book *Imagination is Reality: Western Nirvana in Jung, Hillman, Barfield, and Cassirer.*

3

EDITORS

Neoplatonic Astrological References

Besides Iamblichus, *On the Mysteries*, other Neoplatonic writings with important astrological references include: Sallust, *On the Gods and the World*, and the *Orations of the Emperor Julian*. The seventh book of Proclus' *Theology of Plato* (as reconstructed by Thomas Taylor and included in his translation of *The Six Books of Proclus*) is particularly valuable.

4

EDITORS

Timaeus and *Republic* References

Timaeus 34b and following text refers to the cosmic soul. The sixth book of the *Republic* 508c–511e explains the Divided Line, Plato's metaphor for the levels of knowledge.

5

EDITORS

Stars as Statues of the Gods Reference

For use of the term "statues of the gods" as referring to the stars, cf. Plato's *Epinomis*, 983e–984a, especially the translation by T. Taylor and F. Sydenham, v. 2, p. 400.

6

EDITORS

The Language of Anthony Damiani's Interpretations of the *Enneads*

6A. Reality

For Plotinus there is absolute reality, the One, which is beyond Being or Knowing. Reality means that which is authentic, which never ceases to be itself, and which is not able to be negated. Absolute means the One is self-sufficient, without need of anything.

The relation of the One and the Many can be viewed from two different perspectives: Reality is seen either as exclusive, absolute, and transcendent or as inclusive and immanent. In the transcendent view, reality is equated with the One-Only. All else is either not other than the One or is illusory. On the other hand, an inclusive or immanent view of reality includes degrees or levels. The reality of anything is its participation in the One, or the presence of the One. Plotinus takes this perspective where he says that though the One "is nowhere, nowhere is it not" (V.5.8), and "the Supreme as containing no otherness is ever present with us; we with it when we put otherness away." (VI.9.8) "Degrees" of reality are determined by degrees of participation in the One.

The relation of transcendence and immanence is made clearer if we do not take the Primals (One, Nous, Soul) as completely separate. Thus Intellectual-Principle can be considered as all that is knowable of the One, and Soul as the life-activity of the One. Thus we can distinguish different kinds of reality: Absolute reality (One), self-knowing reality (Intellectual-Principle), living reality (Soul), and manifesting reality (Nature).

This view enables us to know that we really are. Plotinus tells us what we are, what soul is, and that we are soul. Anthony describes the reality of soul as its "inviolable sanctity," for even while it acts here, soul remains itself, unaltered. Yet, when it visits the One, Plotinus says, the soul is now no longer soul. (See VI.7.34–35.) But that is because something can only fully be *what* it is when it emanates. When it returns to the One, soul finds its reality in unity. Soul is the life power of the One. Soul is real because soul is intrinsic to Reality. Our quest is at first a movement towards soul-reality, and later a journey within and through reality, a deepening into reality.

6B. The One

Two questions that Plotinus asks within the *Enneads* motivate much of Anthony's exploration of the One.

> In what sense, then, do we assert this Unity and how is it to be adjusted to our mental processes? (VI.9.6)

> . . . from such a unity as we have declared The One to be, how does anything at all come into substantial existence, any multiplicity, dyad, or number? (V.1.6)

What is the nature of the Absolute—of Reality—and what can be known of it through our relative knowledge?

What kind of answer can we expect? What can be said positively about the nature of the One that will not be wholly negated by that vast Mystery or swallowed up in total relativity? Perhaps we could turn Plotinus' question in VI.9.6 around: How can we adjust our mental processes to the One? Discussion is relevant and helpful to the human situation—both in directing our thoughts and in provoking a positive sense of holiness. Our mind must become acclimated to this rare atmosphere.

Anthony often remarked that there is nothing more important to discuss, at any time or anywhere, than the One. Through this direct focus on the One, thinking can become flexible enough to work within a framework of paradox and allow the words of Plotinus to deliver a hint of the One's nature. Such passages as the following begin to take life within us: Though the One "is nowhere, nowhere is it not" (V.5.8); "Seeking nothing, possessing nothing, lacking nothing," the One overflows. (V.2.1)

Even Plotinus' assertion that *It Is* has tremendous value. Can we expect any further statements about its nature? If the One doesn't have knowledge of itself, how can we expect to gain knowledge of it? Plotinus provides an answer in another context: The One confers what it does not itself contain. (See V.3.15.) The touch of the Alone can evoke specific knowledge that the One doesn't contain as such. And we must be careful to specify what we mean by knowledge. Plotinus says only that the One doesn't have knowledge in the usual sense. (See V.3.13 and VI.9.4.) But it is "not imperceptive" (V.4.2, Deck, p.33) and "it is not to be thought of as having but as being Intellection." (VI.9.6)

It is not that the One is beyond qualities such as knowledge and love because of excluding them. The One is beyond knowledge because it *overwhelms* knowledge—no knowledge or love or any other authentic character can retain its own character in that Infinitude. The One is so infinite that no lesser infinites can survive intact in its atmosphere. It is so infinite that it cannot be divisible in any way. It is so indivisible that it cannot be finite in any way. The One is perfection, infinite power, and super-intelligence.

So we can say something positive about the One, but must clothe the positivity in paradox. On one hand, we must have a framework or we risk being completely swallowed in generalities. On the other, we want to remain as bias-free as possible in order to remain receptive.

6c. Intellectual-Principle or Nous

Plotinus describes Intellectual-Principle (Nous) from its own point of view, and also from the point of view of the other two primals, One and Soul. Our attempts to understand Intellectual-Principle must take into account these different approaches.

Nous is a Greek word for Mind, Awareness, the Knower as immediately united with its content. Nous is both awareness and that of which it is aware. It is inseparable knowing and known. This identity has two sides: It is that which it knows, and it knows completely what it is. There is nothing of Nous which is unknown to itself, and there is nothing known by Nous which is not itself. When you know something completely, you know it by being it. Nous is self-knowledge.

Stephen MacKenna translates *Nous* as "Intellectual-Principle," for it is the state or stasis of being self-knowing. Translating Nous by the terms Divine Mind, or Infinite Mind, or Infinite Intelligence would be useful as well. Nous is a knowingness and awareness, one with its contents (direct and immediate), completely self-knowing throughout. We indicate by "Divine" that the Mind we are approaching is not the ordinary mind. In addition, Nous is Truth, the identity of knower and known, and Beauty, the shining forth of divine Ideas.

Nous is related to the root *noeo,* which means both the contact of the mind and its object, and also the intent or tendance of the mind toward something. *Noeo* means immediate direct experience. The word *Nous* has several related declension words: *noeon* is that which is aware; *noesis* is the act of being aware, or perhaps in-seeing or knowing; *noeto* is that of which Mind is aware—the Ideas. *Nous* means all these aspects together.

Awareness, knowing, illuminating, seeing—all help to capture something of the meaning of *noesis.* Awareness captures the sense of looking, but does not capture the sense of knowing. Nous not only sees its objects, it possesses them. Many places in the *Enneads* also use illumination metaphors for *noesis.* (Many translators use luminosity and knowingness to translate the Tibetan *rig-pa,* which is similar to *noesis.*) *Noesis* is the Seer: both in the abstract sense, and as the truth-seer function in the sage. We may use the term "insight," as Paul Brunton uses it. *Noesis* is both the state of self-knowing and the presence that makes soul insightful. Nous is all there is to know, and if soul is to know itself, it is because it has become the Nous, or knows itself as Nous itself knows soul.

What is the nature of the continuity of the Nous and the One? Does the nature of the Nous tell us anything about the One? We might go either way on the question of whether the Nous tells us anything more than *that* the One is. We may say that there is just no way to express the nature of the One. Even when its object is the One, Nous cannot grasp it as One: ". . . with all its effort to grasp that prior as a pure unity, it goes forth amassing successive impressions, so that, to it, the object becomes multiple." (v.3.11) (See also III.8.8, VI.7.33, 35.) Further, the One is not a *what,* for to be a *whatness* means to be known, which the One is not. Or/and we may say that the Nous is a trace of the One, and expresses it. Nous is the evidence of the One's super-awareness. Nous is a trace of the life and intelligence in its prior. (See v.5.5.) Thus from the Nous we learn about the One (and perhaps this is the only place from which to really learn about the One). The awareness in the Nous (*noesis*) is the presence of the super-awakeness of the One—but as determined. Nous is the *knowableness* of the One. Plotinus himself answers the question in both ways, and we leave it to the reader to ponder the relation (or non-relation) of One and Being. Plato in *Parmenides* (162c) says that the only actual movement there is, is the movement from beyond-Being to Being.

6d. Being

Being is a more difficult word to understand than Nous. Substance (Gk: *ousia*), Being (Gk: *on*), Existence (Gk: *einai*), Is-ness, Essence, and the phrases "That it is" and "What it is" are often confused. Let us consider different senses of Being.

The English verb "to be" is the simplest and most inclusive of all words. "To be" means something IS rather than isn't—the fact *that* it is. The One IS. In fact, the One cannot *not* be. It is as it waked itself to be, as it acted itself to be. (See Plotinus, VI.8.16.) Its *positive* nature answers the question: How come there is rather than isn't?

We may, however, understand degrees or intensities of Being, similarly to degrees of reality. One, Nous, Soul, and manifestation are said to *be* in different ways.

Another narrower sense of Being (*on*) is quite often used by Plotinus to refer to the Nous only. In this sense the One is beyond Being, while Nous is said to have Being or authentic Being. As the One is beyond Being (*on*) and substantial existence (*ousia*) we may ask, What does it mean to characterize Nous as "Being" in the narrow sense of Authentic Existence? In this context, the very meaning of Being is inseparable from Nous. Since we have already pointed out the dual-unity of Awareness and Being, one characterization of Being which is peculiar to Plotinus is that which is Self-aware. Awareness establishes Being, and Being establishes Awareness: To be is to know without images. ("To know without image is to be."—VI.5.7)

But what is the essential nature of Being? Being is the power of the One to emanate. Being is the outward-facing presence of the One, or the contractibility of the One—the fact that something can *be* in the presence of the Absolute. The One does not grant Being to already existent or non-existent somethings—the One's emanation is the act of Being. In order for beings to be, they cannot be only One. Even Self-knowing involves a dual-unity.

Knowing and Being are inseparable. The arisal of the Nous from the One is described in two ways. In one place (V.2.1), Plotinus says that the power of the One emanates. This establishes Being. Turning back to the One, this emanating power is filled and "becomes" Nous. In another place (V.3.11), the indeterminate vision of the Nous trying to know the One is prior. Nous *is* only as the vision becomes determined. The power of the One is so great that its emanant acquires hypostasis, existence (*einai*), and substance (*ousia*).

Existence is literally "to stand out from" and strictly speaking applies to manifestation, where it means that which stands out from the Primals; that which is relational, dependent, temporary, and changing. In a broader sense, Nous stands outward from the One as a distinct existent. So MacKenna used a qualifying term *Authentic Existence*, as that which is eternal, unchanging, not in need of any relation.

For more on the meaning of the term "Being," see Etienne Gilson's *Being and Some Philosophers* and Plato's *Sophist*. Anthony also wrote three essays relevant to the topic of Being in Plato ("Plato's Twofold Being," "Plato's Non-Being," and "The Eight Hypotheses of the Parmenides") which are available in the Wisdom's Goldenrod library.

6E. Hypostasis

Hypostasis literally means standing under, ground or foundation, the support, underlying principle or sub-stance. Hypostasis indicates the status of standing in its own right, taking itself to be and exulting. Plotinus uses the term when he is describing how something other than the One can be Real. What makes something a hypostasis?

a. Hypostasis refers to the act of the One giving stable reality to an emanant through standing firm. Because the One remains what it is, it can grant stasis or stability to what comes from it. It offers a ground for the emanant—Nous/Being.

b. Hypostasis is the condition of the emanant standing as something real in its own right. In order for an emanant to stand as distinct in itself, it has to resist being the original.

c. A hypostasis has the power to give rise to other authentic or outward-facing expressions which have a subsistence of their own.

In order to be itself, anything must stand firm in its own right, finding its own nature in its distinction from its priors, yet having support from the priors. Hypostasis thus involves the movement from the One (the power of the One provides a basis for the Nous), the establishment in the One (the One slips in to hold the emanant), the distinction of the emanant from the One (the fact *that* something exists as distinct from the One, stands in its own right), and the shining out of its nature (the stabilizing of the nature of the Nous through standing as distinct and knowing itself). That which comes to be is not separate from that which was originally "in" the One. (We may describe this with both the loosening and contracting metaphors.)

Strictly speaking the One itself is not hypostasis, since it is too infinite for anything to establish itself in, and too void of things for anything to be or know in it. And the universe is not hypostasis since it is always in a process of change. However, rather than focusing on distinct hypostases, we can take an overall view of reality as hypostatic when we view the circle of reality as having four components: One, Nous, Soul, Body.

7

PAUL BRUNTON

On Describing the Absolute

EDITORS: *In the working draft, Anthony referenced the following passage from Paul Brunton to expand upon how "more harm may be done by* refusing *to speak of the Unity or Absolute God."*

Philosophy understands sympathetically but does not agree practically with the Buddha's consistent refusal to explain the ultimate realization. His counsel to disciples was: "What word is there to be sent from a region where the chariot of speech finds no track on which to go? Therefore to their questionings offer them silence only."

It is certainly hard to capture this transcendental indefinable experience in prosaic pen-and-ink notes. But is it really so impossible for the initiate to break his silence and voice his knowledge in some dim finited adumbration of the Infinite? To confess that intellectually we know nothing and can know nothing about the Absolute is understandable. But to say that therefore we should leave its existence entirely out of our intellectual world-view, is not. For although the exact definition and direct explanation of words are unable to catch the whole of this subtle experience within their receiving range because they are turned into ordinary human intellectual emotional and physical experience, they may nevertheless evoke an intuitive recognition of its beauty; they may suggest to sensitive minds a hint of its worth and they may arouse the first aspiration towards its attainment for oneself.

Why if this state transcends thinking, whether in words or picture, have so many mystics nevertheless written so much about it? That they have protested at the same time the impossibility of describing the highest levels of their experience does not alter this curious fact. The answer to our question is that to have kept completely silent and not to have revealed that such a unique experience is possible and that such a supreme reality is existent would have been to have left their less fortunate fellow men in utter ignorance of an immensely important truth about human life and destiny. But to have left some record behind them, even if it would only hint at what it could not adequately describe, would be to have left some light in the darkness. And even though an intellectual statement of a super-intellectual fact is only like an indirect and reflected light, nevertheless it is better than having no light at all.

So long as men feel the need to converse with other men on this subject, so long as masters seek to instruct disciples in it, and so long as fortunate seers recognize the duty to leave some record—even if it be an imperfect one—of their enlightenment behind them for unfortunate humanity, so long will the silence have to be broken, despite Buddha, and the lost word uttered anew. (*Perspectives*, p. 378)

8

EDITORS

The Dyad

Dyad, or indeterminate Dyad, is a term that tries to capture a sense of the two-ness or paradox of the One's power. At different places in the text, Anthony gives slightly different interpretations of the Dyad: A) the relation of the One and power, B) the complementary nature of the One's power as active and passive, C) the nature of power as abiding and emanating. Some further comments about these different uses may help to give a more complete picture of this elusive concept.

A. One view of the Dyad is that it expresses the metaphysical *relation* of the One and its power. The One, Plotinus affirms, is "infinite not in measureless extension . . . but in fathomless depths of power." (VI.9.6) This relation always remains indeterminate, for power is inseparable from the One; but Plotinus distinguishes aspects of the One such as unity, self-tendance, self-awareness, and power.

B. A second use of the term Dyad is to refer to the twoness or complementarity of power in each of its distinct contexts.

Plotinus refers to the fathomless nature of the One as "at once Act and repose." (VI.8.16) Anthony uses the terms "active and passive perfection." (p. 35) Plato refers to the Dyad in the *Philebus* as Infinity and Bound. Within the One, the Dyad is like matter to the One, and its inter-activity within the oneness gives us a view of a multifaceted and infinitely complex jewel which Anthony calls the Triad. The term "indeterminate Dyad" is appropriate in that any attribution of characteristics of any kind to the absolute and unknowable unity is bound to involve paradox. This paradox is not merely the result of our inability to know through determinate categories; it is the result of the inner nature of the One. Within the One, power represents the glue and the solvent by which the core and content of every authentic real is indivisible from the One. Insofar as the inner nature of the One is fathomless and indeterminate power, without determination, the unities said to be included in it are "swallowed up" into Its vast infiniteness.

Anthony comments, "There probably will be an indefinite number of ways to conceive of the arrangements of the modeless distinctions in the totality of possibilities which exist in the One." (transcript, 4/23/82)

C. Thirdly, we can understand the dyadic nature of power as describing the relation of power abiding "within" the One to that of power emanating. In discussing these points, Anthony said, "The two aspects of power occur when we identify the power with the One and when we think of the power as sheer indeterminacy extending from the One in its super-abundance." (transcript, 4/23/82)

For Plotinus, power is not a separate principle intervening between the Intellectual-Principle and the One. He writes in v.2.1 ". . . the One is perfect, and in our metaphor, has overflowed . . ." (See the discussion of power on page 30 of *Astronoesis,* and refer to *Ennead* v.4.2.) Dyad should be seen as a kind of verb, process, or activity—not a thing. In *Ennead* v.1.5, Plotinus describes the Dyad as a kind of "matter" for the One. MacKenna's first and fourth editions translate the passage differently—taking Dyad as "the underly (or Matter) of The One" (first ed.) and as the "underlie (or Matter) of the Intellectual World [Intellectual-Principle]" (fourth ed.). Anthony saw the validity of both these interpretations. He uses the first edition interpretation on p. 31 but used the fourth edition interpretation in other contexts.

The abiding (transcendent) and emanating (immanent) power are in one sense the same, since the emanating power *is* the power of the One. In another sense they are distinct since in order for the Intellectual-Principle to function, it must be distinguished as itself. The Dyad is the indeterminacy of these two relations of abiding and emanating power as simultaneously same and different.

The power emanating from the One is also dyadic: both emanating from and continuous with the One. It is shaped from within, by the transcendent unities in the One, into Intellectual-Principle (on p. 31). In the context of emanation, power is the corona of the One, indeterminate in itself but determined by the unities within the One. Or it can be said to be determined by the One in one sense, while in another sense it is self-determining.

* * * *

In summary, we understand Dyad as the indeterminacy of the different aspects of power. A similarity to the fourfold Madhyamika Buddhist logic of Nagarjuna and Chandrakirti is evident: Two things are not the same, not different, not both, not neither. But in the case of power, all of the above apply: One and power are identified, distinct, both distinct and separate, and neither.

Focusing on only the sameness of the transcendent and immanent modes leads to a monistic view of reality, while strictly distinguishing or separating them leads to a hierarchical view. The nondistinction of the emanating power of the One and the abiding power allows the continuity of the One and its emanation. This view is emphasized, for example, in the description of the indeterminate impression which the Intellectual-Principle has of the One (v.3.11), and in the intrinsic power of "Intellect loving" (vi.7.35) above intellection, by which it is in contact with its source.

These views of power influence how we understand manifestation. If power remains identical with the One, then reality would be one homogeneous block, and could not be omnipresent to manifestation. If the emanating power were distinct and real, we would have an ultimate dualism in which the One would not be perfect and complete, and the emanant beings would be discontinuous with the One. If power were distinct but not real, we would have an illusory universe. For Plotinus, maintaining the status of emanation of the One as neither illusory nor absolute is achieved through his view of power and essential production.

9

EDITORS

The Double Standpoint: Two Truths

While explaining how the many arise from the One, Anthony goes to great lengths to not reduce the many to the One. He uses the double viewpoint to help explain manifestation without explaining it away, and to account for the realization of a human being for whom both ultimate reality and appearance are simultaneous. The two views of truth are first mentioned on p. 24:

> We must explain how universal relativity is included within the One in a special way. This interpretation will attempt to portray and to account for that possibility and will permit us to expound the two views of truth: the ultimate truth that is the substratum of the appearance, and the truths within the appearance—truths that are not reduced to or resolved into that Ultimate.

The idea of a double standpoint or two truths is stated in different forms in Buddhism, Vedanta, and Platonism. A basic and concise formulation of the two views is found in the following quote from *The Notebooks of Paul Brunton:*

> One of the helpful notions which philosophy contributes to those who not only seek Truth through the intellect alone, but also seek to know how they are to live with that Truth in the active world itself, is the idea of the twofold view. There is the immediate view and there is the ultimate viewpoint. The first offers us a convenient way of looking at our activities in the world and of dealing with them whilst yet holding firmly to the Truth. The first tells us to act as if the world is real in the absolute sense. The second viewpoint, the ulti-

> mate, tells us that there can be only one true way of looking at everything, because there is only one Reality. Since it deals with the Absolute, where time and space disappear and there is no subject to view, no object to be viewed, there is no thought or complex of thoughts which can hold it; it transcends intellect. Therefore it could be said that philosophy uses duality for its practical viewpoint, but it stays in nonduality for its basic one, thus reconciling both. (*Perspectives,* p. 366)

The immediate view, also called practical or relative, seems to be both a human need and an existential fact. We humans need this practical view of objects to live and function in the world. For a world to appear at all, it must necessarily manifest as relative, as multiple, and as distinguished from the unitary nature of the Mind (ultimate truth) by and in which it is manifested.

Anthony emphasizes that the understanding of the two truths depends on the understanding of Power:

> For Plotinus, then, the One-and-*its*-Power stands first and foremost above all and simultaneously sustains and includes lesser levels of reality which are neither self-sufficient nor illusory. Through an understanding of this principle of Power, we can first make available to our minds what we refer to as the double truth . . . (*Astronoesis,* p. 32)

He develops the terms "practical" and "only-truth" in relation to the One and its act:

> We are tentatively establishing this verbal distinction between the One and its issuing act. . . . Consequently, this distinction will ultimately give rise to the practical truth alongside the Only-Truth . . . (*Astronoesis,* p. 31)

Anthony is saying that a careful formulation of the notion of two truths is based on the verbal distinction of the One and its Power. First of all, a notion of truth is a verbal notion. "Verbal" is an appropriate name, as the emanation is related to the Word (*verbum*). Anthony comes back to this theme of the relation of the emanants to the One (see pp. 41–45 in The One: Exposition), and of the presence of the Primals in manifestation (see pp. 67–71 in The One: Symbolization). Although the distinction of One and Power is "only verbal," the emanations of that Power are *not* merely verbal, but are actual principles, or hypostases, such as Intellectual-Principle.

A clarification of this idea of the two truths is needed not only among people living in the world, for whom the sense of reality is superimposed on duality. It is also essential for mystics, who must face the dilemma of reconciling the Void nature of Mind with a manifest world. Any lack of care here will undoubtedly cause confusion and semantic obscurity—for example in statements such as "All is One" or "There is no access to the One from the point of view of the manifest."

It is important that these two truths or views not be sundered. If they are, we will not be able to describe, for example, the realization of the sage, called *sahaja,* in which the two are simultaneous.

> When the sage speaks about the inner constitution of the One, as Plotinus does, the way its Power can rearrange itself or combine itself in any number of ways, that is not acquired when you are in the One. Plotinus is the Nous when he says that. . . . In other words, he is a mouthpiece for the World-Mind. . . . That is what I mean when I say that the intellect ultimately is dualistic, because as high as we can go—the Nous, the point of view of the Intellectual-Principle—there is a subject-object relation. But there the identity is much closer and almost an identity in comparison to what goes on later on. (Anthony Damiani, transcript, 4/9/82)

Paul Brunton also describes the state of the sage in nonduality: "When duality is blended with, *and within,* unity it is the true *jivanmukta* realization." (*Perspectives,* p. 346) Speaking of duality blended *with* unity is equivalent to speaking of the transcendent One and its emanation. Speaking of duality blended *within* unity is equivalent to the inclusion of all four aspects of reality within reality.

Buddhism

In his essay "The Two Truths" in *Kindness, Clarity, and Insight,* His Holiness the Dalai Lama says that the idea of the two truths is a basic structure from which one can view almost the whole of Buddhist doctrine:

> . . . the *bases* are the two truths—conventional and ultimate; the *paths* which depend on them are method and wisdom; and the *fruits* of those paths are the two Buddha Bodies—Form Bodies for the sake of helping others and the Truth Body which is the fulfillment of one's own welfare. (p. 192, emphasis in original)

His Holiness goes on to point out that the two truths are explained differently within different traditional schools of Buddhism, but he emphasizes the teaching of his own Prasangika school. In his book *Buddhism of Tibet and the Key to the Middle Way,* he explains that

> All phenomena that we manifestly perceive have two modes of being. One is the nominal or conventional . . .

> and the other is its final mode of being, its emptiness of inherent existence.... In rough and brief terms, an object found by a valid cognizer distinguishing a final nature is an ultimate truth and an object found by a valid cognizer distinguishing a conventionality is a conventional truth. Therefore, emptinesses and true cessations are ultimate truths, and everything else are conventional truths. (pp. 31–32)

For Prasangika, the ultimate nature of the object or the individual is given the technical description "empty of inherent existence." To see the nature of the world in this new light—as fluid, interdependent, nonsubstantial, and immaterial—requires spiritual vision, or wisdom. So, in terms of the path, His Holiness the Dalai Lama says:

> In dependence on ultimate truths wisdom is developed, and in dependence upon conventional truths compassion and kindness toward others is meditated. These two, wisdom and compassion, must be practiced in union; this is the path of the union of wisdom and method. (*Kindness, Clarity, and Insight,* pp. 198–199)

The result of this practice is the "generation of the two Buddha Bodies":

> The Form Body—the physical manifestation of a Buddha—is the imprint of the yoga of imagining the divine circle that is related with conventional truths. The Truth Body—the exalted mind of a Buddha—is the imprint of the wisdom consciousness realizing emptiness, the ultimate truth. (*Kindness, Clarity, and Insight,* p. 199)

He concludes that the capacity for developing these is inherent in every human, and is called the "Buddha nature."

The Dzog-chen view of Tibetan Buddhism describes ultimate truth as intrinsic self-awareness (*rig-pa*) and the fundamental nature of Mind, distinct from the relative truth—which they call mind-as-basis-of-appearance (*sems*). See Long-ch'en Rb-jam-pa's *Four-Themed Precious Garland,* chapter 3, and the last chapter of the Dalai Lama's *Kindness, Clarity, and Insight* for discussions of this view.

Vedanta

The Brihadaranyaka Upanishad says that "there are two forms of Brahman," (Brh. 2.3.1) which G. Mukhopadyaya translates as "the manifest as well as the unmanifest." (p. 47)

In his work *Studies in the Upanishads,* Mukhopadyaya adds:

> That form which is the active principle in all manifestation is called the *Shabdabrahman* [Brahman which speaks or gives forth the word. —eds.] or in the technical terminology of the later Vedanta, the *Ishvara.* The other which is not so engaged or connected with manifestation is the *Ashabda* or the *Para Brahman.* (p. 47)

And, explaining the nature of the world as both real and unreal:

> The world is not *asat,* non-existent because it is verily a fact of experience.... Its reality is not genuine, because it does not belong to it but is only delegated to it by something else. Hence its reality is a dependent reality and not an independent one.... Dependent reality, borrowed reality, spurious reality is called *mithya,* false, i.e., not genuine. (p. 56)

If one wishes, one may say that the world has some truth, but then

> ... the Absolute or Reality must be called *Satyasya Satyam* the Truth of the Truth, i.e., one must then distinguish between two orders of truth, in which the lower truth is sustained by the higher.
>
> Thus as we rise in the ladder of consciousness we get a clearer and clearer view of Reality but that does not signify that Reality itself is graded in nature.... [I]t is only the power of apprehension that creates the degrees ... in the sphere of *Prana,* which is graded in its very nature, but they do not apply to the Atman, which is single and simple in its being....
>
> Axiologically, Reality exhibits higher and higher values, appears richer and richer in content and being, while logically it remains the same one unchanging Reality. (pp. 56–58)

The Shaivite tantric branch of Vedanta presents a view in which the reality of Brahman includes power or *shakti* as real. This seems close to the Tibetan tantric systems. (See Mark Dyczkowski, *The Doctrine of Vibration,* John Woodroffe, *The Garland of Letters,* and J.C. Chatterji, *Kashmir Shaivism.*)

Platonism

In Platonism the two standpoints of Being and Becoming at times seem like a dualism, if not examined under the light of Neoplatonic thought. These two positions are stated in many works of Plato, such as the *Timaeus.* While inquiring about the fabrication of the universe, Timaeus is compelled to seek out the difference between that "which is always *real being,*" and that which is "*in a state of becoming to be,* but which never *really is.*" (T. Taylor translation of *Timaeus* 27d, emphasis in Taylor) Plato adds that the former is apprehended by intelligence in conjunction with reason, and the latter by opinion in conjunction with sense: Reality and appearance thus require distinct gnostic faculties of apprehension. The distinction of the two standpoints according to the modes of knowledge available is carried further in *Republic* Book VI—first in the analogy of the intelligible Sun and the sensible Sun, then in the application to the Divided Line with its grades of knowledge. And in the famous analogy of the cave in *Republic* Book VII, the philosopher is to be released from bondage to appearances (in the cave) to seek an apprehension of truth. After finding truth, the philosopher returns to the cave to apply it.

10

EDITORS

A Water Analogy

Two Views of the Absolute

In addition to the examples Anthony used in *Astronoesis,* of the gold and trinkets and of the glowing body emanating light, he had also intended to include a further example, an analogy using water, as part of the text. What follows is a reconstruction of the water analogy example, based on fragmentary notes and a class transcript.

To provide an analogy that combines the monistic and multiple views of reality, Anthony modified an analogy traditionally used in Vedanta philosophy: the forms of water as waves, spray, foam, and so on are seen on examination to be none other than water. In Vedanta this analogy is used to illustrate how all forms, attributes, and appearances—even the creative deity Ishvara—are illusory modifications of the pure undifferenced substance, Brahman or Consciousness. (Note: Vedantic philosophy is diverse. Some Vedantins use this analogy to show only that Brahman is the reality of the appearances, just as the water is the reality of the waves.)

In order to expand the analogy, we note that the forms, waves, spray, and so on are descriptions of common human experience. Following modern science, water can be analyzed from a variety of other perspectives—biology, chemistry, physics, etc.—that imply distinct explanations, if not realms of existence. From a sensory point of view, water is wet; from the biological point of view, water is necessary for life; from the chemical point of view, water is made of molecules composed of hydrogen and oxygen; from the point of view of physics, water consists of energy or various subatomic particles, or probability waves. The description of molecules doesn't fully explain the taste of water, nor does tasting it reveal hydrogen and oxygen to the common tongue. These multiple, distinct, intrinsic aspects of water each provide a view of water without being reducible to one another. And also all are water—or non-different from water. Anthony summarized these two points of view as follows:

> Using the above analogy, in the case of the One, the monist point of view corresponds to the view that "all is water" or "all is One." The emanation or hierarchical view is that of the scientists' hierarchy of structures. Both points of view are needed for a complete description of the One, for the total view of nonduality. (Notes and Papers, p. O208)

The view that "all is water" is analogous to monism and applies to the concept of illusory distinctions in the One. The view of distinct stadia of explanation, each science providing a different view of water, illustrates a multiplicity of principles, or aspects, each providing a legitimate irreducible window into reality, as in the view from each of the four quadrants of the metaphysical chart. Together these two views provide a more complete picture of reality.

In discussing this point further, Anthony said:

> If we take Mind to be analogous to water. . . . We have Absolute Mind which is homogeneous and we have Absolute Mind which is hierarchically structured and differentiated. These two positions are both simultaneous and ultimate. Now suppose I say Reality instead of Absolute Mind: Again I would repeat and say that to me Reality has a nature which is homogeneous and it has that nature which stratifies it into various hierarchical principles which are of an eternal nature, and these two together are what I mean by Absolute. If we are going to grasp what Reality is in any way, it will have to be done through this paradox. (transcript, 4/13/84)

It is true that any emanant of the One is already prefigured in the One and is not other than the One. It is also true that each of the emanant principles is a distinct essence that reveals something authentic about the unities. None is reducible to the others, yet all are of the nature of Reality.

11

EDITORS

Gold and Trinkets References

See T.M.P. Mahadevan, *The Philosophy of Advaita,* for a reference to gold and trinkets in Vedanta. See also the treatise on the Golden Lion by Fa-tsang (translated in *A Source Book in Chinese Philosophy* by W.T. Chan and in *Buddhist Teaching of Totality* by Garma C.C. Chang) for an example of the same idea in Buddhism.

12

EDITORS

Emanation

. . . from such a unity as we have declared The One to be, how does anything at all come into substantial existence . . . ?

PLOTINUS, V.1.6

Having obtained some hint of the nature of the One alone, we can ask, "What can it mean for anything besides the One to exist?" If something arises outside of the One, how is the One perfect? But if perfection already includes all, there is nothing to come into existence. In what way is the absolute infinite incommensurable with that which follows? In what way is it continuous with its emanations?

Plotinus answers that perfection overflows, that Reality includes the power of efflux, and that the overflow has the status of authentic existence. The One isn't an abstraction from Being, nor does Plotinus come to a concept of Being by moving from the sensible up to some "authentic" essence. His description starts from the One itself and everything else follows by a contraction, an emanation "from" that perfection.

> All existences, as long as they retain their character, produce—about themselves, from their essence, in virtue of the power which must be in them—some necessary outward-facing hypostasis continuously attached to them and representing in image the engendering archetypes....
>
> Again, all that is fully achieved engenders: therefore the eternally achieved engenders eternally an eternal being. At the same time, the offspring is always minor: what then are we to think of the All-Perfect but that it can produce nothing less than the very greatest that is later than itself? This greatest, later than the divine unity, must be the Divine Mind, and it must be the second of all existence, for it is that which sees The One on which alone it leans while the First has no need whatever of it. The offspring of the prior to Divine Mind can be no other than that Mind itself and thus is the loftiest being in the universe, all else following upon it . . . (V.1.6)

As long as anything is genuinely itself, it emanates or produces an outward hypostasis. For this view of causality Anthony sometimes uses the term *essential production,* because it arises from the essence. Essential means intrinsic—in virtue of its very essence, or nature.

Essential production is clearly related to the nature of Power. In Plotinus' *Ennead* V.2.1, the One is said to "overflow," and the power is so great that the emanant stands as a reality, but is never separate from the One. There are not two realities. The One grants the only reality there is—self-existence, or "itselfness."

The term "engender" gives clues to the meaning of emanation. To engender is to give rise to in such a way that the offspring is produced through the nature of the begetter—naturally rather than artificially, or intrinsically rather than extrinsically. Thus emanation happens without thought, without deliberation. To engender is to unfold what is latent in the source, and what is unfolded carries some essence of the begetter. That which is engendered is distinct from the begetter, yet is continuous with the original. There is a sense in which the emanant is prefigured in the original, and the original is present in the emanant.

Emanation may be understood in terms of a contraction of the absolute infinity of the One. The Absolute is too infinite in its absoluteness for distinct beings to be within it. In order to be, something must be lessened from its status in the Absolute. Authentic beings are simultaneously juxtaposed to the One and have their source and sustenance in the One. The relation is not symmetric: Though the beings must be other than One in order to be (themselves) at all, the One is not "other" to anything. The One merely allows all beings to be themselves. The self-nature of each being is its oneness, the way in which it is rooted in the One. Each being or principle that follows upon the One is both self-determining and dependent upon the One. Being is a self-actualizing of that which already is unified within the One. Anything, in order to be, must first be one: "It is in virtue of unity that beings are beings." (Plotinus, VI.9.1) But in order to be, it cannot be only one.

How is essential production related to other views of causality?

Each hypostasis (quadrant) or level of reality has a different causality. One point of view is that different causal relations—non-causality, emanation, transformation, and manifestation—are primarily appropriate at each of the four levels of reality. Another point of view is that the different forms of causality are all aspects of essential production, and that only by understanding them all will we understand the theory of emanation. We need to allow for various ways of thinking, in which the emanant is a) the original, b) an image of the original, c) a presencing of the original, d) a transformation of the original, and so on.

13

EDITORS

The Maitri Upanishad

An Example of the One as Transcendent and Immanent

Anthony uses the diagram of fourfold reality to illustrate two different views of transcendence and immanence. [*figs.* 121A *and* 121B]

We can examine this contrast in more detail by looking at the end of the Maitri Upanishad:

> He who sees with the eye, who moves in dreams, who is sound asleep and he who is beyond the sound sleeper, these are a person's four distinct conditions. Of these the fourth is greater than the rest. Brahman with one quarter moves in the three and with three-quarters in the last. (Maitri Upanishad, VII.2.11, in S. Radhakrishnan, p. 859)

The "fourth" here means *turiya,* the "fourth state" of consciousness.

The enfolding of principles into one another can be pictured as follows [*figs.* 122A, B, C, *and* D]:

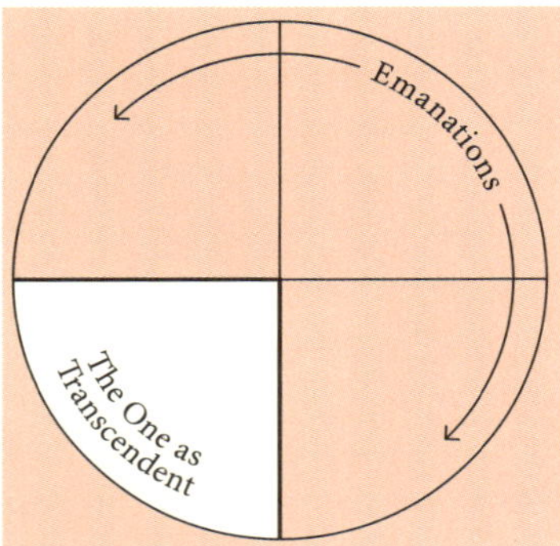

Fig. 121A

The One as transcendent (pictured as the first quadrant) is juxtaposed to the other three levels—Intellectual-Principle, Soul, and Universe. The latter are its emanation, represented by the other three quadrants. In this view, the One prefigures the others and emanates them.

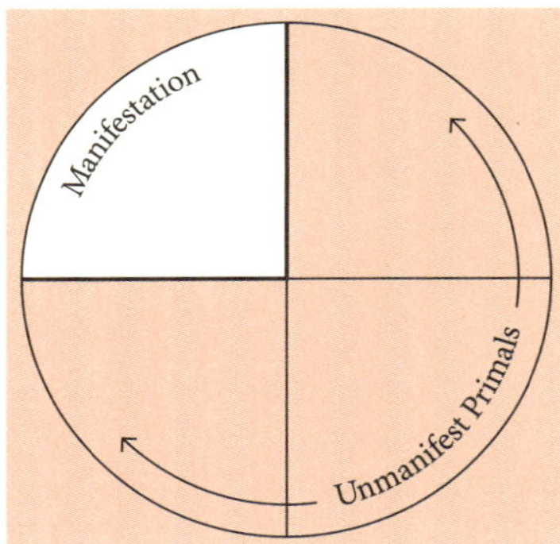

Fig. 121B

From the point of view of manifestation (represented by the fourth quadrant), the three unmanifest Primals are transcendent.

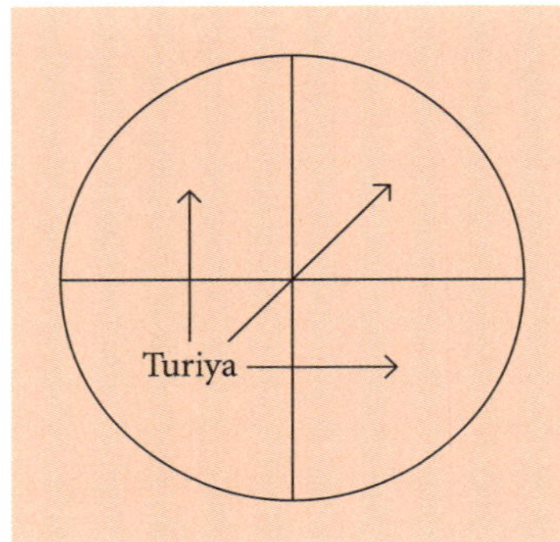

Fig. 122A

In the mandala, turiya *(pictured as the first quadrant) prefigures the other three.*

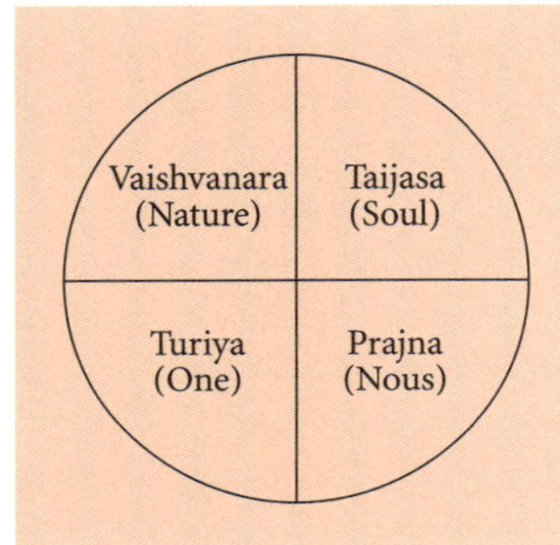

Fig. 122B

Turiya *(as transcendent) may be conceived as the first quadrant and the other states assigned to the following quadrants.*

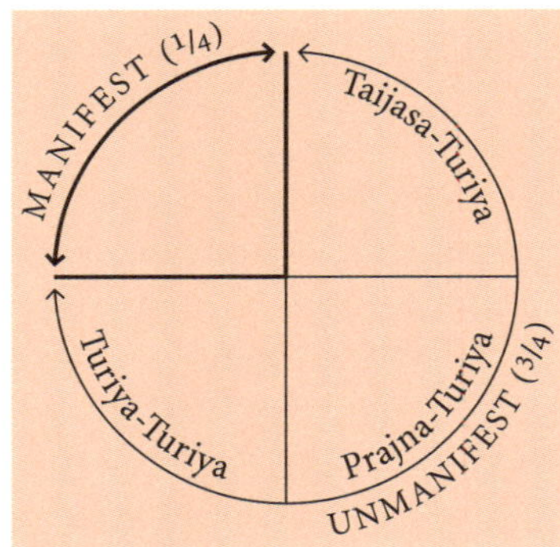

Fig. 122C

Each of the three unmanifest primal principles is conceived as an attribute of pure consciousness ("Brahman as unmanifest") and the three stand in relation to manifestation as the fourth.

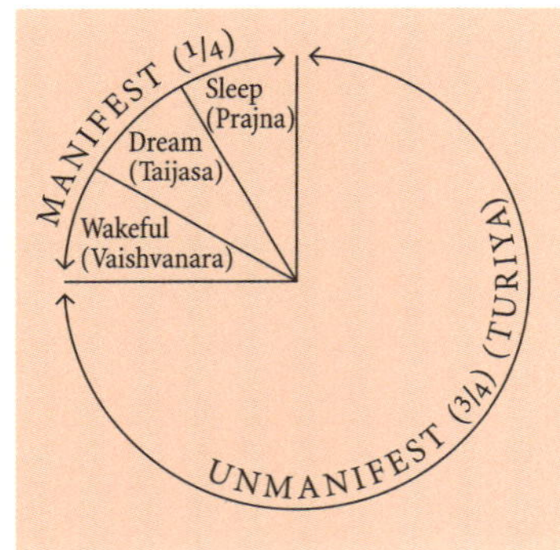

Fig. 122D

Within manifestation, we find a replication of the primal three states. From a point of view within the manifest, *wakefulness, dream, and sleep are conceived as comprising manifestation and the three primal states as unmanifest (Void) are conceived together as the fourth quarter.*

14

EDITORS

Dignities in the First House

The meaning of the 1st house is illustrated by the following quotes from Plotinus:

> This self sufficing is the essence of its unity. Something there must be supremely adequate, autonomous, all-transcending, most utterly without need.... The sovranly self-sufficing principle will be Unity-Absolute ... (VI.9.6)

> One seeing That as it really is will lay aside all reasoning upon it and simply state it as the self-existent ... (VI.8.19) [*figs. 123 and 124*]

Sections 14–19 of Plotinus' *Ennead* VI.8 give a description of the self-existence of the One that can serve as a basis for investigating the meanings of Mars, Sun, Venus, and Saturn in Aries (♂ ☉ ♀ ♄ in ♈). In the following quotes, we have replaced the words He, Himself, etc. with It, Itself, etc.

♂♈ (Mars in Aries) is the self-will of the One. On p. 61, Anthony's text says ♂♈ "represents the power and principle of Individuality."

> God is cause of Itself; for Itself and of Itself It is what It is, the first self, transcendentally The Self. (VI.8.14)

> It is what It made Itself: It is not, therefore, 'as It happened to be' but as It acted Itself into being. (VI.8.16)

> Thus, not 'as It happens to be' is It but as It Itself wills to be. (VI.8.16)

☉♈ (Sun in Aries) is the self-awakening of the One. On p. 62, Anthony's text says ☉♈ "represents the Intellect of the Transcendent One, the author and paradigm of all determined Being which we will find in the Intellectual-Principle."

> Looking toward Itself, so that what we must call Its being is this self-looking ... (VI.8.16)

> A waking without an awakener, an eternal wakening and a supra-Intellection—It is as It waked Itself to be. This awakening is before being, before Intellectual-Principle, before rational life, though It is these ... (VI.8.16)

♀♈ (Venus in Aries) is the self-intention and love of the One. On p. 63, Anthony's text says ♀♈ "represents the Infinite Nature, the uncircumscribed operational activity of the One, the unlived and unpolluted fountain of Life."

> Lovable, very love, the Supreme is also self-love in that It is lovely no otherwise than from Itself and in Itself. Self-presence can hold only in the identity of associated with associating; since, in the Supreme, associated and associating are one ... once more God's being and its seeking are identical: once more, then, the Supreme is the self-producing ... (VI.8.15)

> It is borne, so to speak, to the inmost of Itself in love of that pure radiance which It is, It Itself being that which It loves. (VI.8.16)

♄♈ Saturn in Aries is the placelessness of the One. Anthony connects this idea (on p. 64) to the age-old problem of matter, the Buddhist conception of emptiness, the principle of Difference of the Platonists, and also the Mulaprakriti of the Vedantins.

> ... the Supreme is everywhere and yet nowhere.... If God is nowhere, then not anywhere has It 'happened to be'; as also everywhere, It is everywhere in entirety: at once, It is that everywhere and everywise: It is not in the everywhere but is the everywhere as well as the giver to the rest of things of their being in that everywhere. (VI.8.16)

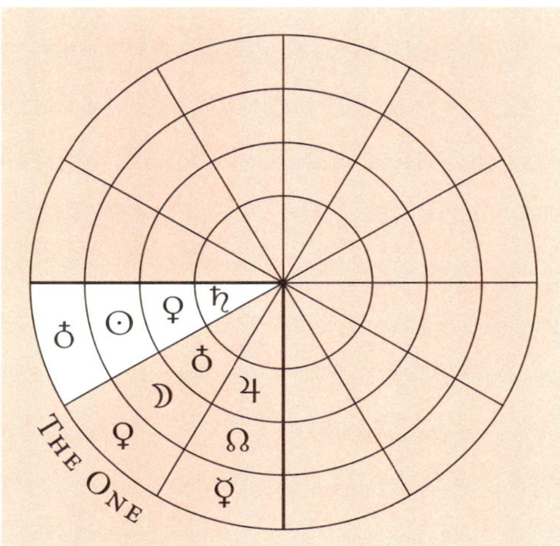

FIG. 123

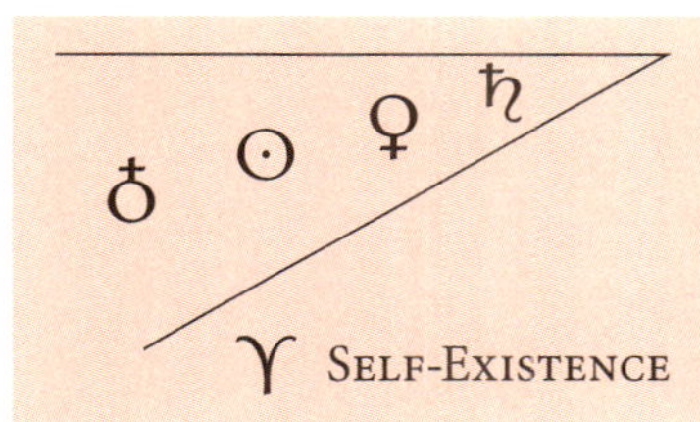

FIG. 124

15

EDITORS

Primordial Matter References

"Mysterium Magnum" is Paracelsus' term for Primordial Matter (cf. Franz Hartmann, *The Life and Doctrines of . . . Paracelsus*). "The Principle of Difference" is one of the ways Plotinus describes matter (*Ennead* II.4.16). In tractates II.4.11–13 he also uses "Alienism," "difference-absolute," "undetermined," and "undeterminateness." T.S. Row, in his book *Notes on the Bhagavad Gita,* describes *Mulaprakriti* as the "veil" thrown over the Absolute (Brahman). His term *Mulaprakriti,* peculiar to Theosophy, means the root source which is beyond the made. Although Vedanta does not seem to use this term, some schools of Vedanta philosophy claim that Brahman is the material cause of the universe (see the *Brahma-Sutras,* verses 1 and 2). For the Samkhya view of *Prakriti,* see the article by K.C. Bhattacharyya "Studies in Samkhya Philosophy" in his *Studies in Philosophy.*

16

EDITORS

Placing Dignities in Rings: Monads

The planets are assigned special positions, called *dignities,* in relation to specific signs of the zodiac. All the dignity positions of a planet represent facets of its deific nature. Placing the dignities in rings gives a dynamic view of the deific pantheon.

We will examine four types of dignities—rulership, exaltation, detriment, and fall—to show how the deific powers represented by the planets manifest the substance (signs) while the signs give creative substance to these functions. For example, Sun ☉ is exalted in Aries ♈, and we write it as ☉♈.

Each planet dignified in a sign is placed according to its dignity within one of the four rings of the chart. [*fig. 125*]

The first kind of dignity, the planet's *Rulership(s),* is placed in the outer ring. A rulership represents the seamless relation of function and substance, planet and sign, one that most efficiently expresses the essence of the sign. Each planet has two rulership assignments. The first of these Anthony calls its *Being,* the second its *Life.* The luminaries Moon ☽ and Sun ☉, which have only one rulership each, he designates as Being and Life, respectively. [*figs. 126 and 127*]

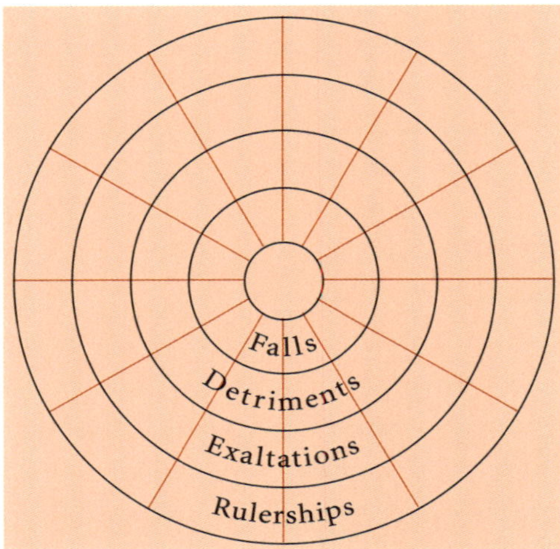

FIG. 125

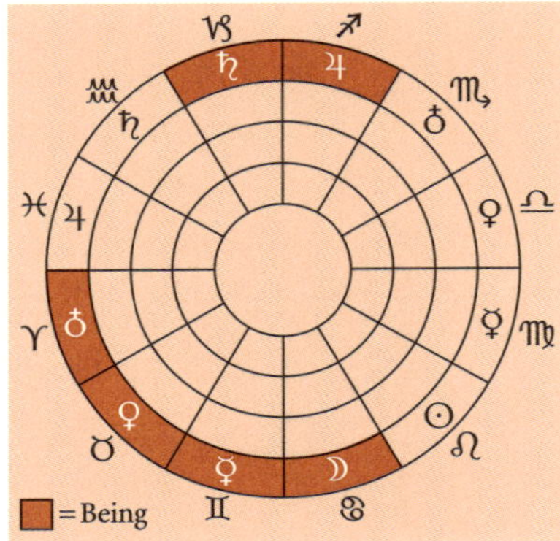

FIG. 126

All the "Being" rulership planets taken together are called the monad of Being.

♅♈ ♀♉ ☿♊ ☽♋ ♃♐ ♄♑

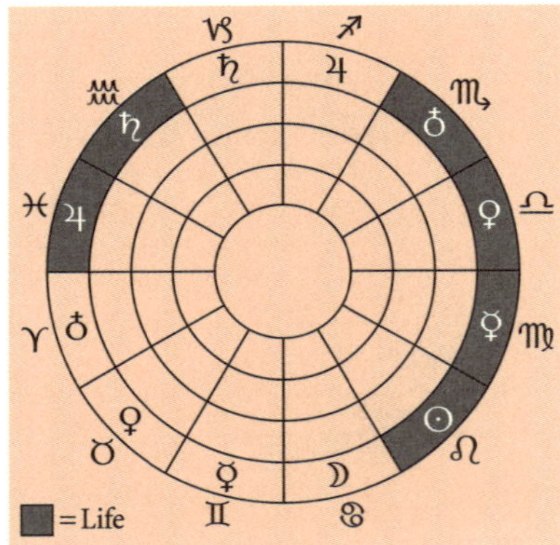

FIG. 127

All the "Life" rulership planets taken together are called the monad of Life.

☉♌ ☿♍ ♀♎ ♅♏ ♄♒ ♃♓

The *Exaltation* dignities are placed in the second ring. These planets are designated *Intellect,* as they indicate the power of intellection within each of the levels of reality. This series of dignities is also called the *separative intellect,* as each planet here represents a divine power by which its Idea is known, and this distinct knowing involves a separation of the knower and known. [*fig. 128*]

Each *Detriment* is, by definition, opposite the rulership of the planet. The detriments in the third ring represent the powers that give rise to Nature and Soul (Psyche). The single ☉ and ☽ detriments are designated *Soul.* For the other planets, which each have two detriments, the detriment opposite a Being rulership is designated *Soul* and that opposite a Life rulership is designated *Nature.* (For example, ♂♎ is Soul opposite ♂♈ Being and ♂♉ is Nature opposite ♂♏ Life). [*figs. 129 and 130*]

The planets that occupy a position opposite to that of their exaltation are said to be in *Fall.* Fall indicates the power to concretize and embody the substance or sign in which these powers function. [*fig. 131*]

In all, Anthony uses forty planetary dignities in four rings to represent reality as function. Each planet has dignities in all four quadrants and all four rings. The Sun ☉ and Moon ☽ each hold four positions in dignity. The other five planets each hold six positions in dignity. The Dragon's Head ☊ and Tail ☋ hold one dignity each. [*fig. 132, p. 316*]

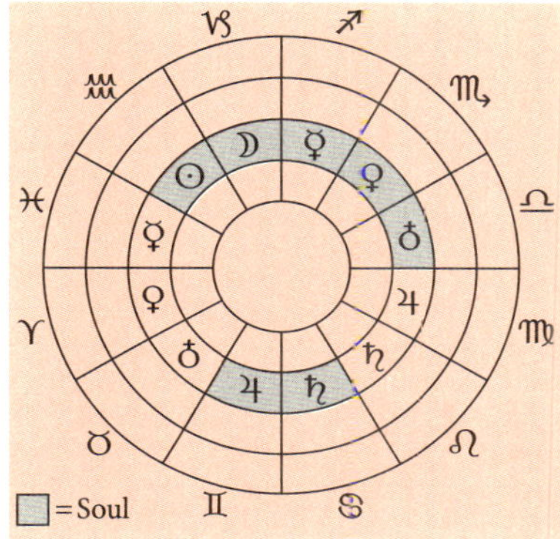

Fig. 129
The planets in detriment above constitute the monad of Soul.

♃♊ ♄♋ ♂♎ ♀♏ ☿♐ ☽♑ ☉♒

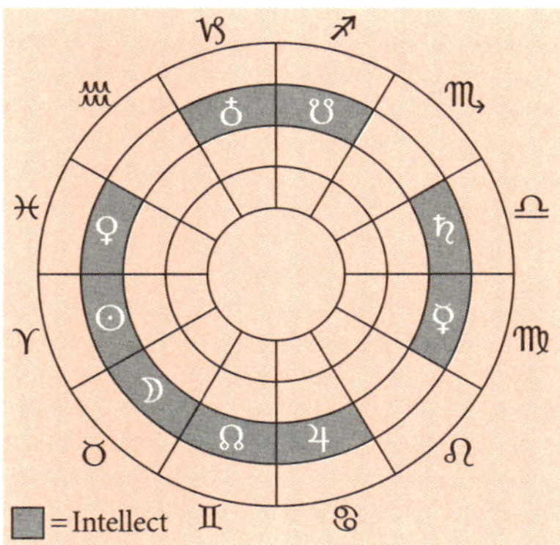

Fig. 128
The exaltations, considered together, constitute the monad of Intellect.

☉♈ ☽♉ ☊♊ ♃♋ ☿♍
♄♎ ☋♐ ♂♑ ♀♓

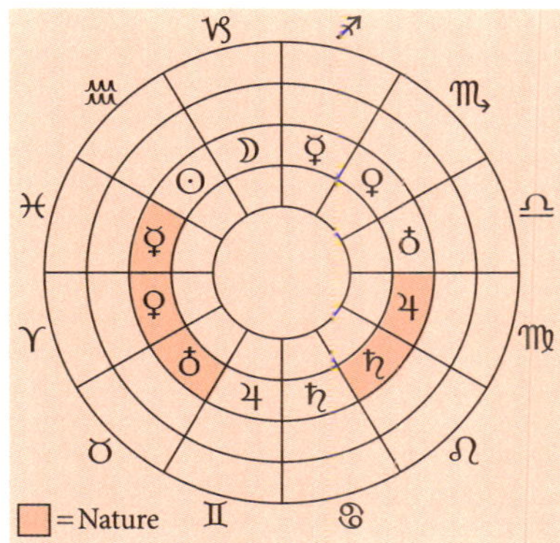

Fig. 130
The planets in detriment above constitute the monad of Nature.

♀♈ ♂♉ ♄♌ ♃♍ ☿♓

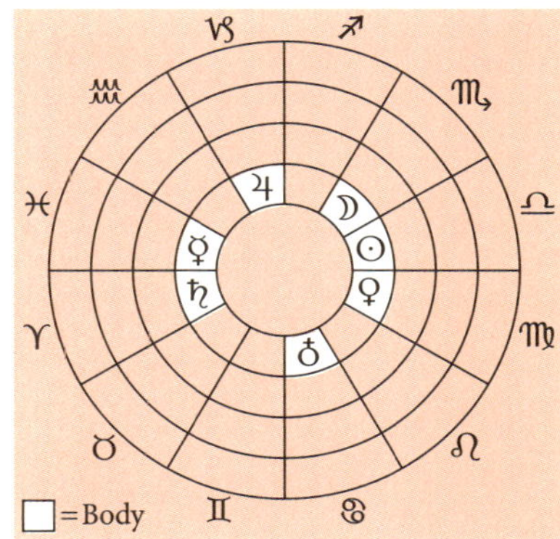

Fig. 131
When considered together, the planets in the seven signs in which they have falls constitute the monad of Body.

♄♈ ♂♋ ♀♍ ☉♎ ☽♏ ♃♑ ☿♓

This geometric mandala of the metaphysical chart with its deities has a significance similar to that of the *thangkas* of Tibetan Buddhism. The mandala of the deity represents a sacred place within which related deities take up residence, while the deity itself is depicted as a divine being in its own right. Not only is the deity located in the mandala, but the mandala is visualized as residing in the heart of the deity as well. In the metaphysical chart, each deity can be conceived as a spherical form, within which we may imagine a version of the entire mandala.

This relation continues inward, leading to an infinite sequence which alters our sense of physical space, and invites us to enter imaginatively into an infinite relation of function and substance. [*figs.* 133, 134A, *and* 134B]

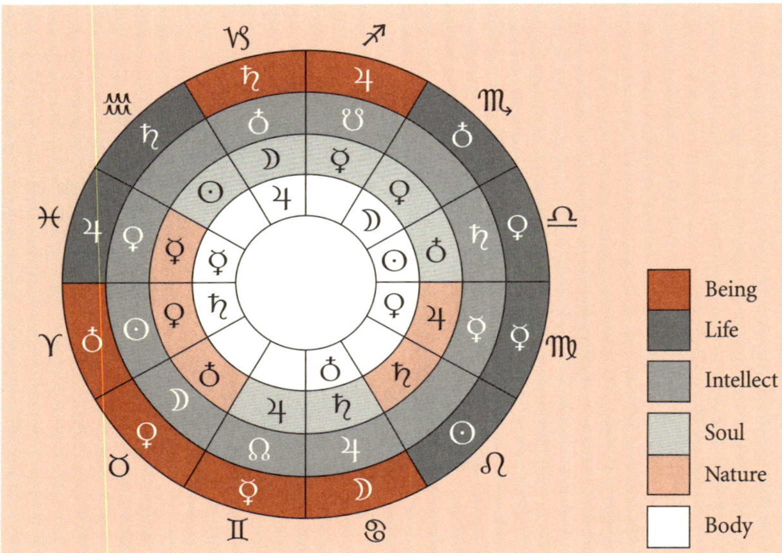

Fig. 132: The chart and forty planetary dignities

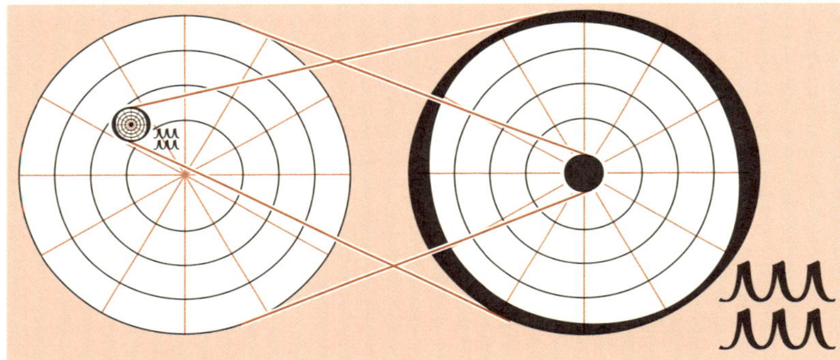
Fig. 133: ☉♒ in the astrological mandala—Astrological mandala in ☉♒

Fig. 134A: Kalachakra mandala with deity at the center

Fig. 134B: Kalachakra deity and consort with mandala hidden in its heart

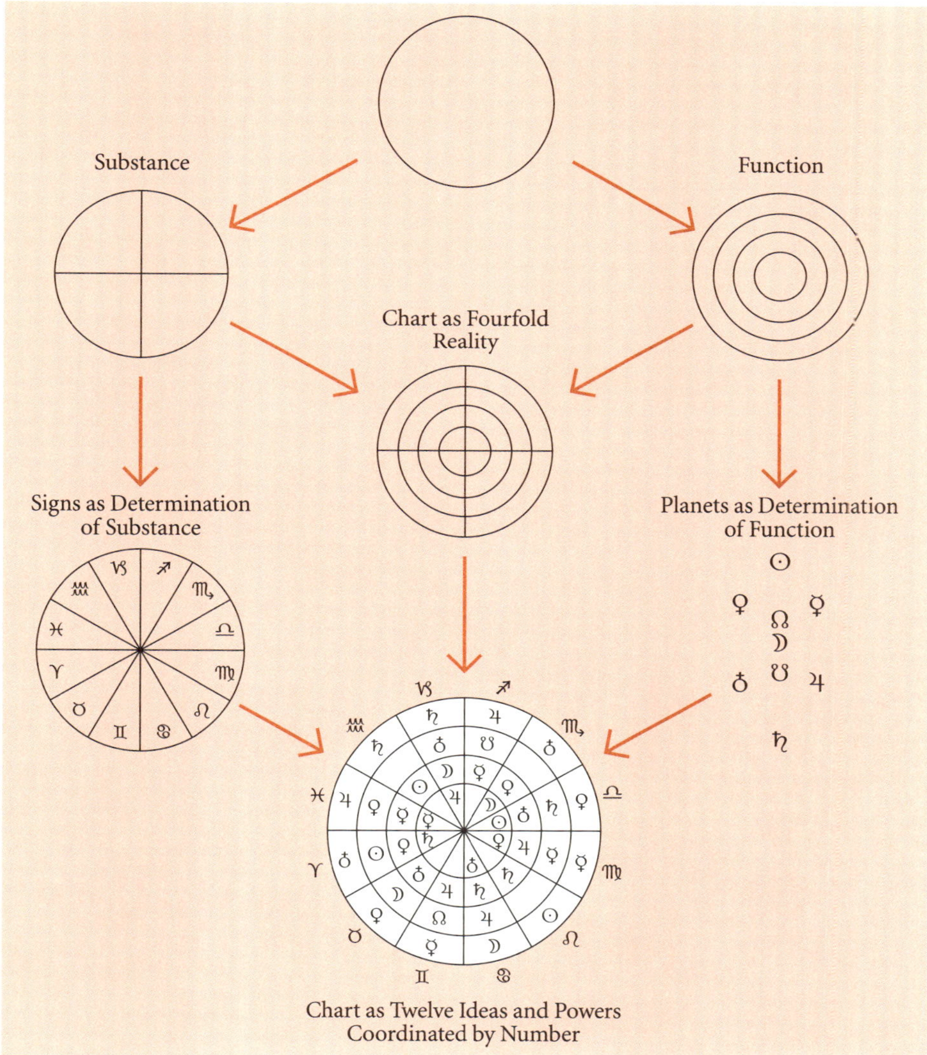

Fig. 135: The dignities complete and complement the geometry of the circle as a means of portraying the nature of reality.

17

EDITORS

Deities in Their Series of Henads

Each planetary deity appears in all four quadrants and all four rings. The complete set of dignities of a planet is called a *henad.* Each henad represents the continuous presence of a divine power or principle, throughout a series of differentiations achieved by functioning through the substances of different signs.

The deity is the specific unity or continuity of the henad. This continuity extends from the One into manifestation, providing a thread that leads souls to remembrance of their origin. The series of differentiations unfolds the nature of that deity and contributes to the establishing of all beings in reality.

The Sun ☉, for example, exhibits a fourfold nature. Its unity is in the first quadrant, its intellect is in the second quadrant, its soul is in the third quadrant, and its body is in the fourth quadrant. [*fig. 136, p. 318*] Wherever the Sun appears, we find the creative principle, the power of light and consciousness.

The levels of the Sun are discussed in several places in Platonism. (See, for example, the "Oration to the Sovereign Sun" in *Two Orations of the Emperor Julian,* Proclus' commentaries in Thomas Taylor's notes to Plato's *Republic,* Dionysius the Areopagite, and Plotinus, VI.7.16.) The fourfold arrangement of the Sun in the chart illustrates Plato's remarkable description of the Idea of the Good (in Book VI of the *Republic*). There, the intelligible Sun and its image, the sensible Sun, together establish four levels of knowledge that Plato conceives as sections of a line. These divisions can be symbolized by the four positions of the ☉ in the chart, where each position of the ☉ establishes both what is known and the power that knows. [*fig. 137*]

By its placement in the mandala, each planet in dignity is likewise woven into the geometry of the mandala in several ways. Consider, for example, Venus ♀ in Aries ♈.

♀♈ *Venus in Aries* (the "detriment" of Venus) represents the self-attraction of the One. But also, as part of the third ring, ♀♈ represents the Nature of the One as outward-facing continuity that endlessly gives without lessening itself. To comprehend both these views is to understand the meaning of harmony, continuity, and all-attractive Love. [*fig. 138*]

As the leader of the third ring, ♀♈ represents the soul's longing for its source, the seeking of the seeker for the Good. It has as its ultimate nature the union of seeker with sought, the dissolution of the soul into the unlived life of the One.

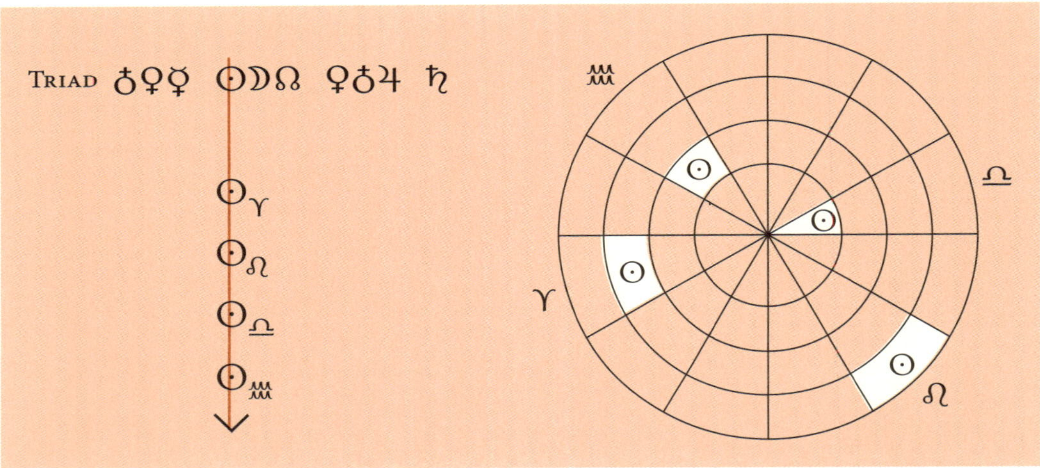

Fig. 136: Henad of the Sun

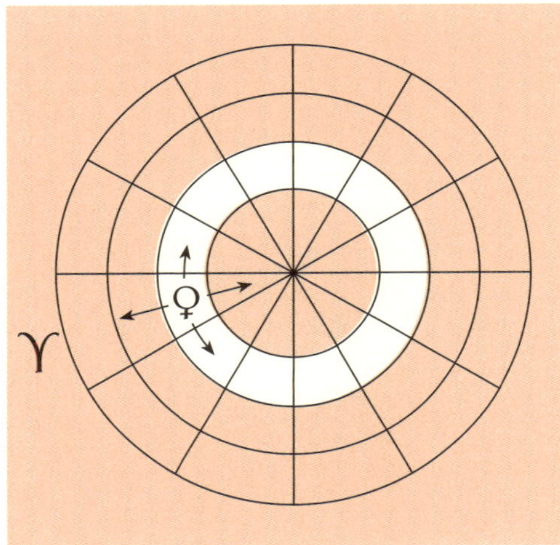

Fig. 138: ♀♈—Soul in the One

Plotinus, iv.4.10: . . . *[soul] has its source in the First and thence, along an unhindered path, it flows into a total of things, conferring grace . . .*

	Manifest Sun		Intelligible Sun	
What Is Known	☉♒	Reason-Principles	☉♌	Self-Existence
Knowing Powers	Perception	☉♎	Contemplation	☉♈

Fig. 137: Henad of the Sun ☉ and the Divided Line

One approach of the soul to the One or the Good is through bliss, *ananda,* love, lover to beloved. The goal, source, or core of this path is represented by ♀♈, which corresponds with what Plotinus calls the self-tendance of the Good. [*figs. 139A, 139B, and 139C*]

Venus ♀ Henad Series

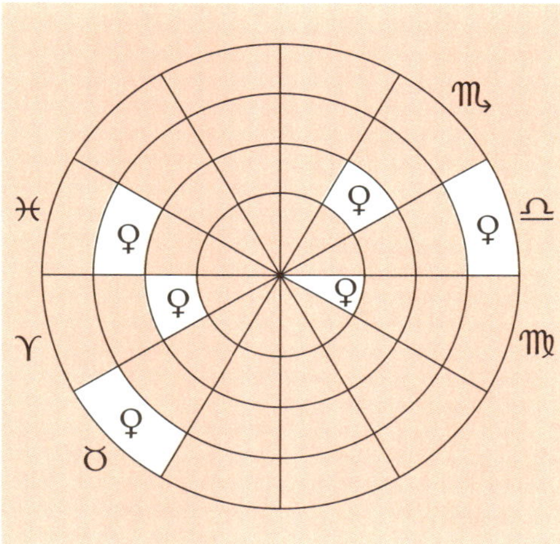

FIG. 139A: VENUS HENAD SERIES IN MANDALA

♀♉	Being	Beauty as the eternal power of the One is the Being of *Pranava Aum*
♀♎	Life	The harmonious hierarchy of Ideas is the Life of the 36 *tattvas*
♀♓	Intellect	Intelligence's desire for or attachment to earthly existence is the Intellect of the five great elements
♀♏	Soul	The love of the soul for its divine image is the Soul of deific imagination
♀♈	Nature	The attraction of all life for the Good is the Nature of the Good as self-love
♀♍	Body	Receptivity to divine power is the Embodiment of the six holy forces

FIG. 139C: VENUS AS THE PRINCIPLE OF SAMENESS

Quadrant	*Rulership*	*Exaltation*	*Detriment*	*Fall*
I	♀♉		♀♈	
II				♀♍
III	♀♎		♀♏	
IV		♀♓		

FIG. 139B: HENAD OF VENUS ♀ BY DIGNITY AND QUADRANT

18

EDITORS

Kashmir Shaivism References

Good sources for more on Kashmir Shaivism are Mark Dyczkowski, *The Doctrine of Vibration,* John Woodroffe (Arthur Avalon), *The Garland of Letters,* and J.C. Chatterji, *Kashmir Shaivism.*

19

EDITORS

Number as Qualitative Language References

Some sources of this meaning of number as the qualitative language of the soul are the *Shrine of Wisdom* article on "Symbols: Number," M.L. Von Franz's *Number and Time,* J.N. Findlay's *Plato: The Written and Unwritten Doctrines,* T. Taylor's *Theoretic Arithmetic of the Pythagoreans,* Proclus' general introduction to Euclid's Elements, M.E. Jones' *Lecture Notes on Pythagorean Astrology,* I.K. Taimni's *Man, God, and the Universe,* M.P. Hall's *Lectures on Ancient Philosophy* and *The Secret Teachings of All Ages,* and R. Brumbaugh's *Plato's Mathematical Imagination.*

20

EDITORS

Two, Five, Eight

The number principles 2, 5, and 8, associated with the dignities of Venus (rulership), Moon (exaltation), and Mars (detriment) in the 2nd house, indicate aspects of power: 2 is the One's inward indeterminate power, 5 is the One's power for infinite self-reflection, and 8 is the One's potentially infinite "outward" power.

21

EDITORS

Transition from the One to the Intellectual-Principle

Anthony envisioned the astrological mandala as a multilevel symbol. Sometimes it represents the One as four aspects of Reality. At other times it represents the Intellectual-Principle and its Ideas. We may see the mandala as a single multidimensional symbol, or several superimposed symbols viewed from different perspectives. In doing this, we focus on various strata of the mandala—as though looking through the lens of a "thought microscope."

Each of the four quadrants, for example, can be represented by an entire mandala. When we focus on the One, we see the fourfold nature of Reality. When we shift to the mandala as the circle of Idea-Intelligences, we see the breadth of the Intellectual-Principle—but the vastness of the One slips out of view. The Intellectual-Principle seems incommensurable with the view of the One, yet the Intellectual-Principle is a trace of the One, revealing something of its immense splendor. [*fig. 140*]

Within the mandala of the One, we emphasize the division of the circle into four quadrants. Within the Intellectual-Principle, we emphasize the twelve Ideas represented by the houses. Rather than to portray the Intellectual-Principle as one quadrant among four, the chart of the circle of Ideas and dignities shows the whole of reality through the vision of the Intellectual-Principle. We no longer have the One itself in focus, but the One-Being. If we look at Intellectual-Principle not as a hypostasis in its own right, but as the emanation of the One, Intellectual-Principle means the entirety of what is knowable of the One.

From Unities to Rulerships

To illustrate a critical link between Saturn in Aries ♄♈ and Saturn in Capricorn ♄♑, Anthony sometimes related the ten dignities of the Triad to the twelve key Ideas of the Intellectual-Principle chart. Figure 141 shows that each of the rings of dignities in the first quadrant—rulerships, exaltations, detriments, and falls—represents the paradigm for one of the four quadrants. Each of the ten first-quadrant dignities determines a rulership in the chart of the Intellectual-Principle. The last three rulers (fourth quadrant) all arise from ♄♈. Figure 142 indicates that the Being (♄♑) of the system of Nature is based on the principle of Otherness (♄♈) in the One.

In the chart of the Intellectual-Principle, the three rulerships of the first quadrant ♂♈ ♀♉ ☿♊ represent a One-Being—no longer the Transcendent One.

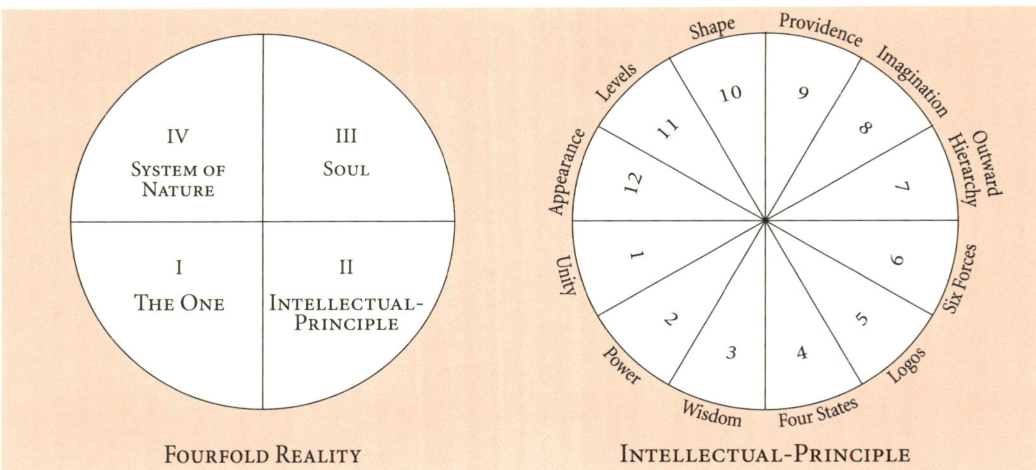

FIG. 140

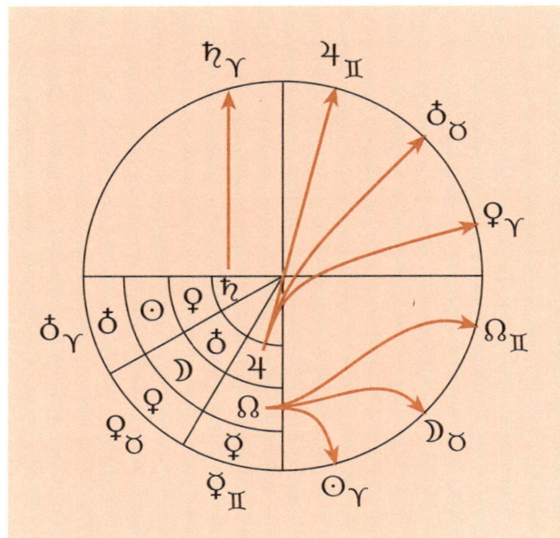

FIG. 141

The 10 and 12: Numbers and Ideas

How do the ten number principles become twelve Ideas? We do not wish to see this change as merely quantitative, adding two to the ten principles we applied to the One. Our search for a more qualitative meaning must distinguish Number as applied to the twelve Ideas constituting the Intellectual-Principle, from Number as the relation of the ten Unities to the One.

As applied to the One, the number principles are absolute, unific, and unique. All the ten numbers together form an inviolable whole. All the glimpses of unity which these numbers give are "dissolved" back into the One. We notice in Anthony's symbolism that there are no glyphs to fill the fourth ring in Taurus ♉ and Gemini ♊. [*fig. 143A*] Subsequently, the transition from the ten divine Unities to the twelve Ideas suggests to us that the last two houses (11 and 12) have a different status than the first ten because their rulerships do not have paradigmatic unities in the One (first quadrant). The last two houses represent the Ideas of "becoming," and indicate something of the indeterminacy of the world-appearance. [*fig. 143B*]

In a note from a discussion of the twelve Ideas, Anthony writes:

> The number sequence tells us that there has been a certain end to our Ideas, for nothing truly new can be produced past 10. Yet if the Intellectual-Principle is to include in its vision particularized subjective entities [in house 11] and their specific appearances [in house 12] we must go further in our analysis.
>
> Universal manifestation, if it is to be truly universal, must include the various levels of individual entities and their own mental fabrications which are the contents of their worlds. The next two Ideas provide for this.
>
> The 11th house is the Form of all the planes of existence in which particular souls may have their experiences, the power of which is Saturn or Difference, but here as Life or motion rather than as the Being of non-Being as in the 10th Idea. All the Life forms that constitute the various levels of existence of the universe are included in the 11th Idea.
>
> Finally, in the 12th house, there is the appearance and dissolution of the sensible universe—the final effect. This immense appearance is within the governance of the Life Power of the Demiurge, and so Jupiter is the ruler in Life here. The dissolution of the life appearances is back into this Universal Solvent of all life, which retains their traces until fresh ones appear to modify these. (Notes and Papers, p. 911)

We may view the last two rulers as the unfolding of the double act of Saturn in Aries ♄♈, the principle of primordial otherness, which has no double

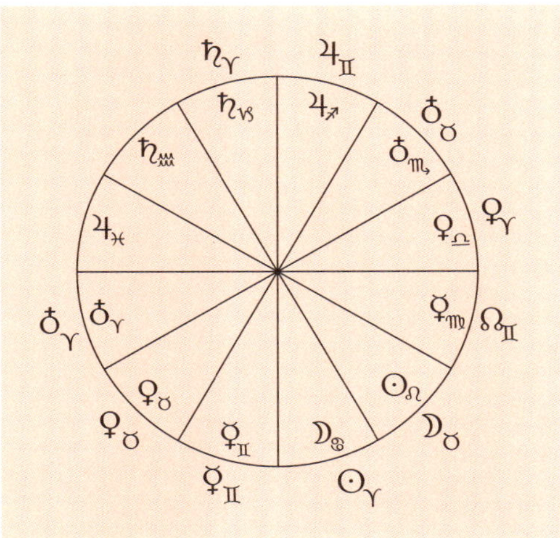

FIG. 142

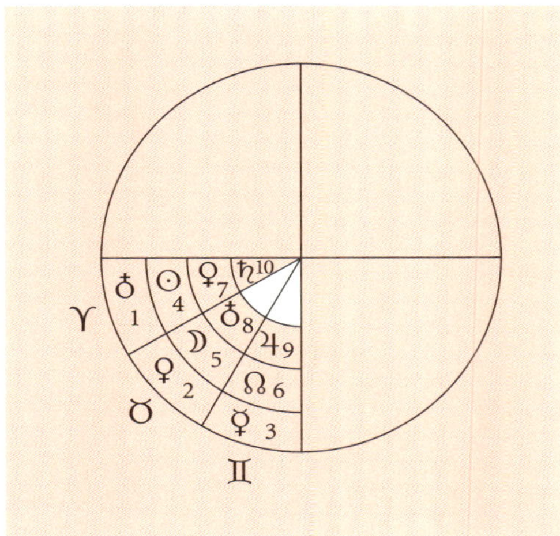

FIG. 143A: DIVINE NUMBER AND THE DIGNITIES OF THE TRIAD

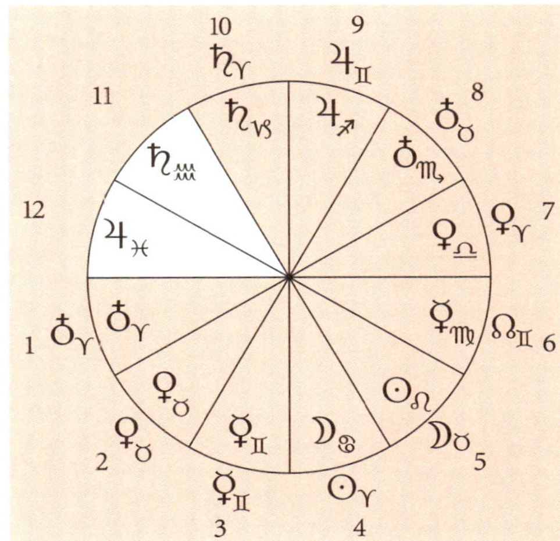

FIG. 143B: DIVINE NUMBER WITH HOUSES 11 AND 12 AS "BECOMING"

act in itself (cf. no 11th and 12th dignities in the first quadrant). Saturn in Aries exhibits its double act only as the receptacle of universal manifestation.

The twelvefold chart allows for the continual instantiation and revitalization of life: Thus there are Ideas "beyond" the 10, which are the unfolding of the 10 principles within the individual and the world. Reciprocally, the presence of the last two Ideas says two things: first, that Number in the Intellectual-Principle (in the view of the 12 Ideas) includes a dynamism and relation that is not present in the One; second, that Number in the Intellectual-Principle has a more determinate role, ordering there what remains "unfigured" in the One.

The One as Substance and Function

Anthony writes that the "One as Triad" prefigures "substance and function [as] the non-dual divine double act." (Notes and Papers, p. 202) [*fig. 144*] Furthermore:

> The Triad, or One and Power, is a balance poised of the double act: on the one hand, the immutable stillness that pertains to passive perfection, and on the other the completing act that is the Power of the One, extending totally and eternally completing each and every aspect of infinity. This is what Plotinus referred to as its completing act, the power which is the underlie or matter for the One and the emanating and informing principle of the Intellectual-Principle. The One revealing itself to itself by means of itself . . . the intrinsic and undifferentiated homogeneity synonymous with perfection may be diagrammed thus as the paradigm of paradigms. (Notes and Papers, p. 211)

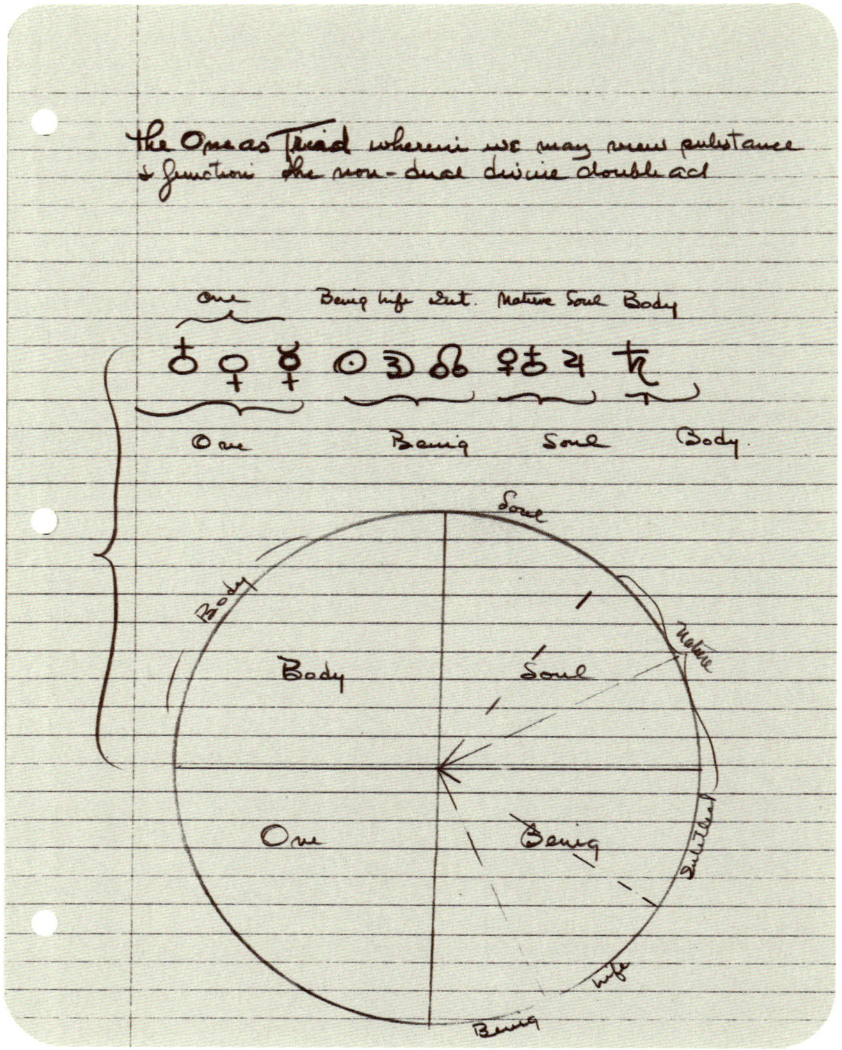

Fig. 144: Anthony Damiani's drawing, Notes and Papers, p. 202

More Diagrams for the One as Transcendent and Emanated

There are several other ways to use the dignities to portray the transition from the Transcendent One to its emanation as the Intellectual-Principle.

A. Each ring in the first quadrant unfolds an entire ring of functions circling through the chart. [*fig. 145*]

B. Each unity unfolds a continuous series of deities (a henad) having positions in both rings and quadrants. [*fig. 146*]

C. The 1st house represents the One as transcendent, and the circle with its four quadrants represents the One as emanated. Anthony says in the text (*Astronoesis*, p. 67) that the four dignities in the 1st house "indicate or suggest that the Absolute Perfection has, in a mode peculiar to itself, a fourfold nature," and that this fourfold nature is unfolded as the four quadrants of the chart. So ♂ ☉ ♀ ♄ represent a) the intrinsic self-existence of the One, b) the unity, intellection, life, and otherness of the One, as well as c) the paradigms or unities of the other three levels of reality which are to be unfolded. [*fig. 147*]

D. Whereas we use the *fourfold* quadrant diagram in our discussion of the One, the *threefold* division is primary in the chart of the Intellectual-Principle. The 4th house together with the first three houses represents the universality of Being. The second section represents Life, and the third section represents the productive thought or Intellect of the universal mind or

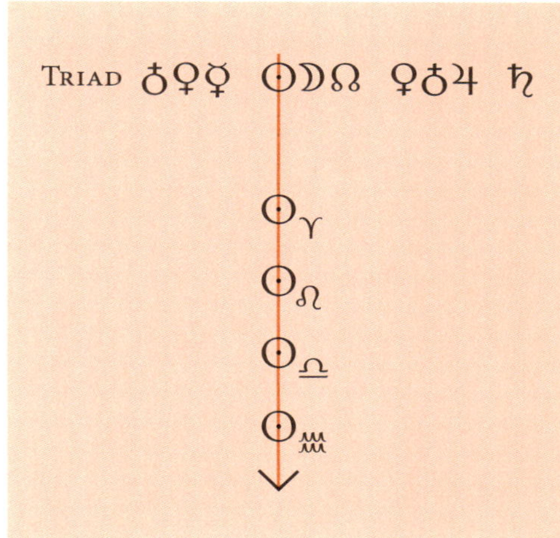

Fig. 146

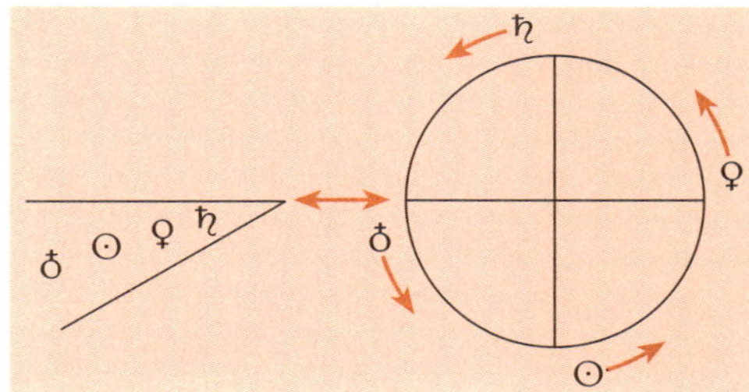

Fig. 147

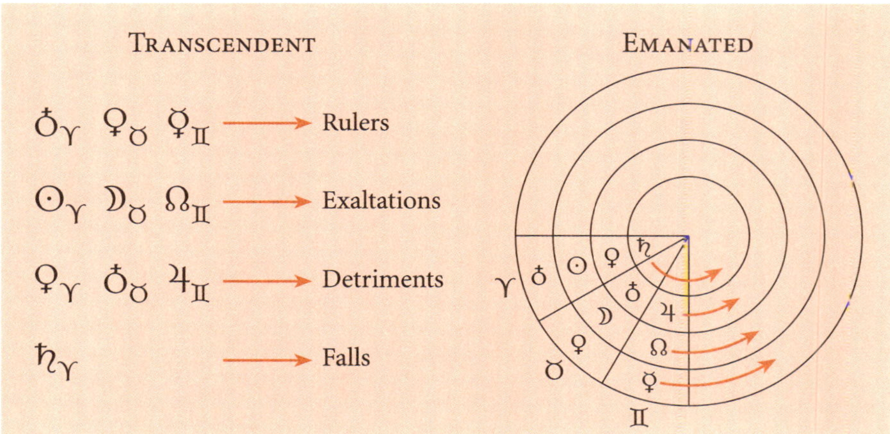

Fig. 145

Demiurge. Looked at in this way, the chart shows Unity assimilated to these three aspects of the Divine Mind. [*fig. 148*]

E. We may show the hierarchical view of the relation between the twelvefold chart of reality and the circle of twelve Ideas. Figure 149 shows that the first three principles of the One are beyond the Intellectual-Principle, while the other principles are subsumed under the last nine ideas.

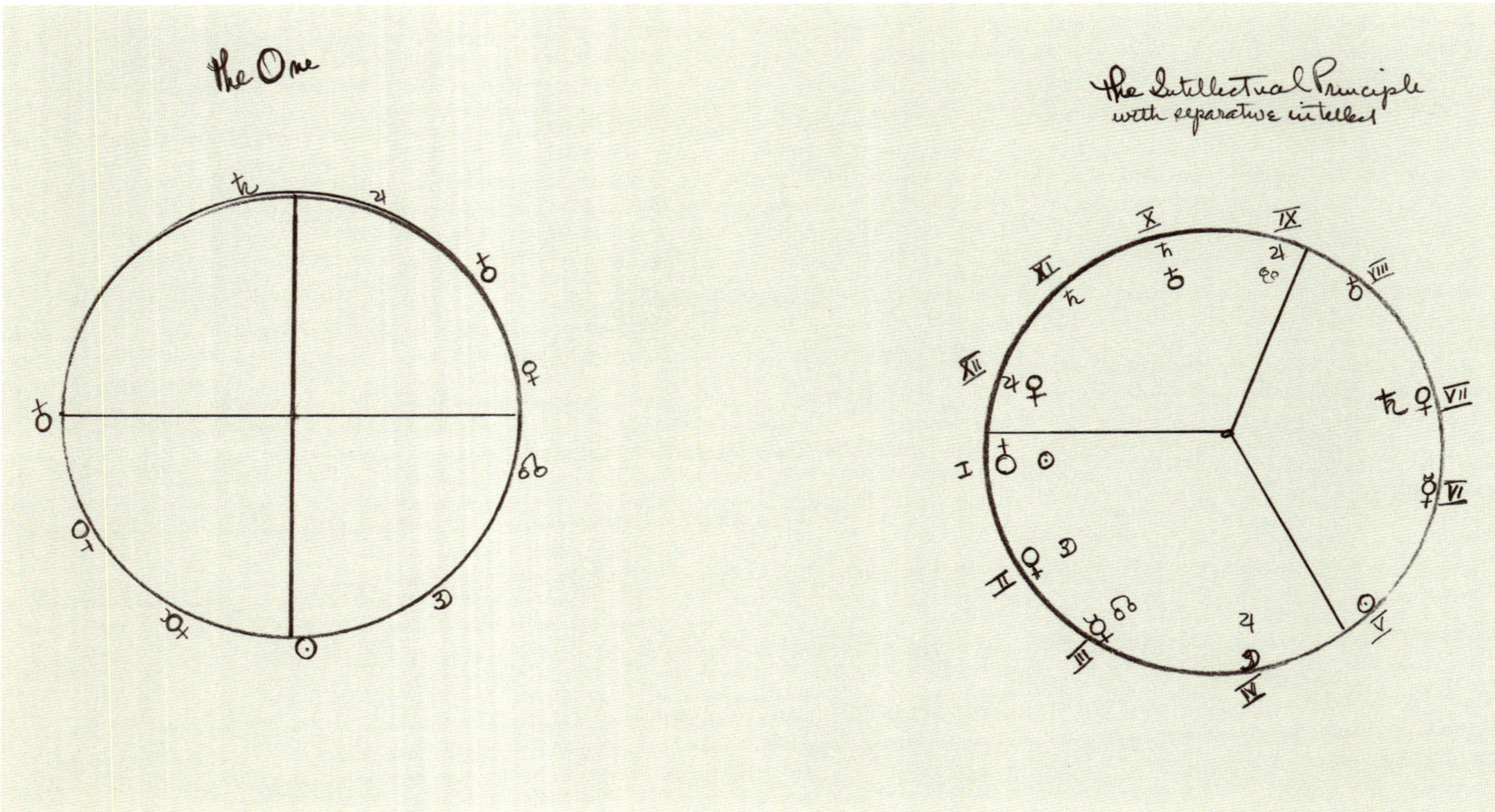

Fig. 148: A drawing by Anthony Damiani

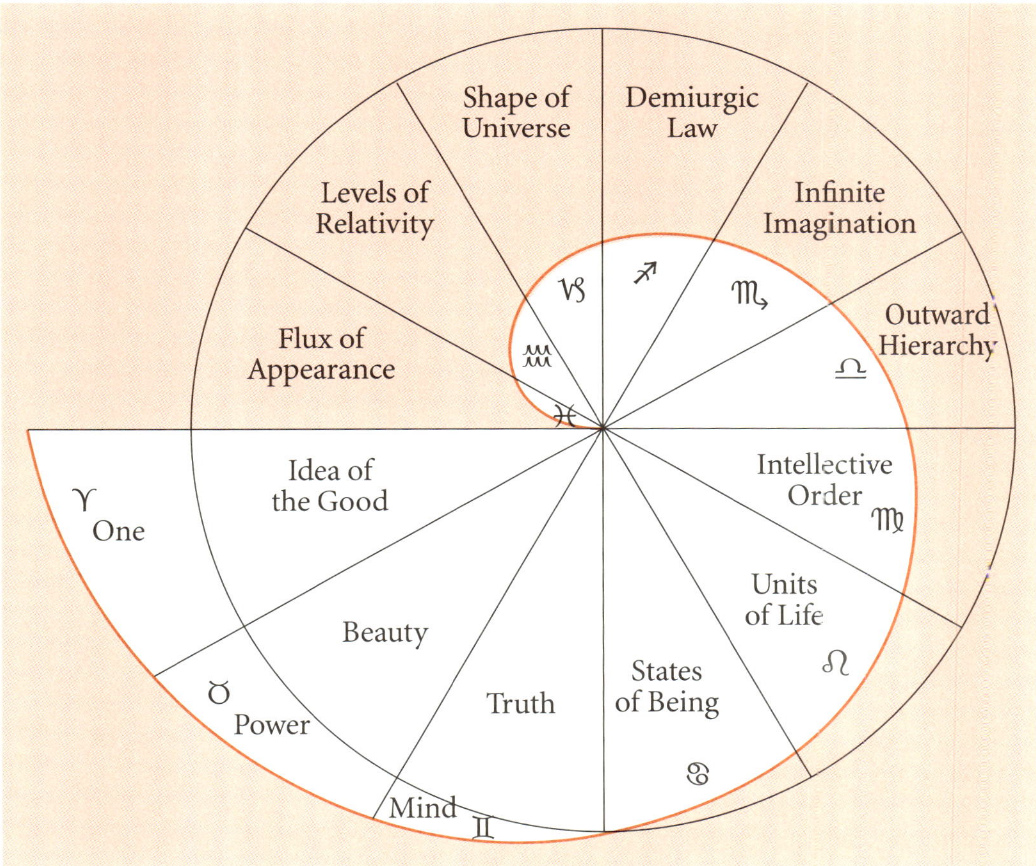

FIG. 149

22

EDITORS

Proclus and the Ordering of Principles or Gods

Proclus Diadocus (*circa* 410 AD–485), one of the last of the Neoplatonic philosophers, showed particular concern in his writings with the orders of gods and associated principles.

The planets in the Triad represent the "ineffable blossoms" or "unities" of which Thomas Taylor speaks. Each is the leader of a series of "deific natures" unfolded within Intellectual-Principle (fig. 45, p. 111). Each planet, representing a divine power, has dignities in each quadrant and ring (fig. 44, p. 110).

23

EDITORS

Vak, A-shabda, and *Pranava Aum* (Taurus)

A-shabda, the unmanifest vibration within the Absolute, is the highest of the four levels of *vak* (sound or speech) in the Shaivite tradition. Sir John Woodroffe discusses these levels in *The Garland of Letters.*

24

EDITORS

Creative Imagination and Vishnu References (Scorpio)

Henry Corbin develops the distinction between imagination and fantasy in *Creative Imagination in the Sufism of Ibn 'Arabi.*

In chapter eight of *Man, Grand Symbol of the Mysteries* (p. 107) M.P. Hall writes, "The word Vishnu is from the root, *vish,* 'to pervade'." The chapter describes the ten avatars of Vishnu as theophanic presences.

25

EDITORS

Prajapati References (Sagittarius)

The Prajapatis, depending on the reference, are of variable number and function, and often associated/identified with the Rishis. T.S. Row, in "The Twelve Signs of the Zodiac," gives their number as nine; he also refers to them in his *Notes on the Bhagavad Gita* (p. 87). Manly Palmer Hall, in *The Adepts in the Eastern Esoteric Tradition, Part One, The Light of the Vedas,* writes of the Prajapatis in one sense that Anthony conceives them, as accompanying the Demiurge (Supreme Lord): "As the Supreme Lord passed from the state of solitary self-existence, by the extension of his own consciousness, to the condition of becoming the cause and sustainer of phenomenal diversity, his objective powers awoke and were embodied as the Prajapatis. Through them he brought forth the cosmos out of chaos, and established the laws which were to govern the material universe." (p. 106) H.P. Blavatsky's *The Secret Doctrine* and Alain Danielou's *Hindu Polytheism* also include references to the Prajapatis.

26

EDITORS

Lokas and *Chakras* References (Aquarius)

The last three chapters of Sir John Woodroffe's *The Serpent Power* interpret the *lokas.* Anthony investigated the relation of the sevenfold functional system of the *lokas* to *chakras* and to the seven planetary functions. Jung's article "Psychological Commentary on Kundalini Yoga" approaches the *chakras* from a psychological perspective. From the point of view of metaphysics, each of the seven *lokas* has seven levels, and all the levels of psychology are all within the lowest *loka.*

27

EDITORS

The Elements (Pisces)

T. Subba Row describes the 12th Idea as the Five Mahabhutas or "the universal solvent of the alchemists." These elements are the basis for various typologies and elemental theories.

Paracelsus describes this "solvent" as the life-energy of the World-Soul, the alchehist into which all things dissolve, and from which all creatures arise. The Nyaya-Vaisheshika school of Hindu philosophy speaks of five eternal substances or *panchamahabhutas* (literally, *panch* = five, *maha* = great, *bhuta* = elements) as space (*akasha*), air (*vayu*), fire (*agni* or *tejas*), water (*ap*), and earth (*prithivi*). These great elements are the ultimate constituents from which the physical world is derived; all the various types of substances in the world arise through combinations of them. K.C. Bhattacharyya, in his essay on "Studies in Samkhya Philosophy," calls these elements "qualitied spaces." (para. 86) In *Man as Symphony of the Creative Word,* Rudolph Steiner describes the elements in the Earth's living imagination; this view is also in Guenther Wachsmuth's *Etheric Formative Forces.* Titus Burckhardt's *Alchemy* discusses the elements in a Western context. The Chinese (and Tibetan) version of the five elements lists water, fire, wood, metal, and earth. (See also the "five agents" in W.T. Chan's *A Source Book in Chinese Philosophy.*)

28

EDITORS

Absolute and Relative *Bodhicitta*

The germinal form of these *bodhicitta* intelligences is fivefold within each living being, eternally present, yet evolving and becoming assimilated through countless incarnations. So His Holiness the Dalai Lama says: "We have within us the seed of purity, the essence of a One Gone Thus (*Tathagatagarbha*), that is to be transformed into Buddhahood." (*Kindness, Clarity, and Insight,* p. 117)

29

EDITORS

Geometry of the Five-Pointed Star

In addition to showing the Pentad of wisdom-insights, the geometry of the pentagon provides an image of the rational intelligence of the human being. In Pythagorean symbolism, the five-pointed star embodies the special relation called the golden ratio which represents mathematically the self-similarity of the whole to the part. This property of self-similarity symbolizes the discovery of meaning as the essential act of life. This is reflected in mathematics as finding a mean between the finite and the infinite. The golden ratio found in the pentagonal star is also the basis of Plato's Divided Line, which schematizes "all that can be known" by the living intellect of the Soul.

30

EDITORS

Chuang Tzu on the Mirror-like Wisdom

Chuang Tzu describes this *mirror-like wisdom* of the sage: "The mind of the perfect man is like a mirror. It does not lean forward or backward in its response to things. It responds to things but conceals nothing of its own. Therefore it is able to deal with things without injury to [its reality]." (quoted in Wing-Tsit Chan, p. 207; brackets are Chan's)

31

ANTHONY

Twelvefold Mandalas for Each Level of the Idea of Man

EDITORS: *Here Anthony illustrates each of the three aspects of the Idea of Man by a separate mandala. The dignities within the four signs of each third of the metaphysical chart are spread around a twelvefold circle. (Editors' Extension II, "The Twelve Idea-Principles," p. 131, provides the paradigm of these twelvefold elaborations. There is also a "reader-friendly" version of the Idea of Man as a unique form of the whole twelvefold Ideas (on pp. 157–165). It is very likely that Buddhism also refers to these three aspects of this Idea when speaking of the* kayas, *or the three ways of being of the Buddha:* Nirmanakaya *is the vehicle of manifestation and form;* Sambhogakaya *is the vehicle of bliss and transformation;* Dharmakaya *is the vehicle of ultimate Truth. [fig. 150]*

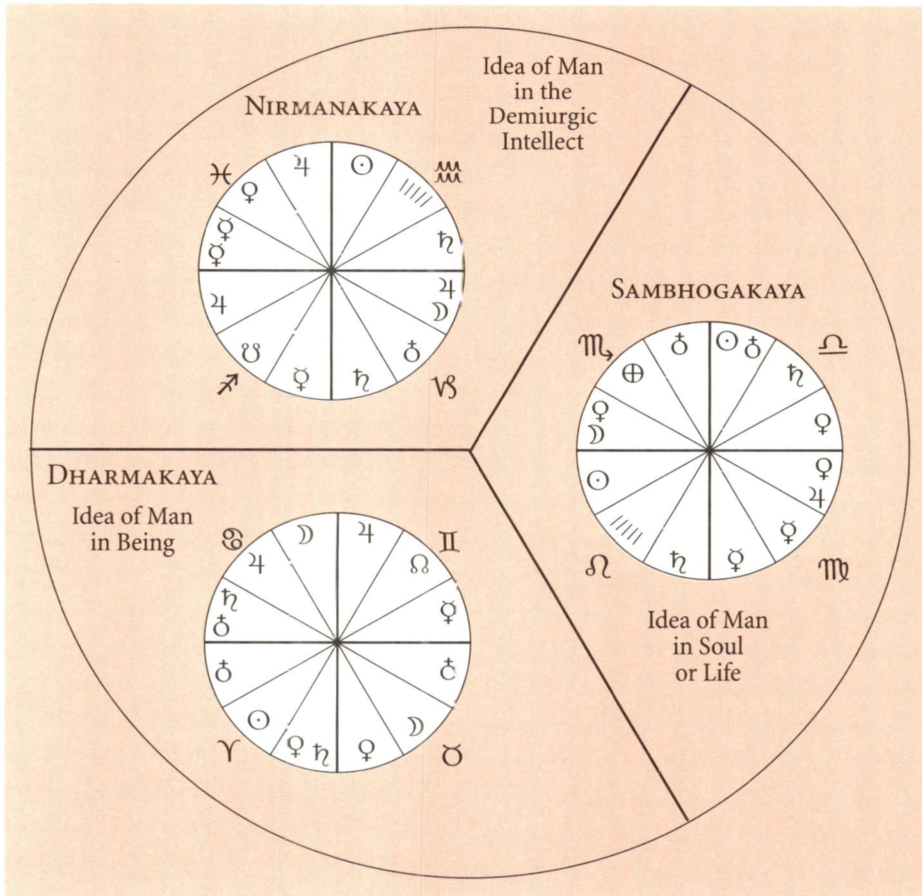

Fig. 150

Mandala of Idea of Man in Being [*fig. 151*]

♈ I QUADRANT *Unity of the Idea of Man is a unique form of the Intellectual-Principle entire*

♁ 1st house • uncharacterizable unity of the Being of the Idea of Man, indistinguishable from the Divine Mind. It is self-subsistent, but one with the Intellectual-Principle.

☉ 2nd house • super-essential essences which pertain to the Idea of Man—the infinite potentialities. Sun = Intellection of Idea of Man which is its primary possession.

♀ ♄ 3rd house • Form of the unity of the Idea of Man as unique—unity and essences are combined to give the peculiar structure of the Idea of Man.

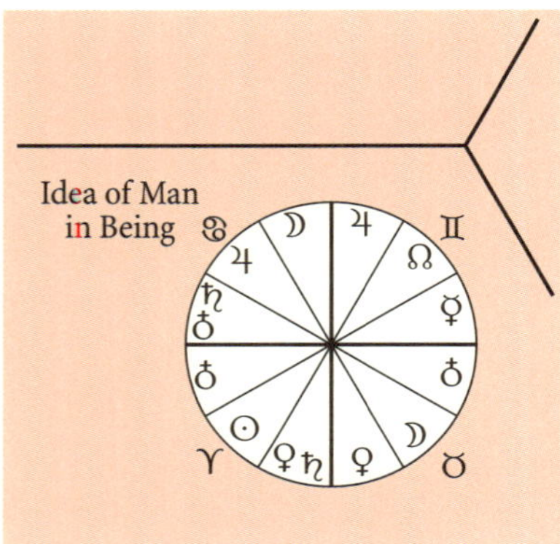

FIG. 151

♉ II QUADRANT *Essences of the Idea of Man in Being are factored out: its power*

♀ 4th house • the four states of Being of any essence in the Idea of Man (the first state is unmanifest).

☽ 5th house • the intelligence inherent in the essences in the Idea of Man.

♁ 6th house • powers inherent in the essences—six forces.

♊ III QUADRANT *Adam Kadmon or the determination of the Idea of Man*

☿ 7th house • the constitution of the intelligible gods—the Grand Man or Being of Adam Kadmon through the first five of the *tattvas* (pure *tattvas*). The later *tattvas* will constitute the Idea of Man in the soul.

☊ 8th house • the subordinate Ideas comprising the Idea of Man—Head ☊ is the unity.

♃ 9th house • the power to manifest these Ideas in Adam Kadmon.

♋ IV QUADRANT *The four states of Being or manifestation of the Idea of Man*

☽ 10th house • shape of the intelligible cosmos in the Idea of Man.

♃ 11th house • prefiguration of the number and grades of beings.

♄ 12th house • prefiguration of all the sensible manifestations of the Idea of Man—this is the basis for the whole appearance of man in the 12th house.

Mandala of Idea of Man in the Soul [*fig. 152*]

♌ I QUADRANT *Unity of the Idea of Man in Soul or the Intellectual Soul*

☉ 1st house • uncharacterizable unity of the human soul as identified with the Idea of Man.

///// 2nd house • essences which will constitute the unit soul's vision of the Idea of Man. [Here the empty position ///// is equated with the planetary rings (the five planets between the Sun and Saturn).—eds.]

♄ 3rd house • the definition of a unit soul—cognitive isness.

♍ II QUADRANT *Essence of the Idea of Man in Soul—its powers*

☿ 4th house • the four levels of operation of each power of soul are defined here, and all the contents of these powers.

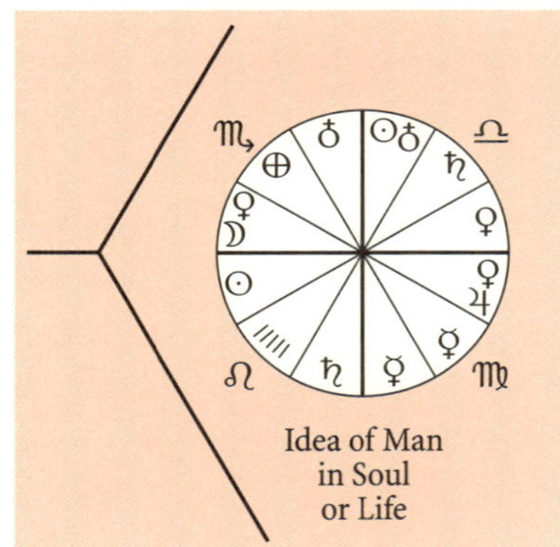

FIG. 152

☿ 5th house • Intellection of the soul—intellectual perception.

♃ ♀ 6th house • differentiation of the Intellection of soul into various powers. The soul perception will embody the essences.

♎ III Quadrant *Definition and outward hypostasis of the Idea of Man in Soul*

♀ 7th house • life of the *tattvas* in the soul—combining of the six pure-impure principles *maya* and *kanchukas* to form a *purusha*—a little knower.

♄ 8th house • soul's vision spread out in thought—essences in ideational form.

♅ ☉ 9th house • power to distribute the reason-principles and embody them in the light.

♏ IV Quadrant *Idea of Man in Soul as existing in a subjective state*

♅ 10th house • soul participates in a total conception of the shape of the universe according to the Idea of Man.

⊕ 11th house • various levels of life grades needed to accompany the experience of the universe. [The Part of Fortune ⊕ is traditionally exalted in Scorpio, although Anthony did not always use it in the metaphysical chart. —eds.]

♀ ☽ 12th house • subjective or soul's *vasanas* in terms of feltness—these underlie the material experiencer, the Idea of Man embodied as soul's congealed feeling states.

Mandala of Idea of Man in *Bodhicitta* [*fig. 153*]

♐ I Quadrant *Unity of the Idea of Man in* Bodhicitta*—participates in the nine Prajapatis [Anthony used the Buddhist term* bodhicitta *(lit., mind of enlightenment) for the embodying soul (or second soul) conjoined with the Idea of Man in the system of Nature. Through its experiences in the cosmos, the embodying soul actualizes the Intelligence of the Ideas.—eds.]*

♃ 1st house • uncharacterizable unity of *bodhicitta* as identical in essence with the higher (first) soul's Idea.

☋ 2nd house • essences which will constitute the total vision of all the sequence of lives that the *bodhicitta* will go through.

☿ 3rd house • Celestial—definition of Man in cosmos—as *bodhicitta.*

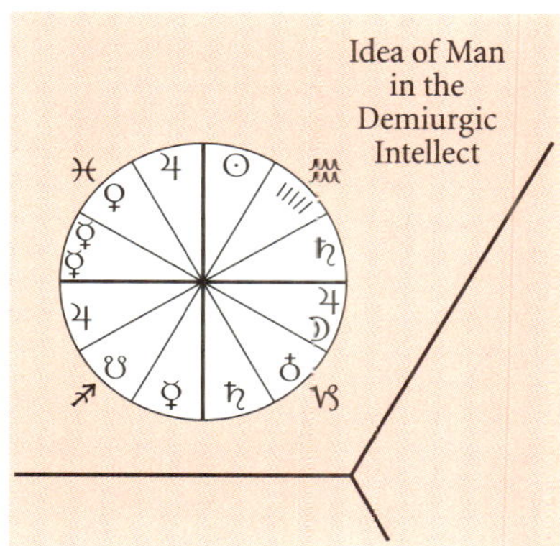

Fig. 153

♑ II Quadrant *Shape of the Intellect of Man in* Bodhicitta *= 360 degrees [The 360 degrees of the zodiac are archetypal essences or reason-principles as experienced by the inhabitants of Earth. See Anthony's Invocatory Prelude, pp. 14 ff.—eds.]*

♄ 4th house • four levels of each reason-principle in the shape of the archetypal realm in *bodhicitta.*

♅ 5th house • intelligence inherent in the reason-principles that constitute humanity in *bodhicitta.*

☽ ♃ 6th house • differentiation of powers of reason-principles.

♒ III Quadrant *Planetary Soul of Man in* Bodhicitta*—Productive Imagination on seven levels*

♄ 7th house • *tattvas* in the productive Imagination—dissipated essences.

///// 8th house • separative intellect organizing various essences into *lokas.*

☉ 9th house • manifests the essences in Imagination—distributes and embodies reason-principles into sensible imagination.

♓ IV Quadrant *Idea of Man in* Bodhicitta *as embodied in physical life elements*

♃ 10th house • shape of individual human life.

♀ 11th house • apparent subject or conscious entity appearing in a certain *loka.*

☿ ☿ 12th house • sense-man as psycho-physical organism.

32

EDITORS

Wisdom and the Heart Sutra

The Heart Sutra spoken by Avalokitesvara is the source of the term "Wisdom which has gone beyond," the Buddha wisdom, that we place in house 3. The sutra begins: "Homage to the perfection of Wisdom, the Lovely, the Holy." (E. Conze translation) We may refer holiness, loveliness, and wisdom to the 1st, 2nd, and 3rd houses, respectively.

33

EDITORS

Fourfold Levels for Houses 9–12

The fourfold of houses 9–12 is the basis for the cosmic unity, intelligence (inerratic), reason (planetary), and physical appearance (sublunary) represented in the 12th house of the chart, as well as the unity of the sage's mind-stream and causal, subtle, and gross vehicles. The System of Nature chapter expands on this relation. Figure 29 (p. 69) in The One: Symbolization, also shows this fourfold.

The unity of the Idea of Man as embodied is the 9th house. In the 10th house, the world is in a state prior to manifestation. In the 11th house, the world is given form by an individual mind. In the 12th house, the world is experienced as a particular kind of world-appearance. We can use the analogy of a dream to understand how this paradigm is imaged in the 12th house. An Idea or archetypal energy in our mind (10th house) manifests through the agency of the mind's power (11th house) as a dream in which the individual and its world arise together (12th house).

34

ANTHONY

On Deepening Consciousness

Comments from a Class Transcript 1/27/84

This emanation, this thought in the Mind of God—our Absolute Self—is what we have to investigate, for this is the key to the mysteries.

Anthony Damiani, *Astronoesis*, p. 173

If you wanted to investigate the nature of consciousness, could you do it by reading books about it? Yes, to some extent. A little bit. An enquiry into the nature of the Self might start off with reading, getting acquainted with some ideas, maybe even thinking you know what your teacher is telling you. But if you *really* want to investigate the nature of consciousness, you would have to *be it.*

And then investigate deeper into it. In meditational practices you succeed in isolating what this I-ness is, what this consciousness is in you. You're identified with this consciousness. You're no longer identified with the psychosomatic. You're no longer dwelling in memories, anticipations, reflections . . . that's all gone. You're in this state of consciousness, you are this consciousness. You are this awareness. Now you can start your investigation into consciousness.

Now that you are this consciousness, you turn this consciousness upon itself. And it deepens, and you find out more about that consciousness. It isn't that when you get a glimpse and you are consciousness—this is it. But there are depths within that consciousness.

An enquiry into the nature of the Self doesn't get off the ground until you yourself have realized that you are consciousness and now you turn that consciousness upon itself and deepen it, inwardize it more and more.

PB [Paul Brunton] makes the remark that when consciousness is inwardized or when insight penetrates into itself, at the first level there's a deepening, the second level contact with the Void, then the third level the Holy of Holies. [cf. *Notebooks,* v. 16, pt. 1, ch. 2, no. 109] Now *that's* an inwardization of consciousness. That's an investigation, an inquiry into the nature of consciousness. You can't do it outside of that nature itself.

Do you remember when PB said that after this state of becoming aware of awareness, there's a further state of being just this awareness? Isn't that consciousness or that awareness deepening? [cf. *Notebooks,* v. 4, pt. 1, ch. 1, no. 128]

For instance, you're meditating and you realize yourself to be consciousness—is this the discovery of the soul? Yes. It is also possible to consider that this discovery may simply be at the fringes, that you're at the outermost, that you've only touched on the surface, that there are depths within it that are treacherous.

When the sage discovers the soul and recognizes he is consciousness, that same sage goes on and discovers that there is what we refer to as Absolute Soul. That sage continuing further discovers the Intellectual-Principle, and investigating and further deepening his consciousness achieves, as Plotinus points out, unity or the Supreme Identity.

Aren't these all stages within consciousness—the deepening and inwardization of consciousness? Think of a sage quietly sitting down and meditating. He has lost awareness of the body, he is this consciousness; now he investigates further into what this consciousness is. And he finds that it is a thought within the World-Mind. He's discovering that it's a point within the World-Mind. Going deeper he becomes aware of a greater principle [Mind]. He can go in deeper and become aware of the Supreme Identity. I'm not saying that he can permanently stay there but he knows that this can happen.

So that's what we are talking about—steps of inwardization of consciousness.

35

EDITORS

Daimons and Angels References

Daimons are divine powers, intermediary between the gods and human beings. Plato frequently refers to daimons. Most notable are Socrates' daimon—who advises Socrates on (in)appropriate action and speech (*First Alcibiades*), and the daimon Love (Eros) in the tale told by Diotima in the *Symposium*. In Plato's *Epinomis* 984d, after the gods come "the daemons and the creatures of the air . . . doing the office of interpreters, and [who] should be peculiarly honored in our prayers that they may transmit comfortable messages." Thomas Taylor writes several extensive footnotes on daimons in Platonism and Neoplatonism; he cites many orders of daimons attached to various gods (see his translation of *First Alcibiades* in his *Plato,* vol. I, notes on pp. 16–25). Apuleius' treatise on daimons, *The God of Socrates,* provides commentary to many of the Plato references as well as to other classical Greek and Roman authors.

Plotinus mentions daimons in tractate III.5 on Love and III. 4 on the Tutelary Spirit, and writes of them as "air beings" in VI.7.11 and IV.3.18. In IV.4.43: "Even the Celestials, the Daimones, are not on their unreasoning side immune: there is nothing against ascribing acts of memory and experiences of sense to them, in supposing them to accept the traction of methods laid up in the natural order, and to give hearing to petitioners; this is especially true of those of them that are closest to this sphere, and in the degree of their concern about it."

In later Neoplatonism, Iamblichus' *On the Mysteries* (an extensive treatise on theurgy) speaks of contacting gods and daimons, the differences between and characteristics of gods, daimons, heroes, their various hierarchies, etc. Proclus writes more conceptually about the hierarchies of gods and daimons (chapters XL–VLV of *The Six Books of Proclus on the Theology of Plato*).

The word "angel" (Gk: *angelos*) means "messenger." In Christian tradition, the concept of the "good daimon" became the "guardian angel"; the "evil daimon" became the "demon." In the early part of the Christian era, Dionysius the Aeropagite (who shows a definite Neoplatonic influence although scholars disagree on who he was and when he lived) wrote about the hierarchies of angels in his *Mystical Theology and the Celestial Hierarchies.* A recent book on angels and their characteristics as related to modern science is *The Physics of Angels* by Matthew Fox and Rupert Sheldrake. It explores the concept of angels in Dionysius the Aeropagite, Thomas Aquinas, and Hildegard of Bingen.

Manly Palmer Hall writes: "The natal daemon is declared by Olympiodorus to be the supreme flower of the soul, for it is the blossoming of soul qualities. To a certain extent, the soul is generated from the interplay of action and reaction in the sphere of sense. The soul is the garment woven from the threads of incident; the natal daemon is consciousness born of experience—the realization begotten of necessity. The natal daemon is the diamond soul, the transmutation of corporeality into incorporeality . . ." (*Lectures on Ancient Philosophy*, pp. 266–267) Or, for Ibn 'Arabi "the Angel represents the essential correlation between the form of a theophany and the form of him to whom it is disclosed. . . . his absolute individuality, the divine Name invested in him." (Corbin, *Creative Imagination,* pp. 62–63)

36

EDITORS

First Four Planets in the Soul Ring

The quadrants/houses represent reality as substance, and the rings represent reality as function. If we take away the dignities, the bare Soul ring represents Soul as a unific function of reality: motion, life, compassion, and so on. This is viewing Soul as a One.

Each substantial intelligence (sign) is unfolded by means of the aspects of function that pass through it, and each function (ring) is differentiated by means of the signs it passes through. In each position of house and ring, the dignity/deity weaves together the substance and function. Each dignity in the Soul ring gives us some insight into the meaning of Soul, while also representing some essential aspect of the house in which it appears.

When viewed as a fourfold, Soul's phases are related to the four quadrants. We will see the reflection of this fourfold in the Unity, Intellect, Soul, and Body of the cosmic soul.

Within the One, ♀♈ ♂♉ and ♃♊ represent the operative power or Soul of the One, and also the unity of Soul.

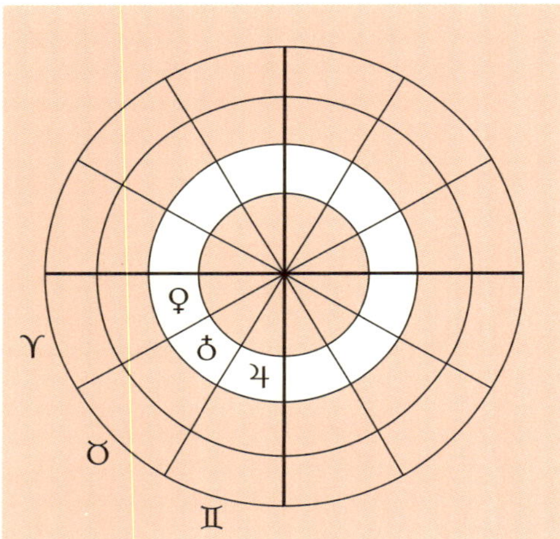

Fig. 154

Soul of the One
♀♈ ♂♉ ♃♊

Fig. 155: Good—Power—Truth

A. First Quadrant ♀♈ ♂♉ ♃♊ [*fig. 154*]

Venus in Aries ♀♈ represents the operational nature or life (third ring) of self-existence (1st house). Plotinus describes it as the self-attraction, self-love, self-tendance of the One. Anthony often spoke of this as "the unlived life of the One."

At the same time, ♀♈ is the core essential nature of Soul: the inner motion which always tends toward union with the Good. This dignity represents the absolute inviolability of Soul. Placed in the 1st house, Self-Existent Brahman, it stands for the Vedantic conception of *Atman:* the self-ness of the soul.

Next, Mars in Taurus ♂♉ is the infinite divisibility (♂) which is intrinsic to the One's power (♉) (which is also called the Word, and referred to by Anthony as the infinite potentialities or *samskaras* of the One). Through this deific aspect, the One can be entirely present to each of an infinite number of Ones. In terms of the Soul ring, ♂♉ represents the essential nature of the Soul as divisibility, the nature of Soul to individuate the infinite potential of the One, or infinitely spread out the potentialities which abide in the One Reality.

Thus ♀♈ loving motion toward the Good and ♂♉ infinite divisibility of power characterize Soul in its most unific nature.

From the loving union and infinite divisibility arises the operational power of the Mind of the One. Anthony describes this as an intellectual energy void of name and form. Jupiter in Gemini ♃♊ represents the operation of the Infinite Void Mind. This is the power of Thought, or the wisdom which can measure out the Void into Thought. As a power of the Soul ring, ♃♊ represents Soul's intrinsic wisdom as a principle of manifestation, capable of the transformation of Void into form. This wisdom not only manifests the Ideas, but also returns all things to the One. It is therefore both wisdom and compassion: It represents the quality of Soul that "provides for all things."

All three of these dignities ♀♈ ♂♉ ♃♊ thus indicate something about the nature of Soul, and Soul's participation in Unity: in Good, Power, and Truth. [*fig. 155*] In different places, Plotinus gives descriptions of the soul's return to the One which may be represented by these three dignities.

In VI.7.33–35, for example, he describes the soul's ascent through love, putting aside even the self-knowing of the Nous to participate in the One through Loving—the path of ♀.

In V.3.17, Plotinus describes the ascent to the One as an approach through which we are to "Cut away everything." This points to the path of ♂.

In VI.7.36, he describes the "grand learning" which the soul must acquire—the path of ♃. Perhaps better, in VI.9.3, he says that the One is always present, but we are present to it only when we turn to it, or recognize it as such.

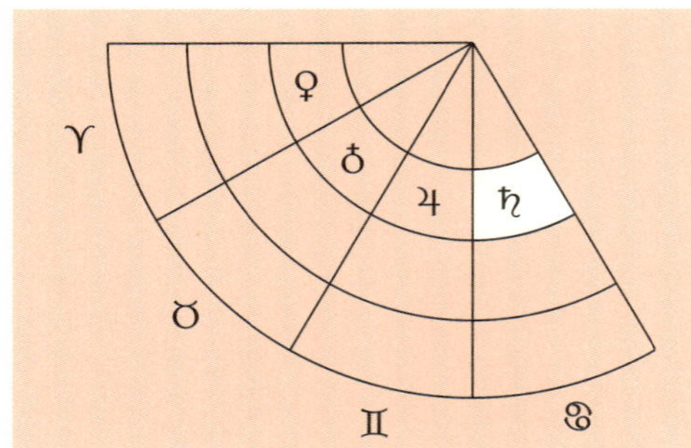

Fig. 156

B. Saturn in Cancer ♄♋

We may also view Soul according to the number 3, as related to the three thirds of the circle. In this view, ♄♋ completes the first third of the Soul ring, in which the Being of Soul is established. [*fig. 156*]

In the 4th house, Saturn in Cancer ♄♋ is the operative nature of the four states. Here, the principle of Difference (♄♋) is the divine monad needed for reality to be hypostasized. At the same time, ♄ is the principle in Soul which allows Soul to have different levels of consciousness: *turiya, prajna, taijasa,* and *vaishvanara*—or, unity, intellect, soul, body.

With the four detriment dignities in houses 1–4, ♀♈ ♂♉ ♃♊ ♄♋, we have a view of the intrinsic nature of Soul as a living essence. It is now able to be permeated by the Ideas in the Nous.

37

EDITORS

Soul Ring Planets Cross References

For commentary on the first four powers of the Soul ring (Venus in Aries ♀♈, Mars in Taurus ♂♉, Jupiter in Gemini ♃♊, Saturn in Cancer ♄♋), see Note 36, p. 331. For a brief technical commentary on the last Soul ring planet, Mercury in Pisces ☿♓, see Note 50, p. 344. See Editors' Extension IV (p. 191) "Seven Soul Powers" for a discussion of Soul ring dignities: Saturn in Leo ♄♌, Jupiter in Virgo ♃♍, Mars in Libra ♂♎, Venus in Scorpio ♀♏, Mercury in Sagittarius ☿♐, Moon in Capricorn ☽♑, and Sun in Aquarius ☉♒. For commentary on the seven planets as psychological archetypes and instinctual functions, see Note 54, "Planets as Rational and Instinctual Functions," on p. 351.

38

EDITORS

The Henad of Jupiter ♃

Jupiter, or Zeus, is that which contains all. In one myth it is the sky god, and is therefore a metaphor for Mind, which like the sky contains all forms in its vast expanse. In another myth, Jupiter swallows Phanes (the Idea of the world) and comes to contain within itself the entire universe. It is thus the ground for all manifestation. A third meaning of Zeus is "that through which life is given." (cf. *Cratylus* 396b) It provides the living medium through or within which the intelligence of the Ideas can manifest.

Jupiter finds the common basis for the Ideas. It dissolves the separateness of the Ideas without destroying their uniqueness. It preserves differences within a common medium, provides a common medium in which differences can emerge. Jupiter is the outer recognition of appropriateness, or correspondence. It provides the outer synchronicity with inner events. Jupiter is also the faith in the guidance of the sages, World-Idea, the Wisdom of the cosmos, and so on.

As can be seen in figure 157, ♃♐ contains the energy of the previous Jupiter positions in ♊ ♋ ♍. Thus it measures out the void, knows the distinct levels of Being, and incorporates the six forces of nature. ♃♐ represents the Demiurge, which faces both the void and manifestation. It combines in itself the vision of the intelligibles and the manifest appearance. Jupiter manifests Thought, and returns all thought to the nature of Mind. Jupiter assimilates the Ideas, which are infinite substances, and transmutes them into a world. Jupiter is the medium, combining in itself the extremes of void and appearance, or the *double standpoint.* It has within itself the power to Think, not deliberative thinking

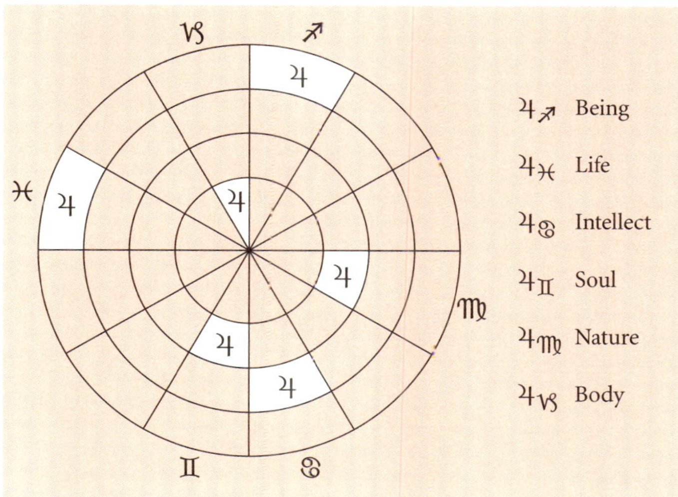

FIG. 157: HENAD OF JUPITER ♃

but Thought as the power to manifest: to expand, imagine, combine into a common medium and make a whole of different Ideas so they can manifest. We don't really know what the power of Thought is. In Hinduism it is called *Shabda Brahman*—that which has the power of the word or speech. Paul Brunton describes this power as the World-Mind.

We can relate the inner and outer aspect of the Demiurge to ♃♐, the only Being planet in the third (Soul) quadrant. [*fig. 158*] We can also expand the relation of outer and inner to the fourth quadrant and second quadrant. [*figs. 159A and 159B*]

♃♑ is the wisdom of the underlying matrix or pattern of body. The shape of space is ultimately the nature of Geometry, the structure of the space which is the real stuff of the universe.

♃♓ is that through which Life is given to the Ideas. It is a kind of living medium or solvent for Intelligence, and in Pisces it dissolves the traces of life into a common imaginal substance. This is the "primordial soup" of *vasanas* (tendencies built up in prior states of manifestation). It is the matrix from which all things grow. Or, rather, it provides the energy and ground for the self-determination of the growth of the universe.

♃♍ is the matrix of the six forces of nature present in Soul. This is the wisdom that appropriates the vision of Ideas and relates the Ideas to a common substance. Jupiter here knows all the interrelations of all forms which can manifest.

♃♋ is the intuitive knowledge of the states of Being or consciousness, for it is the solvent that knows the matrix of all the states as consciousness. This Jupiter holds the kaleidoscope: viewing the interrelation of the states. It is called "all-ground cognitiveness" in the Nyingma Buddhist tradition. (See Herbert Guenther, *Tibetan Buddhism,* p. 118.)

Another way to represent the "twofold" nature of Zeus is to use the ♊–♐ axis. ♃♐ is the principle of the Soul's ideation as a power of the One, which sees only the paradigm represented by ☊♊. ♃♐ represents the actualization of the Void Intellectual Energy: the Mind of the Demiurge, which has "swallowed" a vision of all universes, represented by ☋♐. [*fig. 160*]

♃♊ is the power within the Void Mind to measure out the infinite. Anthony calls this power "an intellectual energy void of name and form." It holds the matrix of all intelligence in the Absolute. This Jupiter is part of the androgynous nature of

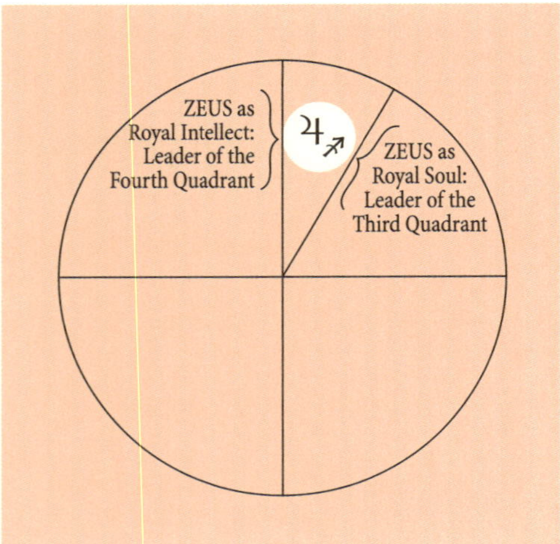

Fig. 158

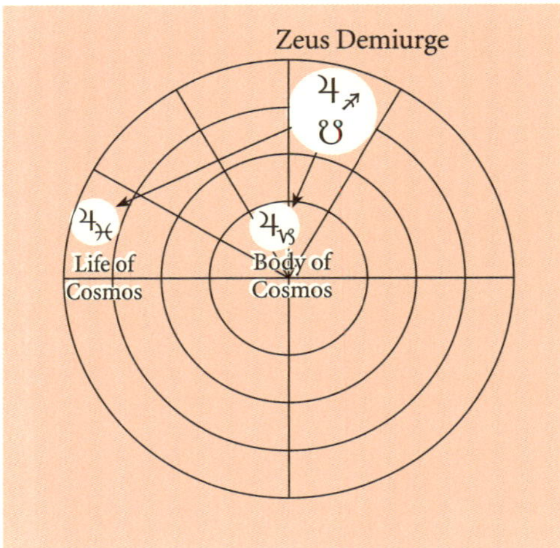

Fig. 159A

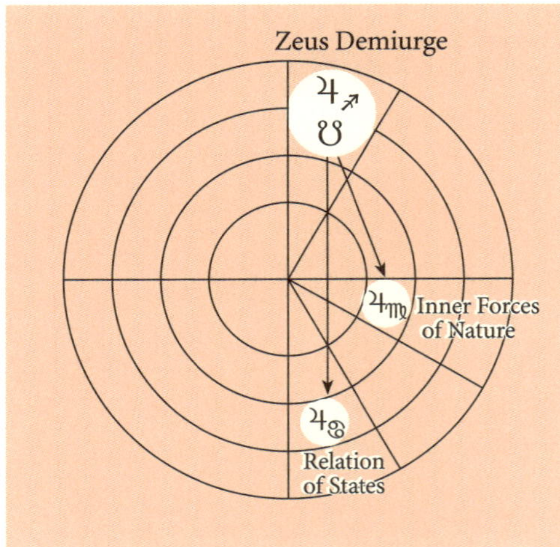

Fig. 159B

Mind, combining wisdom and power. In another symbol system, Jupiter holds the *damaru*—a Tibetan double-headed hand drum that signifies that the only real change is that between beyond-Being and Being. At all levels, the paradox of Jupiter seems to be the paradox of the One and Many. ♃♊ has the Word which stays within the Void, and ♃ speaks this Word in house 9 as ♃♐.

If we put together the two views of Jupiter in ♊ and ♐, we get an insight into the relation of the dignities in the third ring, led by ♃♊ (Soul of the One), and the dignities in the third quadrant, led by ♃♐ (Being of Soul). Perhaps these two represent the function (ring) view and substance (quadrant) view of the Universal Soul. The ring view shows Demiurgic thought productive at every level of

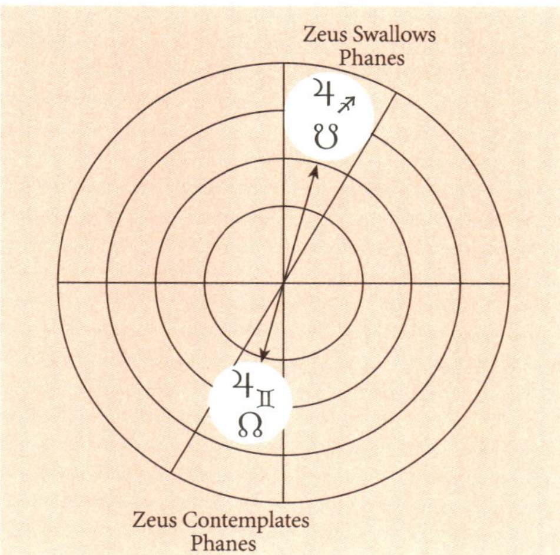

FIG. 160

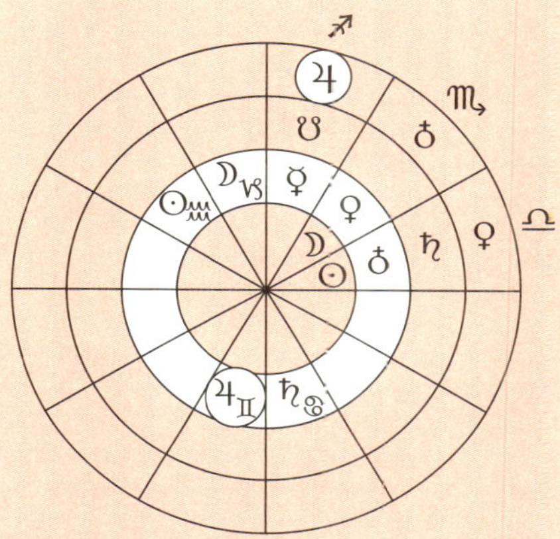

3RD RING • *FUNCTION IN SUBSTANCES*

Unity	♃♊			I Quadrant
Being	♄♋			II Quadrant
Soul	⛢♎	♀♏	☿♐	III Quadrant
Nature	☽♑	☉♒		IV Quadrant

3RD QUADRANT • *SUBSTANCE WITH FUNCTIONS*

Rulership	♃♐			Being
		⛢♏	♀♎	Life
Exaltation	☋♐		♄♎	Intellect
Detriment	☿♐	♀♏	⛢♎	Soul
Fall		☽♏	☉♎	Body

FIG. 161: JUPITER AS UNIVERSAL SOUL—UNITY OF FUNCTION AND BEING OF SUBSTANCE

reality, and the quadrant view is the assimilation of Ideas to Life. [*fig. 161*]

39

EDITORS

The Relation of Jupiter ♃ and Virgo ♍

Soul's life essence is the third ring, running through all the quadrants. There are two ways to view the relation of Jupiter ♃ and Virgo ♍.

- The Idea of Virgo ♍ radiating into Soul evokes the power of reason as Jupiter in Virgo ♃♍. In this view, Virgo ♍ is differentiating the pure life of Soul to actualize its reason, and Jupiter ♃ is the presence of the six forces in the mode of Soul.
- Jupiter in Virgo ♃♍ is intrinsic to the Soul ring, and its act is to incorporate or appropriate the six forces to the soul's use in fulfilling its vision of the Nous as the cosmos. In this view, Jupiter in Virgo ♃♍ is the soul's access to the six forces.

40

EDITORS

Plotinus References on Soul and Memory

For Plotinus on soul and memory, see *Enneads* IV.3.31, the first part of IV.4, and all of IV.6.

41

PLOTINUS

Souls as Logoi of the Intellectual-Principles

But what place is left for the particular souls, yours and mine and another's?

May we suppose the Soul to be appropriated on the lower ranges to some individual, but to belong on the higher to that other sphere?

At this there would be a Socrates as long as Socrates' soul remained in body; but Socrates ceases to exist, precisely on attainment of the highest.

Now nothing of Real Being is ever annulled.

In the Supreme, the Intellectual-Principles are not annulled, for in their differentiation there is no bodily partition; each separate phase possesses identical being within the diversity. It is exactly so with the souls.

By their succession they are linked to the several Intellectual-Principles, for they are the expressions, the Logoi, of the Intellectual-Principles, of which they are the unfolding; brevity has opened out to multiplicity; by that point of their being which least belongs to the partial order, they are attached each to its own Intellectual original: they have already chosen the way of division; but to the extreme they cannot go; thus they keep, at once, identification and difference; each soul is permanently a unity (a self) and yet all are, in their total, one being.

Thus the gist of the matter is established: one soul the source of all; the many founded in that one, on the analogy of the Intellectual-Principle; those many are at once divided and undivided; that Soul which abides in the Supreme is the one expression or Logos of the Intellectual-Principle, and from it spring other Reason-Principles, partial but immaterial, exactly as in the differentiation of the Supreme. (IV.3.5)

42

ANTHONY

Notes on the Solar Logos/Cosmic Soul

Compiled from Astronoesis Working Drafts and Class Transcripts 1/27/84, 2/3/84, and 9/19/84

Interpenetration of Solar Logoi: *Asmita* or I AM of the Sun as Infinite Light

As we ascend to consider the inner intelligence of the Sun soul, the immense vision of Plotinus begins to appear to our receptive minds. The star souls (cosmic souls) are emanations of the Nous; the infinite life-intelligence of the Sun soul is interpenetrating with all the other souls. Through their mutual omnipresence, each cosmic soul's intellection or contemplating constitutes its World-Idea. In its inward ground, each soul is the receptacle of divine beings (Ideas)—for these logoi are the divine radiations of Nous.

The *asmita* principle, or the soul of the Sun, is prior to this sensible Sun that you see, and it is the presupposition of the sensible Sun. Now the soul of this Sun is an infinite formless light. There are other Suns, also infinite formless light; consequently they all interpenetrate each other. This would be the *asmita* or principle of manifestation for any Sun we speak about, any Solar Logos or Ishvara. It is the God that you know—that you hope to know. The soul of the Sun is also referred to in the Buddhist literature as Amitabha—Buddha of infinite light. Here every Sun interpenetrates with every other Sun.

> The light of our world can be allocated because it springs from a corporeal mass of known position, but conceive an immaterial entity, independent of body as being of earlier nature than all body, a nature firmly self-based or, better, without need of base: such a principle, incorporeal, autonomous, having no source for its rising, coming from no place, attached to no material mass, this cannot be allotted part here and part there: that would be to give it both a previous position and a present attachment. (Plotinus, VI.4.8)

If you think of each Solar Logos as a vast entity of light, then you would be speaking of many Solar Logoi in that subtle state where their infinite light extends throughout—to use a spatial metaphor—so that all Solar Logoi interpenetrate one another. Even though we think of their life as infinitely extending and each interpenetrating the other, yet each Solar Logos remains distinct from other Solar Logoi, but includes them within itself. A Solar Logos is a universal Idea and each Idea participates in all the other Ideas and vice versa.

Because of this interpenetration (of souls), the Sun knows the Ideas that each one of these formless lights is conveying to it. The glow, as it were, of all the other pure intelligences interpenetrating with our Sun is the aura emanating into our Sun; it is also the aura emanating into any one of them. We will represent the Sun soul's intellection of Ideas by the selection of logoi that constitute the sidereal zodiac.

A Solar Logos enjoys a bliss of all the other universal Ideas within itself, but remains itself. And the same applies to all the other universal Ideas. They also enjoy the bliss of all the others but remain themselves. If they didn't remain themselves, that is, numerically identical with their own selves, then we would have only one soul. The Solar Logos, or the I AM of any Solar Logos, is infinite light that interpenetrates with all other infinite lights. This interpenetration of all the lights with one another is like the soul is One-and-Many. That results in each Solar Logos having its interpretation of the World-Idea—of what the World-Idea is—by a reflex.

Inerratic Intellection in the Solar Logoi

When the Sun soul interprets to itself its infinite inner vision, through its own universal reasoning, it brings about within itself the archetypal form of the World-Idea—a unique form of those intelligences among the infinite outpouring of Nous—with which it has a particular affinity. This celestial wisdom is symbolized by the Sun's sidereal zodiac and constitutes its intellectuality. Symbolically, the Sun's travel through the cosmos represents to us the universal reasoning through which our Sun too is assimilating facets of the unfathomable divinity of Nous. Within its principle of manifestness—boundless light—arises the formless intellection of the World-Idea.

This infinite light of each Solar Logos assimilates the light from all the other Solar Logoi, and what results from that is its version of World-Idea. The World-Idea at this level is pre-sensible. There is a steady contemplation, and it is like a stabilized ecstasy and rapture that is uninterrupted; but that is the only place you will find this.

A vision of the pure Soul in the Nous is held within the boundless consciousness of the I AM of a Sun soul, and is represented as the inerratic sphere—a version of the Ideas in the Soul. A certain selection of the star intelligences is required in order for this individual *asmita* principle to form the World-Idea for itself. That is, the star intelligences represent the availability to the individual soul, in a separative mode, of the Ideas which are in the transcendent Nous. At that level, there is no sensible world; there is only the sheer presence of these Ideas—the combination of Ideas necessary that is going to produce the sensible world much later.

The Sun's intellectuality is represented in the way *it* interprets the starry system. Each Sun's vision of the Intellect is peculiar to it—includes all the Gods in a unique way. So its participation in the intelligence of the universe (and the other logoi) is a unique zodiac (termed the sidereal zodiac or celestial dragon). Plotinus says that here the soul is the recipient of divine beings. The inerratic sphere represents the intellectual space (locus) for the Sun soul, the ordered plan of the cosmos for the Sun. Its intellectual vision (of the Nous) is represented sensibly by its singular motion about its axis—the pole of Vesta—and the sidereal zodiac is the sphere resulting from the interaction of the universal reason-principles with the soul's own intellectual vision.

The first Reflex, this level of Intellection, already has a threefold plan. The World-Idea is already structured according to this threefold plan. Once that becomes effectively differentiated, it is conjoined with these particular souls—Uranus, Neptune, and Pluto. Uranus rules over the Intellective field, Neptune over the Reasoning phase, and Pluto over every Sublunary phase. The triple mode of knowing, willing, and feeling is derived from (or has its origin in) the contemplation that is being attributed to the Solar Logos—its outer vision of the World-Idea.

The Idea of the world that is to be derived from the contemplation of those Ideas is done by the reasoning power of the soul, not by its intellection. Its intellection grasps the Nous; its reasoning power is going to manifest that Nous as sensible. Again, this is not so far away from your own being—that you have the power to contemplate something, to reason about the inspiration or intuition when it is given to you, and then, so to speak, manifest it.

Overself: Sun Soul and Individual Soul

Each (unit soul) Overself, conjoined with the Sun soul, may also enter into the intrinsic knowing-by-being, and find this divine aura as its innermost essence. There the soul finds itself to be consubstantial with (the Radiation) the All-Soul—Brahman. Only the intelligence of the Nous (God) touches the Soul in its still ground, radiating into it the mystery of Creation.

Associated with the soul of the Sun, this infinite formless light, are many individual Overselves. Remember what Plotinus said: "beings by the side of Being." (VI.4.4) All these beings are by each others' side; there's no sense at this level in saying Overselves. At this level it's one Overself, one Soul from the point of view where light interpenetrates light and all souls are conjoined with one another. But if you speak from the point of view of manifestation, then you've got to say an individual Overself. The contiguity of all beings is prior to manifestation.

The Hindus would call this (Solar Logos) Brahman. And the individual souls are called Atman. And they say Atman is Brahman. At that level we speak of the Overself conjoined to the cosmic soul. Individual soul may later have attributes (*upadhis*), but this never has *upadhis*.

Here [at the level of the soul of the Sun] we speak of one soul or many souls, or one and many. That is the nature of real Being. Similarly, your mind will not exclude my mind and my mind will not exclude your mind. Each mind will remain numerically identical with itself and not be the other.

To define the meaning of Soul as One-and-Many, I have to use the context that the self of the Sun is your soul, and it's also my soul. Now in that case I could see that since that self is one, these many souls are one. But insofar as each and every one of us is projected forth from that self of the Sun, and each one has a historicity that is different from the others, and that has intrinsic powers which are developing which separates it from the others, then it's a many. In this context, and only by using contextual language, can that be understood.

> Again, it is admitted that the particular soul—this 'part of the All-Soul'—is of one Ideal-Form with it, but this does not entail the relation of part to whole . . . (Plotinus, IV.3.2)

When we speak about your Overself and cosmic Overself as conjoined together, your Overself is not a part of the cosmic soul. There cannot be any such relationship of part or whole between souls.

If you experience yourself as this kind of consciousness—which is absolutely immaterial, sizeless, dimensionless, all light—then you can easily understand that the Sun also has something like that and that principle in you is conjoined to the principle which we are calling the soul of the Sun. It doesn't take great ingenuity to see that, but if you don't know yourself as that kind of being, then you could never understand a passage like that.

Each unit soul, Sun soul or human soul, is likened to a crystalline *sphere* of infinite consciousness. In this state of union, all the unit souls enjoy each other's *bliss*. Also, each soul receives the intelligence of the Nous as the pre-manifest World-Idea.

The Idea of the world as it is in the Overself is different from the way we see it here. In the Overself, you have the intellectual Ideas, the intellectual cosmos. There the Overself is robed in the glory of the universal Ideas. The experience of the intellectual world or the cosmos in the Overself is the world in a pristine form. It's the Idea of the world to be—not the way we see the world, but in its pristine form. There the archetype of everything exists: There's no coming, no going, no creaturely existence—none of that is there in the intellectual cosmos. This is the experience of the World-Idea within the Overself. But that's the same as saying that the World-Idea as the Overself experiences it and the World-Idea of the Solar Logos are the same. So if the way the human soul, the Overself, experiences the World-Idea in this pure golden hue is the same as how the Solar Logos does, this is what is meant by saying that the Overself has contemplation or intellection of the World-Idea. The way the sage perceives the World-Idea is not the way we do. So the sage can separate the sensible manifestation of the World-Idea from the archetypal Idea. He can actually separate them. He can ignore the sensible and gaze or contemplate on the intellectual cosmos within the soul.

"The God in the sun is the 'I' in me"—this put tersely is the essence of man's relationship to divinity. A whole book may be needed to explain it, a whole lifetime to get direct experience of its truth as insight.

PAUL BRUNTON, *NOTEBOOKS*, V. 16, PT. 1, P. 1

43

ANTHONY

The Animate and the Alien

Within the cosmos — the vast thought-form in the cosmic soul—a particular psychosomatic organism comes to existence, and seeks for Soul. A body and mentality organized by the cosmic circuit, composed of the life traces, reason-principles, and elements of the Earth, draws to itself this projecting ray of Soul. It seeks to perpetuate itself, to appropriate what it can of Soul's life. The unit soul in its true nature is impartible, but what this intruder (cf. Plotinus, VI.4.14–15) receives is this phase, a divisible life or ray which is some form of communion with Soul. Reciprocally, the nature of the understanding phase of the soul has an affinity for the forms of experience which can be gotten through this particular body. The natural reason-principles which constitute the organism, and the reason-principles which the understanding of the soul has made its own, have an inherent linkage within the intelligence of the World-Idea. In this union of the souls' projected powers—that phase which is becoming divisible—with the organization of a body, we have the coming to birth of the living Animate, or what we term the "coagulated ego complex." When this phase of Soul has been associated with a body, the two are so intertwined as to be almost indistinguishable, and we are said to have become the dual thing—organism and light. As part of the constantly reconstructed World-Idea or world image, the body/mentality is a receptor for the world image, while the light of understanding projected into the entity interprets the meaning (i.e., reason) which is inherent in that manifestation. Through this constant transformation of the understanding, the understanding is itself evolved and is "withdrawn" to the higher soul when the body/mentality has run its course.

As a further consequence of the Soul's omnipresence, we have to realize that even now "we are not apart." The sense in which we cannot become the authentic Mind is that we are always that Mind and no addition can make Mind more authentic. The infinite and unfailing power of Soul is ever-present, but the compulsion which focuses our attention on the image keeps us ignorant of what we truly are. What is called non-Being is the appearance of the world image which includes the ego, which is always becoming, never truly Being. The compulsive tendency of the appearance—a thought construct—to appropriate some image of the soul's awareness, to live and call itself "I," is coupled with the corresponding desire for a separate existence which is lodged in the soul's divisible phase. The identification of these two forms the ego, [and] results in the alienation from the True Soul; thenceforth the "We" includes not only the higher, but [also] this other. We must make a clear distinction here, [and say] that the divisible tendencies and manifesting faculties of Soul *cannot* be destroyed, for they are intrinsic to the nature of the I AM, but the compulsive desire within it to separate from the Real (i.e., egoism) can and eventually must be destroyed. The desire in the reproductive soul, and the corresponding compulsive, attractive habit-energies which are lodged in the organized body, have to be "put aside." When this is achieved, appearance of the cosmos is not thus annulled, but the relation to it changes; and as Plotinus points out, for the sage, side by side with the Authentic Intellection can occur the image of that. Lessening of the alienation, the separative egoity in us, is a primary means to realize the God within us—the I AM—and the reality of Mind. "By the lessening of the alien in you, you increase. . . . engaged in the alien [i.e., the ego—eds.], you will not find the All." (Plotinus, VI.5.12)

44

ANTHONY

Soul and the Idea of Man

We speak about this Soul as projecting from itself a unit of life or a stream of life which we will call the unit of life. Now this is what reaches all the way down to the Earth and starts out as an amoeba and goes through all the various kingdoms of nature, rising up to the level of man. When this stream of life reaches the level of human, it has entered the reasoning phase. These two together, the transcendent Overself and that emanant from the Overself which we're now calling that individual ego, is the "We." (transcript, 3/9/84)

When you say "unit of life," you're not speaking about man, you're speaking about soul. The term is anonymous. A unit of life could be any sort of life. It could be the idea of an angel. It doesn't have to be the idea of man. But once the unit of life marries itself, or weds itself to the Idea of Man, then that soul becomes Man. Then you have to go down that road. Remember Plato's remark? Unless you've heard of the Idea of Man your soul will not become it. It must at some time have heard of it—that means the Nous. (transcript, 9/5/84)

45

EDITORS

Universal Man

Now just as these three [Unity, Intellectual-Principle or Being, and Soul] exist for the system of Nature, so, we must hold, they exist for ourselves. I am not speaking of the material order—all that is separable—but of what lies beyond the sense realm in the same way as the Primals are beyond all the heavens; I mean the corresponding aspect of man, what Plato calls the Interior Man.

PLOTINUS, V.1.10

Anthony wrote: "Considered together, the last four houses also represent the macrocosmic principles of Universal Man. They are an image of the Idea (of Man) in the Nous." (Notes and Papers, p. N4) He tentatively correlated the last four houses (Sagittarius, Capricorn, Aquarius, Pisces) of the system of Nature with the celestial, empyrean, aerial, and aquatic Man, with the terrestrial Man as the final effect. (See figure 98 on page 236, which is based upon a drawing in Anthony's Notes and Papers.) Thomas Taylor cites Proclus as his authority in these orders of "universal man":

> It is necessary, says Proclus, that immovable causes of all things which have a perpetual subsistence in the universe should presubsist in the intellect of the fabricator of the world. . . . Hence, of man so far as man, and of every individual form in animals and in plants, there are intellectual causes; and the progression of all

things from thence is not immediately into these material genera.... From *man itself,* therefore, or the ideal man in the demiurgic intellect, there will be, in the first place, a certain celestial man; afterwards an empyrean, an aerial, and an aquatic man; and, in the last place, this terrestrial man. All this series of form is perpetual, (the subjection proceeding into that which is more partial,) being suspended from an intellectual unity, which is called *man itself.*" (note 1 to Plato's *Parmenides,* p. 42)

Anthony also correlated houses 9–12 with the four worlds of the Kabbalah—the World of Emanation (*Azilut*), the World of Creation (*Beriah*), the World of Formation (*Yezirah*), and the World of Effect (*Asiyyah*). [*fig. 162*] The three unmanifest Primals (One, Intellectual-Principle, Soul) he correlated with the unmanifest principles, *Ain, Ain Soph, Ain Soph Aur.*

In his Nature material, Anthony had noted the following passage from S. Halevi, *Adam and the Kabbalistic Tree:*

> According to Kabbalist tradition God generates, out of the Void of Non Existence, wherein He is All and Nothing, the first state of Unmanifest Existence. From this World without End, crystallizes a realm of Limitless Light, in the midst of which, there emerges a point of no dimensions, called the First Crown. These three states of Unmanifest Existence become the negative background to the positive universe that streams through the primal point of the First Crown to evolve into the archetypal world of Emanations. This manifestation on its completion is known as Adam Kadmon, that is, the Primeval and Universal Man. Although composed of the ten aspects of the Creator, Adam is the symbol of a unified working whole, the first man being a living image of the relative universe. (p. 15)

The "World Without End," or *Ain Soph,* is the Intellectual-Principle, while the "realm of Limitless light" or *Ain Soph Aur* is the inner realm of the Universal Soul, which we have symbolized by the third quadrant. The first point, the Crown of the World of Emanations, is the Demiurge—Jupiter in Sagittarius considered as the one Logos or principle of manifestation. As such, the Demiurge is Divine Thought, that divine attribute whereby the Void Mind is brought into Form. Whether we say that the Demiurge is thinking, imagining, or manifesting the formless Ideas, the universe is a direct reflex of its inner Vision. In Shaivism, the Demiurge is called *Shabda Brahman,* the primordial

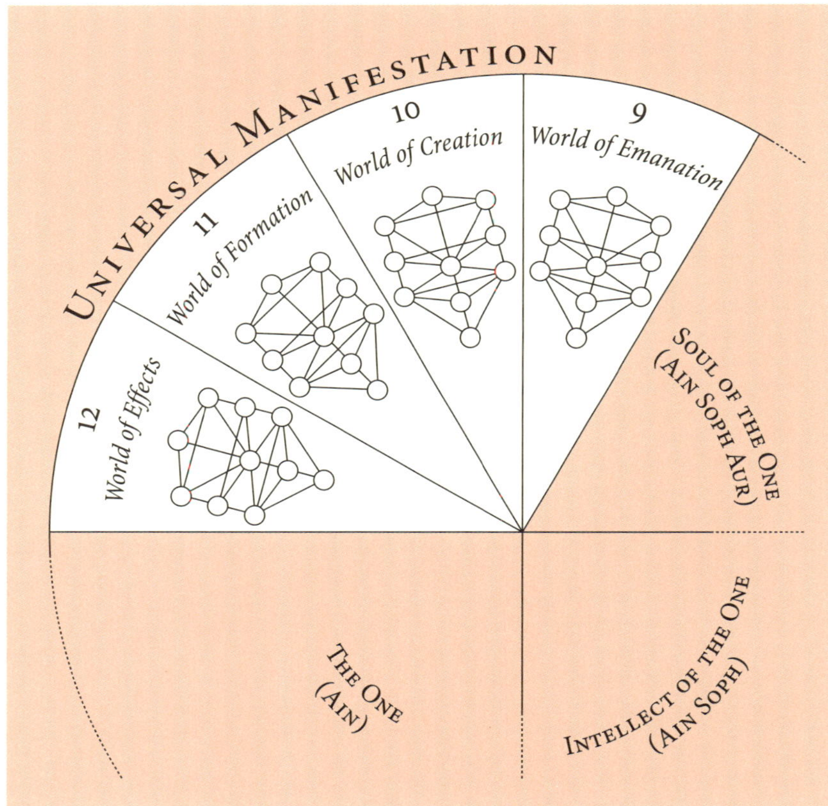

Fig. 162

point (*bindu*) from which arises the manifest levels of the Word (*vak*). (Note that the Celestial Unity of the Grand Man, the 9th house, the house of the Demiurge, is opposite, and thereby rooted in, the 3rd house Idea, that of the androgynous Adam Kadmon.)

Continuing with Halevi on the four Adams:

> 'In the beginning,' that is, when Adam Kadmon had been *emanated* [editors' emphasis] into existence, God created heaven and earth, time, the firmament and the elements. He also brought forth plants and living creatures. On the sixth day of Creation God said 'Let us make (a) man in our image' and so a second Adam came into existence who was the perfection of Beriah, and its steward. Later, in Gen. 2:7, 'The Lord God formed man' and this third Adam became the synthesis of the World of Formations. When the yeziratic Adam fell with his Eve, he descended into Asiyyah, the world we live in. These four Adams are an expression of the four levels within the azilutic Tree of Adam Kadmon. (*Adam and the Kabbalistic Tree,* p. 33)

46

EDITORS

House 10 References

Anthony comments further on this substratum of universal manifestation: "We can also equate this with the *Prakriti* of the Samkhya system and the seven grades of matter evolving from it, with the exception that in the Plotinian system, all the differentiations which evolve out of *Prakriti* are or were originally in-formed by the soul, for the whole world is convolved in and by the soul." (Notes and Papers, p. N7)

The 10th house indicates the causal state of universal manifestation, the first principle of existence of the World-Idea. Ideas do not need a locus in order to be, but they do need one in order to exist or to manifest. Manifestation requires a Place, but this is not Matter in the materialistic sense: Matter is the externalizing power of the Mind, or the infinite locus of mental space. As Corbin (*Creative Imagination*) puts it, "Creation is an act of the divine imaginative power" which is "essentially a theophanic Imagination." (p. 182) Every moment there is a recurrence of creation likened to a "sigh of Divine Compassion." (p. 201)

This first state of Ideation is characterized as the infinite geometric structure and stratification of every universe, the essential and unchanging principles of manifestation. At this level the World-Idea is geometrically infinite, shape without extension. Shimon Halevi (*A Kabbalistic Universe* and *Adam and the Kabbalistic Tree*) describes this second "world" as the World of Creation. The Bible describes this separation of heaven and earth, the Abyss (the Void nature of primal matter) being the receptacle for the imposition of order. (cf. Genesis 1:2)

There is a stratification of the unmanifest Reason-Principle of the All as it is superimposed on the principle of Difference. Direct mystical experience of this stratification is the purpose of Buddhism's famous four formless absorptions: infinite space, infinite consciousness, emptiness, and the peak of cyclic existence. (cf. J. Hopkins, *Meditation on Emptiness,* p. 207)

47

EDITORS

Applying Three Contexts to Saturn ♄ as the Principle of Matter

Exploring the placements of Saturn ♄ in the metaphysical chart illustrates how the chart can be used to map out a philosophic problem. This particular example also provides an opportunity to discuss one of the most difficult notions in the Perennial Philosophy—Otherness or Matter. Earlier discussions of this notion in *Astronoesis* include the discussion of Saturn in Aries on pp. 64, 72–75 in the One: Symbolization; p. 143 on Capricorn in "The Twelve Idea-Principles"; Note 15, p. 314, "Primordial Matter References"; and Note 46 "House 10 References."

In the Plotinian view, matter is "empty": absolutely without determination, quality, substance, or actuality. It provides the "negation" necessary for anything to be. To be is to stand out from the One, to *ex-ist* (Latin, literally: "to stand out"). A principle of Otherness is required to differentiate Being from the One. This presence of Otherness *within* the One-without-a-second is what makes the One truly remarkable.

Anthony equates this principle of Difference or Otherness (Saturn in Aries ♄♈) with pure undetermined space or "the formlessness of emptiness." (*Astronoesis,* p. 74) Otherness or Emptiness in this sense is the primary focus for the Madhyamika system in Buddhism. That there is nothing which exists isolated in and for itself, that "all self-nature is empty" (*svabhava sunya*), is the teaching of the Heart Sutra.

Three contexts in the chart illustrate three views of "matter."

A. The whole series of Saturn ♄ dignities in different signs (the henad series) illustrates different meanings of the principle of Difference or Otherness.

B. Led by Saturn in Aries ♄♈, the body monad ring provides meanings for matter as pure receptivity to Forms at every level.

C. As the ruler of the fourth quadrant, Saturn in Capricorn ♄♑ provides a symbol for the Being of the fourth quadrant system of Nature. Here Saturn ♄ represents the Being of non-Being, matter as the underlie of all appearances in manifestation.

A. Saturn Henad Series: The Principle of Otherness

The ♄ henad portrays the meaning of "matter" as the principle of Otherness or Difference. Unfolding the hierarchy of positions of ♄ as Body, Nature, Soul, Intellect, Life, and Being shows the central meaning of difference in a spiral pattern from the first to the fourth quadrant. [*fig. 163; also fig. 33, p.75*]

Being: ♄♑ represents the place of manifestation, becoming, momentariness, separateness, discreteness. ♄♑ is the receptivity of matter to be informed by the Ideas, the principle of discreteness or distinct boundaries within which the universe can appear.

Life: ♄♒ represents the function of ever-becoming: the laws of relative causality, what Buddhism calls "combined dependent origination," the order of the life of the universe. Saturn here provides the conditions by which consciousness can appear as a center of experience. However, to mistakenly take the sensible appearance as something in its own right is to be separated from reality.

Intellect: ♄♎ as exaltation is the leading planet in the Intellect ring of the Soul quadrant. Its presence here indicates that the Soul contains within itself a form of intellection based on a differentiation and hierarchy of reason-principles. This intellection in Soul is *dia-noia*—literally divided Nous—which is the function of discriminating the Ideas.

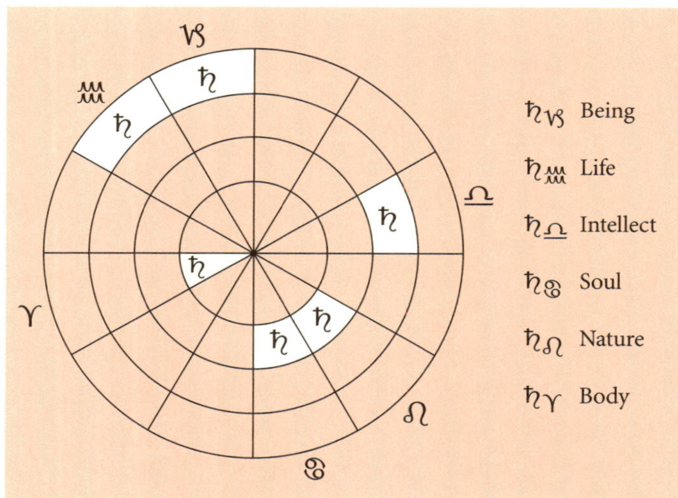

Fig. 163: Henad of Saturn ♄

Soul: ♄♋ in the third ring represents the principle of multiple states of the Soul, and the universal power of Soul that differentiates the unific consciousness of the One into multiple states.

Nature: ♄♌ is the vision of the Soul in the Nous, Soul as living receptacle of the Ideas. Here Saturn is the nature of Soul that remains inviolable through all its differentiations.

Body: ♄♈ indicates the principle of Difference or Otherness found in the One. Here it does not produce any actual difference, so we speak only of the principle of illusory distinction. Anthony writes of this principle in the chapter on the One as "pure Space" (p. 72) or "the 'ground' for the emanation from the Unity." (p. 73)

B. Saturn and the Body Monad

Following the Body ring offers another view of matter. Each of the planets in the Body monad represents a mode of receptivity, a place or embodiment for its substance or sign, and at the same time a "densification" of the primal absolute "space" of ♄♈. Continuing by quadrant with the dignities in the Body ring, ♂♋ and ♀♍ represent the "intelligible matter" of the Intellectual-Principle, ☉♎ and ☽♏ concretize the light of the Soul, and ♃♑ and ☿♓ are the principles of Body within the system of Nature. [*fig. 164*]

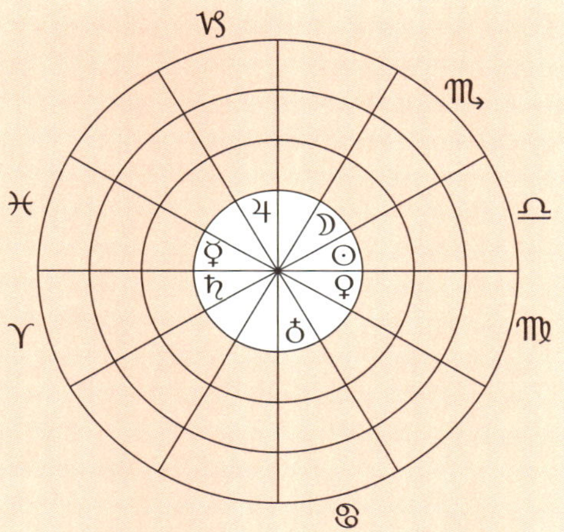

♁♋	Separates particular kinds of concrete manifestations of the four states of consciousness
♀♍	Receptacle for divine power, six holy forces
☉♎	Embodiment of reason-principles or Forms in Soul
☽♏	Reflector and container for images in Soul
♃♑	Geometric laws governing the form of space
☿♓	Embodying individual

Fig. 164: Body monad

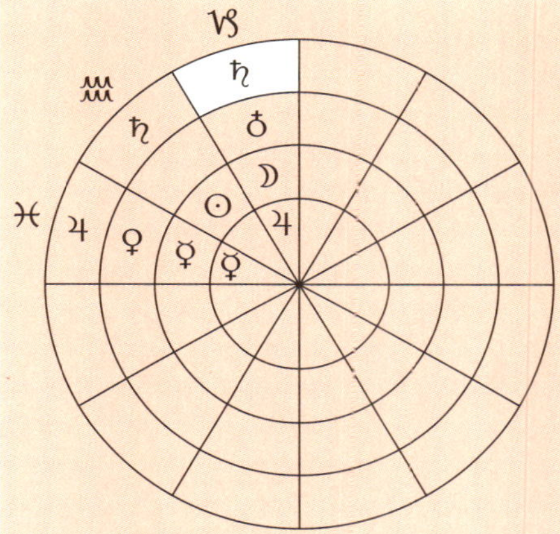

♄♑		The fundamental *Being* of all manifestation is becoming
♄♒	♃♓	*Life* activity of manifestation: actual function of manifestation and beginningless matrix of causes and conditions
♁♑	♀♓	Objective *Intellect* of universal World-Idea and inerratic *Intellect* of every soul in the cosmos
☽♑	☉♒	*Soul* of universe: karma (memory) and ideation (imagination)
☿♓		Unit of life perpetually evolving within *Nature*
♃♑	☿♓	*Embodiment* of universe and individual as Ideas in World-Mind

Fig. 165

C. Saturn in the Fourth Quadrant: The Functioning of Manifest Otherness, or Becoming

♄♑ as the ruler of the fourth quadrant gives a third view of matter. It represents the 10th house "tentatively illumined matter" (cf. Plotinus VI.7.7), or a delimitation of the absolute space of ♄♈. We may view the dignity of ♄♒ as equivalent to the principle of combined-dependent origination. The other dignities in the fourth quadrant are all principles that appear in the context of this primal matter. [*fig.* 165]

48

EDITORS

72 Powers and 360 degrees

Late in his life, Anthony was trying to derive the 360 degrees of the zodiac from the interaction of the 72 powers of the central Sun with the Five Great Elements. As a context for 72, the intelligence of the Universal Soul or Demiurgic Mind (World-Mind) is said to include 12 deific aspects, called the Liberated Order of Gods in the Orphic theology. Each deific aspect includes or has available all of the 6 forces (*shaktis*) which Proclus termed (in *On the Theology of Plato*) Being, Life, Intellect, Soul, Nature, Body. The combination (12 x 6) gives 72 aspects of God, or 72 Divine Names. These 72 represent the (objective) powers of the Demiurge, and are the powers of the central Sun, and thus are non-individual. The 72 are the powers of the God-Called-Names, the God whose divine attribute is to transform the unmanifest into form. The individual soul draws upon these powers from the Demiurge in order to realize the Idea of Man.

In the metaphysical chart, there are two streams associated with 5. First stream: 5 is associated with the 5th house Idea of the 5 Dhyani Buddhas or the Idea of Man. Insofar as the soul includes the combined Wisdom and Power of the 5 Dhyani Buddhas and the 72 gods, there is an intelligible cosmos within the Soul—the Idea of Man refers to this intelligence, and this is connected with the inner nature of the central Sun, the Sun in Aquarius. Second stream: The 5 Mahabhutas or Great Elements is the 12th house Idea. The Universal Soul also manifests the Ideas into the realm of the 5 great energy fields or Mahabhutas, bringing about a sensible expression of Wisdom and Compassion. Each individual unit of life participates in both these streams, the intelligible and the sensible in the cosmos. The 72 powers combined with the 5 Wisdoms give the idea of the 360 in the soul; the 72 powers combined with the 5 Mahabhutas gives the 360 in the sensible zodiac.

49

EDITORS

The Four Levels of the 12th House and the Divided Line

The four levels (rings) of the 12th house represent four levels of knowledge established by the 11th house Sun as cosmic soul. Anthony often made a correspondence between these four levels and the four levels of Plato's Divided Line as discussed in "What Is an Idea?" (p. 125).

50

EDITORS

Mercury in Pisces ☿♓

Nature Mercury in Pisces ☿$_{♓N}$

If we represent the individual soul as the ☿$_{♓N}$ our diagram [*fig. 166*] shows that included in its constitution are three components. First, ☿$_{♓N}$ is part of the Soul ring, and participates in the nature of Soul. It is an inviolable consciousness, and all the powers along the Soul ring can be present to the individual. Secondly, ☿$_{♓N}$ is part of the series of ☿ planets, and as such is a logos of Ideas, representing the coordination and instantiation of Ideas. Thirdly, ☿$_{♓N}$ is an aspect of Pisces, the continual flux of appearance, and has a historical relation to the evolving World-Idea.

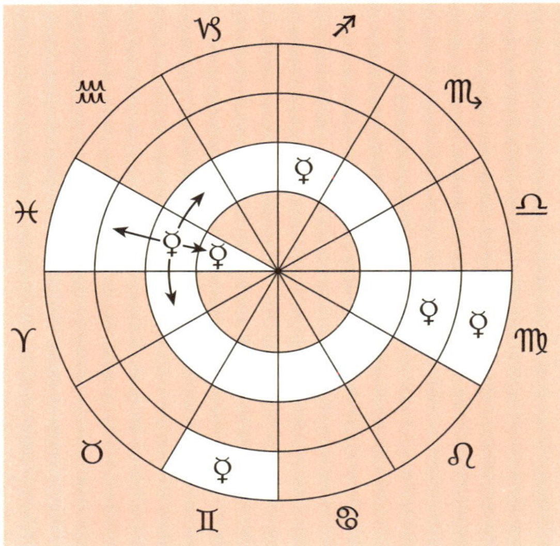

FIG. 166

The Double Mercury in Pisces ☿$_{♓N}$ and ☿$_{♓B}$

From the point of view of soul, the cosmos is the soul's own activity, and the planetary spheres are experienced as the rational powers of the Sun soul. These unembodied rational soul powers are symbolically contained in the ☿$_{♓N}$ at the end of the Soul ring. The rational powers of the Sun soul can be called "functional images," for they are the presence and transformation of the metaphysical principles represented in the Soul ring. These powers function as a harmonious unit, an undivided mind in continuity with the Divine. Plotinus says in VI.4.9: ". . . where these powers appear, their . . . source must itself be omnipresent as an undivided whole." Each planetary soul expresses a divine power, and is a soul in its own right. The Earth, and each of the other planetary souls, participates in the integral and seamless functioning of the whole planetary system.

From the point of view of body, the cosmos is a manifest body. The unified functioning of the cosmos is duplicated within the Earth's own planetary imagination. These embodied powers, operating as instinctual and archetypal functions, are represented by the ☿$_{♓B}$ in the Body ring. When we split the body from the soul we have materialism or Descartian dualism, and astronomy as other than "Astronomia." To put these two perspectives represented by the double Mercury in Pisces (☿$_{♓N}$ ☿$_{♓B}$) together, soul and body, is the point of view of philosophic astrology. In this context Anthony points to the sensible universe as a symbol for the mind which thinks it into existence.

51

PAUL BRUNTON

Evolution and the Unit of Life

EDITORS: *"The Third Phase of Soul" (p. 252) is a commentary on the following excerpt from Paul Brunton's* Notebooks.

What is the use, ask many questioners, of first, an evolution of the human soul which merely brings it back to the same point where it started and second, of developing a selfhood through the long cycles of evolution only to have it merged or dissolved in the end into the unselfed Absolute? Is not the whole scheme absurdly useless? The answer is that if this were really the case, the criticism passed would be quite a fair one. But it is not the case. The unit of life emanated from the Overself begins with the merest glimmer of consciousness, appearing on our plane as a protozoic cell. It evolves eventually into the fullest human consciousness, including the intellectual and spiritual. It does not finish as it began; on the contrary, there is a grand purpose behind all its travail. There is thus a wide gulf between its original state and its final one. The second point is more difficult to clear up, but it may be plainly affirmed that man's individuality survives even in the divinest state accessible to him. There it becomes the same in quality but not identical in essence. The most intimate mental and physical experiences of human love cast a little light for our comprehension of this mystery. The misunderstanding which leads to these questions arises chiefly because of the error which believes that it is the divine soul which goes through all this pilgrimage by reincarnating in a series of earthly forms. The true teaching about reincarnation is not that the divine soul enters into the captivity and ignorance of the flesh again and again but that something emanated from the soul, that is, a unit of life that eventually develops into the personal ego, does so. The Overself contains this reincarnating ego within itself but does not itself reincarnate. It is the parent; the ego is only its offspring. The long and tremendous evolution through which the unit of life passes from its primitive cellular existence to its matured human one is a genuine evolution of its consciousness. Whoever believes that the process first plunges a soul down from the heights into a body or forces Spirit to lose itself in Matter, and then leaves it no alternative but to climb all the way back to the lost summit again, believes wrongly. The Overself never descends or climbs, never loses its own sublime consciousness. What really does this is something that emanates from it and that consequently holds its capacity and power in latency, something which is finited out of the Overself's infinitude and becomes first, the simple unit of life and later, the complex human ego. It is not the Overself that suffers and struggles during this long unfoldment but its child, the ego. It is not the Overself that slowly expands its intelligence and consciousness, but the ego. It is not the Overself that gets deluded by ignorance and passion, by selfishness and extroversion, but the ego.

The belief in the merger of the ego held by some Hindu sects or in its annihilation held by some Buddhist ones, is unphilosophical. The "I" differentiated itself out of the infinite ocean of Mind into a distinct individuality after a long development through the diverse kingdoms of Nature. Having thus arrived at consciousness of what it is, having travelled the spiral of growth from germ to man, the result of all this effort is certainly not gained only to be thrown away.

Were this to happen then the entire history of the human race would be a meaningless one, its entire travail a resultless one, its entire aspiration a valueless one. If evolution were merely the complementary return journey of an involutionary process, if the evolving entity arrived only at its starting point for all its pains, then the whole plan would be a senseless one. If the journey of man consisted of nothing more than treading a circle from the time of his emergence from the Divine Essence to the time of his mergence back into it, it would be a vain and useless activity. It would be a stupendous adventure but also a stupid one. There is something more than that in his movement. Except in the speculations of certain theorists, it simply does not happen.

The self-consciousness thus developed will not be dissolved, extinguished, or re-absorbed into the Whole again, leaving not a trace behind. Rather will it begin a new spiral of evolution towards higher altitudes of consciousness and diviner levels of being, in which it will co-operate as harmoniously with the universal existence as formerly it collided against it. It will not separate its own good from the general good. Here is part of the answer to this question: What are the ultimate reasons for human wanderings through the world-process? That life matters, that the universe possesses meaning, and that the evolutionary agonies are leading to something worthwhile—these are beliefs we are entitled to hold. If the cosmos is a wheel which turns and turns endlessly, it does not turn aimlessly. Evolution does not return us to the starting point as we were. The ascent is not a circle but a spiral.

Evolution presupposes that its own possibility has always been latent within the evolving entities. Hence the highest form is hidden away in the lowest one. There is development from the blindly instinctive life of animals to the consciously thinking life of man. The blind instinctive struggles of the plant to sustain itself are displaced in the evolutionary process by the intelligent self-conscious efforts of the man. Nor does this ascent end in the Vedantic merger or the Buddhistic annihilation. It could not, for it is a development of the individuality. Everywhere we find that evolution produces variety. There are myriads of individual entities, but each possesses some quality of uniqueness which distinguishes it from all others. Life may be one but its multitudinous expressions do differ, as though difference were inherent in such expression.

Evolution as mentalistically defined by philosophy is not quite the same as evolution as materialistically defined by Darwin. With us it is simply the mode of striving, through rhythmic rise and fall, for an ever fuller expansion of the individual unit's consciousness. However, the ego already possesses all such possibilities latently. Consequently the whole process, although apparently an ascending one, is really an unfolding one. (*Perspectives*, p. 368)

52

ANTHONY

The Sheaths/Subtle Bodies and Plato's Divided Line

EDITORS: *This note introduces many terms with which readers may be unfamiliar. Anthony integrated ideas from the esoteric traditions of East and West into his understanding of the soul's vehicles or sheaths. Sources for more information on these terms include: G.R.S. Mead,* The Doctrine of the Subtle Body in Western Tradition *and also his essay "The Soul-Vestures" in* The World-Mystery.

Within the cosmos—the combined effect of Universal Soul's intellectual illumination or blueprint of the Universe, and a Sun soul's rational elaboration of its portion—we place the operation of the individual soul which is "becoming divisible." The individual human soul, itself having phases of intellectual vision—Saturn in Leo ♄♌—and rational powers—Jupiter in Virgo ♃♍—puts a part of itself into the realm of manifestation.

The third or the divisible phase of Soul, which we might also refer to as its principle of manifestation, is the emanation of a subtle light which is the link of the higher phase of individual soul with its lower phase. The cosmos or the World-Idea which is pictured without is a reflex of that which is received within, i.e., Nature. Nature is the medium through which our soul, or rather the light from it, is associated with various vehicles in the cosmos and our personalities are bound up with the will of cosmocrators. In the three phases of the individual soul, we symbolize this light of the soul—whether pertaining to a cosmic or an individual soul—as the Mercury in Pisces ☿♓. The Mercury in Pisces phase of an individual soul is a dual product of the cosmic circuit and what is evoked from the inner nature of the individual soul. This lower soul is provided with various sheaths, (i.e., Moonlike, etc.) that are developed in and by the cosmic soul's activity. This light must be conceived in a way as the permanent emanation of the higher soul with a reflex of its pure reasoning and intellective phases, and also as evolving through a vast series of lives to a state where it is a perfected image of its prior, endowed with all the excellent qualities of the Ideas, and finally actualized as a sage. This is the imitation of God spoken of by Plato.

From its first association with the heavens, the soul descends to more and more gross vehicles—or, rather, contracts its light in its association with grosser and grosser vehicles. Or, we may say that the vehicles are the participation in the various levels of the cosmos *by* the light of Soul.

As the ray of Soul descends, its participation in the various levels of the cosmos gives it these denser and denser mediums of experience. So this second "soul" is formed in part from an impartible essence—the light of higher Soul—and that which is becoming divisible in the "sheaths" or vehicles. This, we feel, is the real reference of the famous passage in the *Timaeus* where the soul is fabricated of the impartible and partible essence, sameness and difference. The impartible is pure Soul. The becoming divisible is the cosmic circuit's imaginal production. The participation and blending of the one with the other is the second soul—that which is evolving *in* the cosmos.

The ray of our soul first finds its place within the universal logos, and then, at the second level, with the fixed sphere of this Sun or Solar Logos. This is spoken of as the descent of the spark or light through the various spheres. The participation in each successive sphere is denser and denser, and will constitute the various vehicles of the Soul. These levels of vehicles correlate to the five aspects of universal manifestation. [*fig. 167*]

PLOTINUS, IV.3.15: *The souls peering forth from the Intellectual Realm descend first to the heavens and there put on a body; this becomes at once the medium by which as they reach out more and more towards magnitude (physical extension) they proceed to bodies progressively more earthy. Some even plunge from heaven to the very lowest corporeal forms; others pass, stage by stage, too feeble to lift towards the higher burden they carry, weighed downwards by their heaviness and forgetfulness.*

PLOTINUS, IV.3.17: *Various considerations explain why the souls going forth from the Intellectual proceed first to the heavenly regions. The heavens, as the noblest portion of sensible space, would border with the least exalted of the Intellectual, and will, therefore, be first ensouled, first to participate as most apt; while what is of earth is at the very extremity of progression, least endowed towards participation, remotest from the unembodied.*

All the souls, then, shine down upon the heavens and spend there the main of themselves and the best; only their lower phases illuminate the lower realms; and those souls which descend deepest show their light furthest down—not themselves the better for the depth to which they have penetrated.

Augoiedes/*Anandamaya Kosha*

The Mercurius phase of Soul as the pure form of manifestness—that is, universal consciousness as our "I AM"—is first identified with the highest reaches of heaven. There resides the intelligence of the cosmos in its pure (*sattvic*) aspect, called the reason-principles or Ideas variously combined to form the universe. Here the Mercurius assumes substantiality as the pure joy of existence called the *anandamaya kosha,* or the Robe of Glory. However,

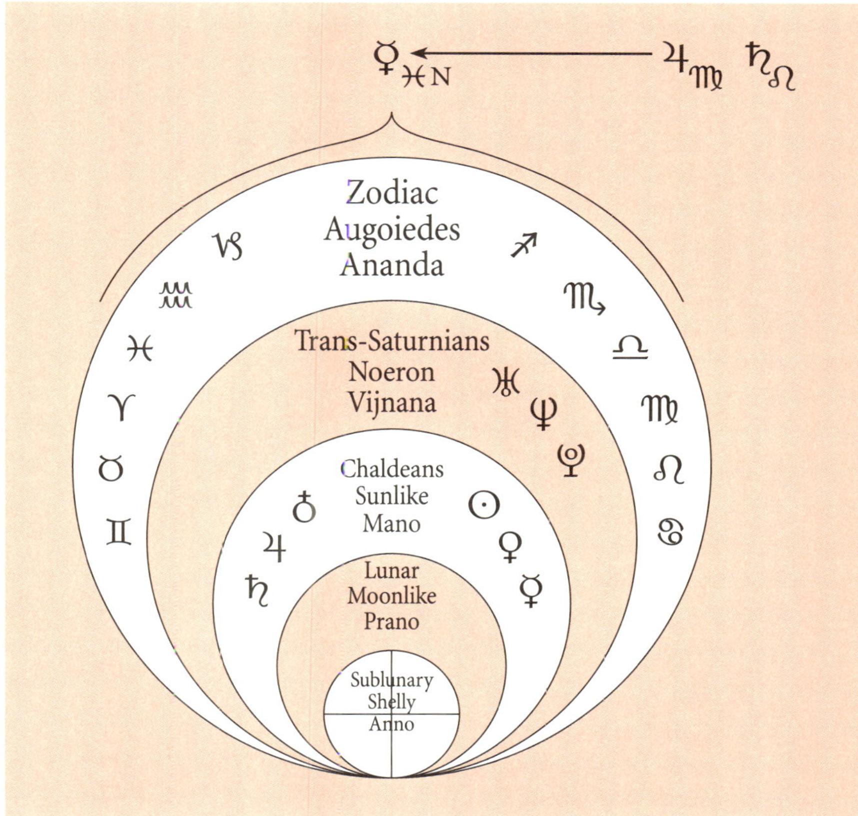

FIG. 167

the principle of pure light can also be considered *a priori,* that is, before it manifests a cosmos. Through the function of manifesting, the soul takes itself to be embodied in the reason-principles. The I takes itself to be this mind, the self-affirming certitude of "I am the World-Idea." At this level, the I or *asmita* is the substratum or basis for the differentiation of the appearances within itself, i.e., the lower principles which evolve from Reason. The soul that still has its wings is associated with a vestment of luminous reason-principles that form the "Robe of Glory" or Augoiedes, the radiant body, and may truthfully say, "The world is my idea." For here, the intelligence of the stars is the locus for the soul's first association with the cosmos, and it is infused there with the intelligible form of the cosmos, or World-Idea. Within the realm of manifestation, this universal consciousness is the higher knower in us. At the station of Reason and its act—that is, the clear shining of the intelligence which contains the Ideas for the formation of the world—the highest phase of the Mercurial soul is in harmony with the operation of the cosmic soul and Universal Soul in unfolding the World-Idea for itself.

The first vehicle needs to be explained in more detail. Here we are speaking of a soul associated with a particular star or group of stars. (From above, we are speaking of the intellectuality of the star.) Each star, as a part of the universal reason-principles, represents the immutability of a particular version of the Idea of Man of which there is some finite number. The intellectuality of each star is peculiar to it, but in their totality these stars are the archetypal illumination of the cosmos, the reason-principles of the Universe emitted by Universal Soul, or the 360 degrees of existence. This is the sensible analogue to the Intellectual-Principle *per se*—in which each Idea is a unique form of the whole. These stars are the intelligences that govern the world. Each star is itself a soul, the soul of a god. Its body is the body of a supercelestial god without any hiatus or separation. When our soul returns to its star, it participates again in the intellectual godhead of that star. The peculiar star with which the ray of our soul associates must be in accord with the peculiar vision of Nous which our soul has in its highest phase (the image to which it tended). (cf. Plotinus IV.3.13) The light of the individual soul, associated with that version of the Idea of Man represented by the reason-principles of that star, becomes the authentic essence of the second soul called the Celestial or Augoiedes.

Noeron Body / *Vijnanamaya Kosha*

The ray of Soul in its universality, dressed with its robe of reason-principles, participates in the spiritual powers of absolute knowing, willing, and feeling, through which it modalizes or particularizes itself as this individual soul at the level of *vijnanamaya kosha* or the Noeron body. This Nous or intelligence of the cosmos as manifested through an individual soul includes that particular soul's role in the World-Idea, and constitutes its spiritual knowledge. Consequently, the soul functioning at this level of intelligence is a spiritual gnosis that includes the spontaneous knowledge and activity of the next level, the planetary spheres or the *manomaya kosha.* This self-knowing within the soul through the modes of knowing, willing, and feeling is symbolized in cosmology/astrology by the three trans-Saturnian functions of Uranus, Neptune, and Pluto, the absolute powers of knowing, feeling, and willing, respectively.

As we have said, the "I principle" or *asmita* identified with the *vijnanamaya kosha* includes the knowledge of the lower realms. It is that power to know by which the Ideas are variously combined to form the World-Idea for an individual soul. This latter is received by the functioning of the *manomaya kosha.*

Sunlike Body/*Manomaya Kosha*

The *manomaya kosha* is the realm of the cosmocrators, the powers of the cosmic soul symbolized by the sensible Sun and the planets, which are elaborating the sublunary worlds. The individual soul conjoined to the cosmic soul participates in the functioning of its powers. The soul or mind receives the World-Idea, or rather, intuits it and at the [previous] level of *vijnana* makes a representation of it for itself. Inherent in this production is the intellectual nature of mind, including the fabricative and gnostic powers of the Soul that simultaneously imbue its representations with significance.

In order to understand the soul's manifesting of the World-Idea, it is necessary to describe the activity of the cosmic soul at this level. The mind of the cosmic soul is receiving the World-Idea, and, with its powers or representative faculty, organizes the Earth's memory accordingly. This organization is the given content of the world of sense or the subtle image of the world, which is still prior to our reflective consciousness.

This imaginal essence or the Earth's memory then becomes the medium through which that prior mind can become reflectively aware of its prior unconscious activity. At this level of our unconscious mental functioning, the World-Idea is being incessantly instantiated and transmitted. That is, the soul's image-making faculty is constantly recreating the World-Idea. This subtle recreated image of the World-Idea is not optical, but rather an abstract schema or synthesis, and is prior to our reflective consciousness; it is an alogical whole which nonetheless contains all the features that will later on become available to us through our murky, reflective self-consciousness. The mind's prior activity of organizing the "matter"—or more precisely, the Earth's memory or the elements—into clusters of sensations which constitute the functional images that become the phenomenal subject and object of our knowledge, *is* our actual empirical experience. These clusters of sensations or the organization of the Earth's memories into a functional image, is the medium through which the soul or mind incarnates and therein ultimately becomes self-reflectively aware of its prior activity.

The functioning of the psychosomatic organism—which is, as we have said, an organization of the Earth's imagination—may be referred to as imitative being. Here are located the presuppositions or basic attitudes that will be included in the manifesting of an individual's mental powers. Here and only here, within the organization of the memory or imaginal essence which is the depositing of mankind's life, may we speak of Jung's archetypes as the precondition for an individual psychological experience, and say that philosophy is but a mouthing of these archetypes.

Moonlike Body/*Pranomaya Kosha* and Elemental Body/*Annomaya Kosha*

The sublunary system of tropes or life traces has been organized by the planetary spheres into a variety of species and their environment, in conformity to the World-Idea, as far as this is possible. This imaginal realm of the Earth's dragonic belt constitutes the Moonlike and Shelly bodies, the lowest vehicles of the soul's experience. That our mind is really cosmic in extent is obscured in this final identification with the lowest (imaginal) sheath, and the apparent bifurcation of the soul into an empirical subject and a phenomenal object occurs here. A simplified picture of the vast cosmological presuppositions must be delineated in order for us to be able to understand our sensible experience. It is approximately in this way that the ancients conceived what conditions were necessary for the arisal of knowledge. The entire panorama of the sensible world which was fabricated by the gods was a necessary prerequisite—the stage upon which the actors made their entrances and exits.

Correspondences with Plato

These five vehicles line up with the Vedantic and Buddhist terms, with the levels of the cosmic manifestation, and with Plato's Divided Line. [*fig. 168*] They are *all* conceived as in the context of the light of Soul—the third phase of the Nature Mercury in Pisces ☿$_{♓N}$, the efflux of the Soul.

Divided Line	Sensible Sun		Intelligible Sun	
	1	2	3	4
5 Vehicles as the Microcosm	Moonlike and Shelly Bodies	Sunlike Body	Noeron Body	Augoiedes or Radiant Body
Bodies / Koshas (Vedanta)	Pranomaya Kosha / Annomaya Kosha	Manomaya Kosha	Vijnanamaya Kosha	Anandamaya Kosha
Skandhas (Buddhism)	Earth and Water	Fire	Air	Aether
Corresponding Macrocosmic Level	Terrestrial / Aquatic	Aerial	Empyrean	Celestial
Level of Cosmic Manifestation	Sublunary = Elements and Appearance	Chaldean Planetary Spheres	Uranus, Neptune, Pluto	Realm of Stars and / or 360 Degrees
Samadhis	——	Nirvitarka Samadhi	Nirvicara Samadhi	Ananda Samadhi

Fig. 168

The two lowest sheaths—*mano* and the combined *prano/anno*—can be portrayed as the sensible world of Plato's Divided Line whose author is our Sun.

Conjecturing upon the realm of shadows, the individual soul ultimately participates in true belief as to the nature of those images. Here the soul is participating in a body formulated or fabricated according to the sensible Sun's understanding of the reason-principles, that is, its version of the Ideas contained in the inerratic sphere.

Once the awareness of the Sunlike is realized and actualized—which may be spoken of as living the rational life, freed from the overwhelming compulsion of the memories and habits—there is the further need to participate and evolve within us the clearer intuitive vision and efficacy of the Ideas through the triple modes of super-reflective Willing, Knowing, and Feeling. That each individual human soul must apply these powers to the understanding of the World-Idea and their [own] role within it is to understand at the same time the universal reason-principles that are manifesting themselves or working themselves out through and within the World-Idea. Here we must think not only in terms of developing the rational soul powers, but must align these to the intelligence of the Ideas which are manifesting themselves. At this level we make available the spiritual function of Reason, or the Noeron, which informs the functioning of the rational mind and recognizes the higher or universal knowing principles, those principles which are the higher knower or the intelligence of the sphere of the inerratic Ideas. These spiritual Presences are brought into manifesting by the functioning of Uranus, Neptune, and Pluto in relation to our natal chart. Similarly the transits of the Chaldean planets can be used to illustrate the intrusion into the ego of the rational mind. (cf. Plotinus VI.7.6)

The intellectuality within us (that is represented by the inerratic spheres), particularizes itself (becomes available to each individual soul) through its un-unifiable triple function of Knowing, Willing, and Feeling. This could be conceived of as an image of the multiple organ of Intellect's self-vision in the Nous. The soul through these transcendent functions is the recipient of the Ideas—a spiritual gnosis often referred to as the *vijnanamaya kosha*—that includes the rational soul's content. After the full development of Knowing/Willing/Feeling and their being brought into coalescence arises the possibility of insight or true self-cognition of the Overself. The reason-principles that robe the pure I—or light of the Soul—are referred to as the *anandamaya kosha,* the joy of existence and the inherent Intellectuality of the Ideas available to the I AM or Overself.

In order to become aware of our higher nature, various systems have been devised. For example in the Yoga system the transition to the Sunlike body (which they refer to as the *manomaya kosha*) and the subsequent subtler levels as a conscious attainment cannot be accomplished without the technique of meditation. They speak of it having a threefold aspect: There is concentration on the thought, resulting in the stilling of the empirical mind, which leads to an inwardization of consciousness and then silence. This technique is called *samayana* in yoga terminology. This meditation leads us out of the empirical mind into the mind of light, the first of a series that leads finally to the formless contemplation of the pure Soul and Nous. The first level of meditation, if successful, would lead to the disidentification with the phenomenal or lowest sheath, to the viewpoint of the Sunlike body. The successive application of the threefold technique on this and the following levels leads to a series of transitions through the sheaths or forms of thought, leading up to the I AM or Overself. Apparently, although the transition between levels does involve a cessation of thought at each level, the pure light of the Soul, or the I AM principle, does not become immediately available in normal practice because of the immediate appropriation by the next subtler level of thought with which it becomes identified.

EDITORS: *The bottom line of figure 168 (p. 349) shows the levels of* samadhis *(absorptions) in yoga terminology. As the vehicles develop, the individual acquires the powers needed to enjoy the states of knowledge described in or by Plato's Divided Line. (See I.K. Taimni,* The Science of Yoga, *p. 31 ff.)*

53

ANTHONY

On the Topography of the Subtle World

EDITORS: *In his Invocatory Prelude, Anthony establishes a context in which to explore what he refers to as "the topography of the subtle world." This "topography" suggests gradients, contours, and deeper levels of the cosmic mind. There are no markers in the depths of Mind, but we can delineate broad grades of the cosmic soul and World-Idea using, as Anthony says in his Prelude, the sensible cosmos as a symbol for philosophic principles. Page references below are in* Astronoesis.

- The ancients spoke of the realm beyond the constellations as the World of Ideas. (p. 12)
- These . . . Ideas were mythologically represented by the constellations. . . . the signs . . . are the paradigms of all living forms or species symbolized by the archetypal zodiac or inerratic sphere, which represents the immutability of the Ideas. In this sense the stars are a theophany of radiating intelligences, distributing their presence and informing the universe through patterns of intellectual energy. (p. 12)
- An individual soul in its own right, our Sun comprehends the nature of these intelligences through its intellect and proceeds to unfold this prodigious intelligence. Each part of the whole (the solar system)—that is, each planet or rational power—may be considered in its own right, or may be considered as part of the whole. (p. 14)
- Our Earth is included within the functioning of the rational soul of the Sun. . . . the Earth is also a recipient of the Ideas, and through the intermediary of the rational powers of the Sun gradually there is organized and manifested on Earth a variety of species that will partially . . . reflect the Ideas or animal intelligences. (pp. 14–16)
- The traces of life left behind by these creatures are retained by the soul of our Earth as the content of its memory, and this is . . . the "Dragon." . . . This is what the degree symbolism of the chart refers to (p. 16)

54

EDITORS

Planets as Rational and Instinctual Functions

Every individual human soul as related to the Earth soul participates in the unembodied rational soul powers (the Undivided Mind of the Earth) as well as in the embodied functions imaged in the dragon. The rational powers correspond to the soul's virtues or perfections (Plotinus writes about these in 1.2), while their presence in the embodied ego takes the form of archetypal determinators (corresponding to Jung's archetypal functions) and of instinctual (unconscious) functions.

Another way to say this is that the undivided rational mind is creative and gnostic, organizing and transforming the noumenal World-Idea into experience. The embodied ego functions are transformed by these powers, and are available to the individual person in experience.

The planets are the presence of the Soul powers or functions. They function at a spectrum of levels, corresponding roughly to the main divisions of Plato's Divided Line:

> Natal planets in houses are the physical/instinctual functions.
>
> Natal planets in signs/elements are the psychic functions.
>
> Natal planets on natal degrees are the mental body powers.
>
> Planets as transiting are the rational soul powers, virtues. [*fig. 169, p. 352*]

Only as reflected within the individual ego (what Plotinus calls the living Animate) are these powers fragmented. The actualization of these powers is illustrated in astrology, where the rational powers of the soul are represented by the transiting planets and the embodied powers by the planets in the natal chart. The individual functions represented by the natal positions are both continuous with and distinct from the "transiting" powers. Thus there is a spectrum of levels of function for each of these powers, from the instinctual to the rational.

Delineations for individual planets through the spectrum from rational to instinctual levels follow. These delineations are by no means complete; they simply point toward key identifying characteristics of each planet's activity. How an individual planet will function depends on its degree in the zodiac, as well as its complex aspect pattern with the other natal planets, and its position by sign and house. In addition, when reading a natal planet on a degree we have to consider the level of maturity of the ongoing unit of life, its power and character, where and how it is identified with the body, etc.

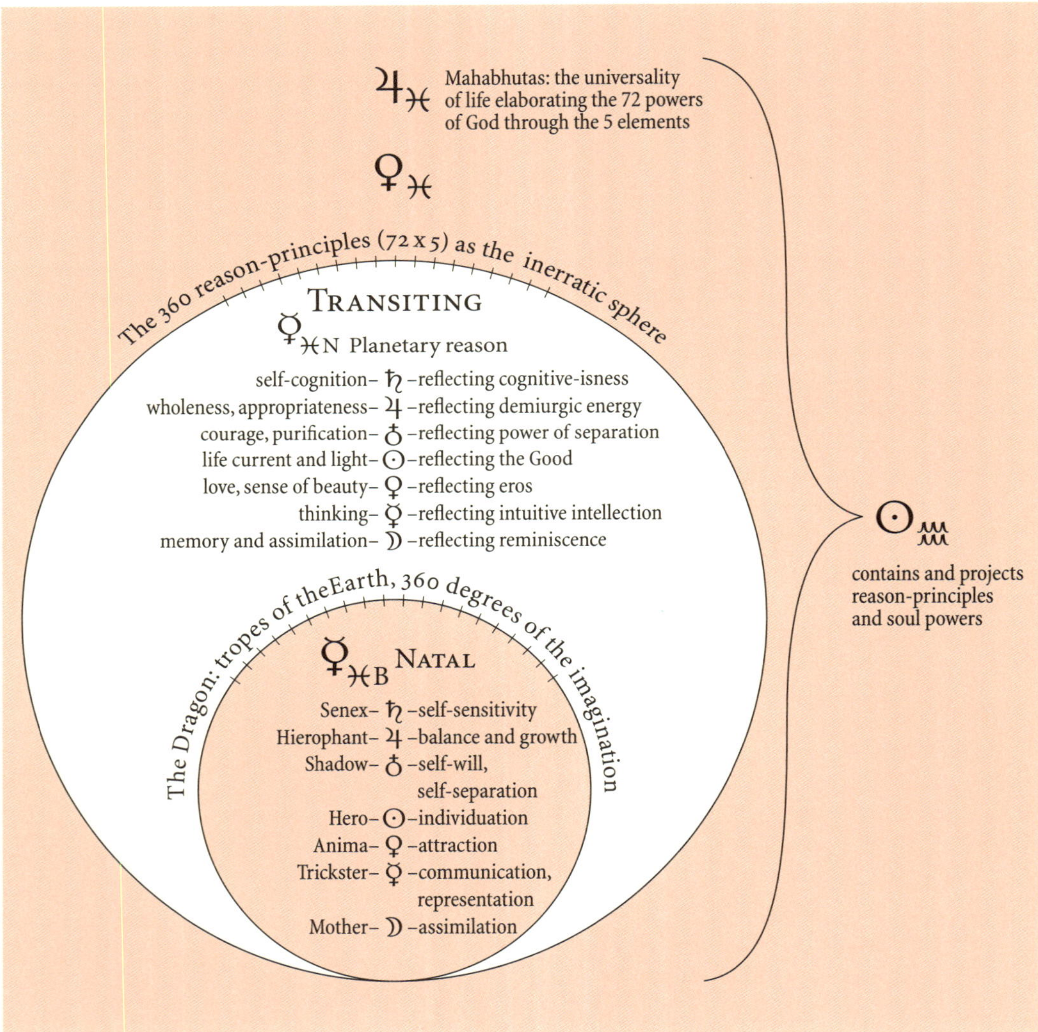

Fig. 169

Sun ☉

The Sun essentializes life according to the Good and is the manifest symbol of the Good. The Sun is not bare life, but *meaningful* life, the Good for the individual. As Anthony remarks, "Soul is free when it moves toward the Good." (*Astronoesis,* p. 199) The Sun guides you on the path to the Good. This meaning carries through into the natal chart as well.

> We have to look at the Sun as life. . . . No one lives just life, he lives *a* life. As soon as you put in the word "a," "a life," you're speaking of an idea that dominates an individual life. In other words, it is this idea which is dominating that life and which must be expressed in and through that life. . . . Jung always refers to the Sun as the principle of individuation. (Anthony Damiani, transcript, 1/30/84)

In the individual, the Sun represents the central life force of the incarnation. It is the fluid power of imagination to manifest and perceive the creative idea of the life. The Sun ☉ is the life energy of the chart, the center or feeling of I, the life current, or heart current. All the planets are articulating this light, like a prism disperses white light into colors. Anthony says: "The Sun in the solar system is the central Force and individuality of the Cosmos. Likewise the Sun as the central unity is present in the individuality as the central point of a person's ego." (Notes and Papers, p. S58)

> The Sun degree in the chart may be used to symbolize the individuating quality of that life, being brought to self-reflective awareness through the function of the other six powers within the organism. The resulting

functioning of this psychosomatic organism brings about the appearance which is the field upon which the reasoning soul passes judgement. (Anthony Damiani, Notes and Papers, p. S128)

The Overself or soul, which is like the Sun for us, emits a ray of light, or stream of light, which is united with the vehicles. This light or life is not a pure light, but has a very specific structure. It carries the idea which the individual is going to work out in this life, and this idea is the Sun degree. This degree gives us some idea of the quality of the living consciousness communicated continuously to the individual, and gives us some idea of the form in which meaning will be found in the world.

Moon ☽

As the rational soul power of "Recollection" or "Reminiscence" (the Platonic *anamnesis*) the Moon recalls the soul to its true place in the Divine Mind.

Through the power of memory, the rational soul assimilates its experience of the World-Idea and provides a continuity through the discontinuous or moment-to-moment manifestation of the world. "Memory [at the rational level] is primarily a power, and not [mere] passivity, and memory is even impeded by the body." (Anthony Damiani, Notes and Papers, p. S57; cf. Plotinus IV.6.3, IV.3.26)

Through the Moon's cyclic nature (which we see as its phases) an individual assimilates experience, retaining traces of experience (i.e., *vasanas*) as memory. The lower memory is that "associated with the psychosomatic organism, the convolutions or engrams left in the brain—mythologically, Minerva's snakes." (Anthony Damiani, Notes and Papers, p. S57) We can think of eating all kinds of food: physical food, emotional food, or mental food; at all these levels we are nourished and nurtured by the inner Mother.

> ... this is the imaginal structure of the Mother archetype, for it is these memories which are nourishing and feeding the everyday assimilation of experience. Through this the soul is bound by the limitations of time and the sequence of events. (Anthony Damiani, Notes and Papers, p. S57)

> I think of it as one uninterrupted process. The thought of the body is an organ through which the perception of the sensible world could take place so that it could be brought back into the person. I think apperception sums it all up. . . . We have to posit the existence of the body within the primal image, and there we think of it as part of that primal image. We are thinking of the body really as a thought, as part of that huge thought. . . . The body is the organization of all these *vasanas* unfolding themselves. In that very process the soul is assimilating, through the Moon, the *vasanas*. . . . The seed thoughts go into the heart center, reach the brain, exfoliate, and now there is a perceived world. . . . Then this has to be brought back in—this is the function of the Moon. (Anthony Damiani, transcript, 1/30/84)

Mercury ☿

As a rational soul power, Mercury (the Greek Hermes) makes available the inner meaning through which we can interpret experience, and also communicate our inner knowledge outwardly. Mercury weaves together the objective reason-principles and subjective feeling-states at each level of experience. This communion of inner and outer reason is the soul's language.

> For Understanding, the true, is the Act of the Intellections: in many of its manifestations it is the assimilation and reconciliation of the outer to the inner. (Plotinus, I.1.9)

Mercury/Hermes rules the crossroads, places that are in-between, and the journey as a place in itself. Hermes is the guide of souls, for it is the messenger of the gods, and communicates to us the inner guidance. And it is Hermes who gives us the Hermetic arts, such as astrology. "[Mercury] is the psychopomp—guiding souls and leading them to the higher." (Anthony Damiani, Notes and Papers, p. S56)

The individual participates in Mercury as the power of communicating, naming, connecting, classifying, articulating. Mercury is also "fascination." Mercury makes meaning, or "means" things through ratio, thought, and image. When the World-Idea comes into the individual mind, it is relativized, connected with concepts, overlaid with personal opinion. Thus Mercury helps us make sense of the world indirectly by means of images, symbols, and representations. Mercury is the Trickster, and may represent the world in a way conformable to the primal essences, or not.

> People feel that they have magic when they can name something. The Mercurial function is evolving a language in which it can reflect itself. (Anthony Damiani, transcript, 1/30/84)

Venus ♀

Venus indicates the desire of the soul to fulfill its vision of the Nous through union. The soul is where it loves. Venus is love, the power to seek and to see beauty, and to unite with the beautiful. Here Beauty seeks to unite the soul in Love to its prior.

> And the Soul includes a faculty peculiarly addressed to Beauty—one incomparably sure in the appreciation of its own, when Soul entire is enlisted to support its judgement. (Plotinus, I.6.3)

In the individual, Venus represents the aesthetic sensibility to value what the soul loves and to desire the objects of its vision. "This evaluation [provides] the subjective feeling states needed to accompany the objective principles in the soul." (Anthony Damiani, Notes and Papers, p. S55)

Venus is the sense of beauty, evaluation, union, desire, attraction—what attracts a person's attention to one thing or another. When the World-Idea comes in to the individual mind, it cannot be grasped as a whole: The Venus function selects what the attention will focus on, what part of the scene will be emphasized. Some function decides how this world will be portrayed, what forms will best manifest how one feels the World-Idea's energy pattern.

> Venus is *evaluative.* It will take an image that bears the traces and nuances of certain types of feelings and say this will do better than that, not in a cognitive way, but in a feeling way . . . [as a] felt significance. (Anthony Damiani, transcript, 1/30/84)

> . . . in Venus [is] the basis for the feelings of the Anima archetype. This archetype binds you through feeling to a certain kind of experience . . . (Anthony Damiani, Notes and Papers, p. S55)

Mars ♂

As a rational soul power, Mars also represents the courage and fortitude needed by the soul in its journey toward freedom. Mars indicates the desire of the soul to become an individual through separation and activity. Mars is will, the power to actualize the soul's potential. Mars not only propels the reason-principles into life, but provides the energy needed to purify experience so as to distill and reclaim these reason-principles for the soul's return journey to itself.

> . . . purification is its [soul's] awakening from the baseless visions which beset it, the refusal to see them; its separation consists in limiting its descent towards the lower and accepting no picture thence, and of course in the banning of all that it ignores when the pneuma (finer-body or spirit) on which it is poised is not turbid from gluttony and surfeit of impure flesh, but is a vehicle so slender that the Soul may ride upon it in tranquillity. (Plotinus, III.6.5)

Anthony writes: "The ancients speak about the Mars amputating base matter—the base matter being the appearances taking place in the dragon. . . . It is a particular requirement to bring about the purification and separation of the soul essences from their association with impure images. . . . It requires fortitude to bring a halt to the compulsion of the phantasy, or—to put it conversely—to choose which essences you will experience and will yourself not to be overcome by." (Notes and Papers, p. S54) The embodied individual experiences this presence as craving and will.

In the individual, Mars is "work done" or "energy actualized." It is the psychic energy or empirical will needed to get things done. Mars must take initiatives and build up courage. To have new experiences, you have to let go of the old ones. Purgation, purification, and the practices for the quest all require Mars. One person commented that Jung calls the Shadow "that which you didn't want to be . . . but find you are anyway."

> In other words, this energy is grooved, it keeps flowing in that direction. If you say, well I want to alter it a little bit, then you will find out what the Shadow is . . . that lower aspect of the energy which has already been directionalized for a long time refuses to make a change. (Anthony Damiani, transcript, 1/30/84)

Ultimately, the instinctual energy of separation and individuation represented by Mars has its source in the soul's desire for separative existence, and thus complements the Venus as soul's longing to fulfill its desire through union.

Jupiter ♃

Jupiter represents the power of rectitude and appropriateness and balance, which mediates the wisdom appropriate for the embodying soul. As a rational power of the soul, it makes available the metaphysical principle of the Six Forces of Nature. Jupiter envisions the whole of the I and the world, sees it as it is before separation. It operates instinctually or intuitively, without deliberation.

In the individual, Jupiter balances and coordinates all the instinctual functions. Jupiter as a psychological function rules the homeostasis or harmony of the planetary functions as a whole.

> I was thinking of it [Jupiter] as the homeostasis of the instinctual operation within the psyche. There is an instinct toward homeostasis, that is to maintain a balance, and this is instinctual in any psychosomatic organism. (Anthony Damiani, transcript, 7/11/83)

Jupiter determines the unfolding of the sequence of the archetypes. Because Jupiter has an intuitive view of which Ideas are needed for the soul, it often appears as the Hierophant, the guide. In Hindu astrology, Jupiter is the Guru: It guides you on the path to wholeness.

> The interrelationship of the psychic functions, for better or worse, is what Jupiter is indicating. That's why they refer to it as the guru of the chart. . . . If you consider very carefully how the psychic functions are related to each other, you're speaking of the functioning of the entity as a whole. (Anthony Damiani, transcript, 1/30/84)

> Hence Jupiter often appears as the Hierophant, the guide, which is our reason, but in the lower organism becomes the wisdom or the correctness or the incorrectness of the instinctual functioning. (Anthony Damiani, Notes and Papers, p. S52)

> Isn't it nice to have someone inside you like a little compass, telling you you're going the wrong direction? (Anthony Damiani, transcript, 7/11/83)

Jupiter indicates how the psyche will grow and open up to the world of experience in new ways.

> Jupiter represents expansion. . . . If the interrelationship of the psychic functions to each other is a fitting kind of relationship, we have the expansion of the individual. (Anthony Damiani, transcript, 1/30/84)

This energy of Jupiter (in Latin Jupiter is Jove, hence Jovial) is both kingly and expansive and enthusiastic. It decides what is needed for the next phase of growth, what is "appropriate," and what is right to appropriate in accord with the soul's need to be more whole and complete. Since the new experience is always unknown, Jupiter relates the existing ego to the unknown.

Saturn ♄

Saturn is self-reflection. Soul knowing itself "by the Intellectual-Principle with which he becomes identical" (Plotinus, v.3.4) is Soul's access to the light of the Nous, the higher knower. There is also a self-knowing which accompanies all reasoning, and this is Saturn: "the one that takes cognizance of the principle in virtue of which understanding occurs in the Soul or mind." (Plotinus, v.3.4) This refers to Soul's own reflective cognizance of itself as a "subject," which accompanies its other functions at every level.

In the individual, Saturn sets the boundary between the I-thought and the world-thought. Saturn is sensitive to what is in me and what is not: self-reflection, self-awareness, the mass of *vasanas* constituting a sense of I, holding together as a fixed entity.

> Every living being is a mass of sentiency. It is a reflection of the principle that wherever you are, you are self-aware, self-sensitive, a reflection of the principle of the soul's omnipresence in the psyche. (Anthony Damiani, transcript, 1/30/84)

Saturn is the power that allows us to reflect and turn back to the center of self, and is the wisdom accumulated through experience. "This power is psychologically manifest as the function of self-apperception and forms the imaginal category of the Senex, the wise old man." (Anthony Damiani, Notes and Papers, p. S53)

In order to reflect on experience, you must stop for a second, stand where you are. Saturn is this stoppage. It is also often associated with Death, for in order to be born anew, there must be death. As the slowest moving of the seven inner or Chaldean planets, it indicates the cycles of karma, the taking account of the whole life experience. Saturn leads us on the path to self-knowledge (small "s").

Saturn may be paired with Jupiter. The growth of experience brought by Jupiter leads to the building up of Saturn's self-cognition. And in the reverse way, Saturn building a self-reflective center of experience invites further expansion and wholeness through Jupiter.

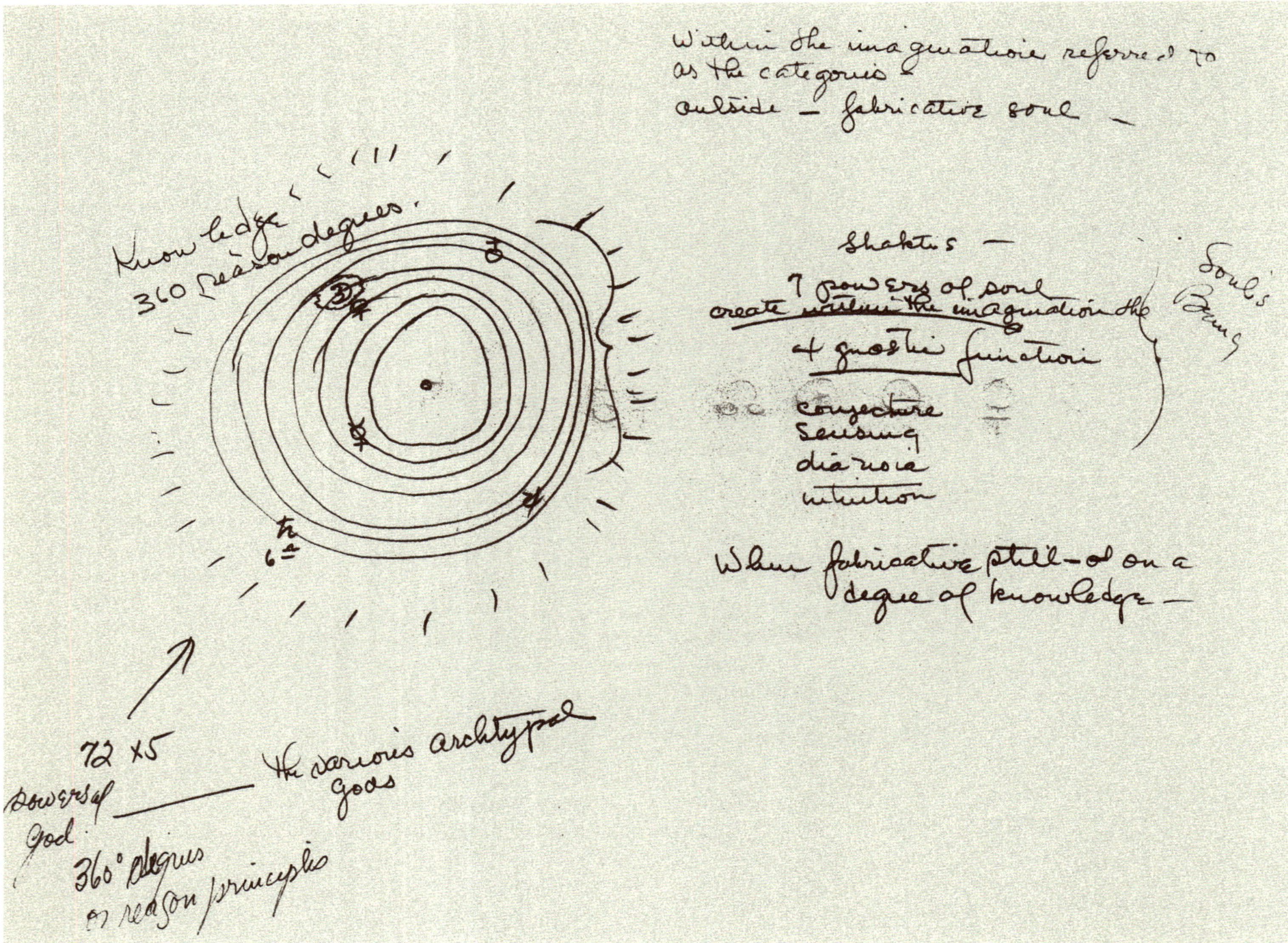
Within the imagination referred to
as the categories –
outside – fabricative soul –
Knowledge
360 reason degrees.
Shaktis –
7 powers of soul
create within the imagination the
of gnostic function
conjecture
sensing
dianoia
intuition
Soul's
Being
72 x5
the various archetypal
gods
powers of
God
360° degrees
or reason principles

Glossary

THIS GLOSSARY INTRODUCES readers to Anthony Damiani's sometimes unique meanings for key terms that appear in *Astronoesis.* Many of these terms have a variety of meanings for different schools of thought. Unless stated otherwise, the explanations given here confine themselves to a) Anthony's use of an original term, or b) his most common interpretation(s) of terms borrowed from other systems or thinkers and adapted for his own purpose in context.

Terms that appear in SMALL CAPS indicate cross references within this glossary. An asterisk (*) indicates a term related to soul's "modes" with body. Please consult the General Index for references to where these glossary terms are used, and often explained, in the text itself.

Absolute Soul Usually, the source of Soul in the Divine Mind (NOUS), prior to any individuating activity.

Ahamkara* Literally, "I-maker." Generally correlates, as used in SAMKHYA, with REASONING or the WITNESS-I.

Animate/Couplement* Plotinian terms for the individual union of "organized body" and a "ray of the soul." (cf. I.1.7) Usually it refers to the organism of only one life, not the ongoing life-to-life continuum. Other roughly equivalent terms are EGO or PSYCHOSOMATIC organism.

Terms that appear in SMALL CAPS *indicate cross references within this glossary. An asterisk (*) indicates a term related to soul's "modes" with body.*

Aseity Unique characteristic or essentiality (literally, is-ness). *Oxford English Dictionary* has "underived or independent existence."

Asmita* Sanskrit term meaning the (mis-)identification of the soul or self with anything objective: any thought or bodies. Anthony uses this term to refer to the individual evolving unit of life, or UNIT SOUL, that goes from life to life. This term should be considered in relation to the terms WITNESS-I and I AM. Yoga Sutra II.6 says: "*Asmita* is the identity or blending together, as it were, of the power of consciousness (*Purusha*) with the power of cognition (*Buddhi*)." I. Taimni comments: ". . . in its wider and more intelligible meaning it may be considered as the identification of consciousness with the vehicle through which it is being expressed. . . . The Samskrita word *Asmita* is derived from *Asmi* which means literally 'I am.' 'I am' represents the pure awareness of Self-existence and is therefore the expression or *Bhava,* as it is called, of pure consciousness or the *Purusha.*" (*Science of Yoga,* pp. 142–143)

Aspects Significant angular relations formed between the PLANETS in their movement through the ZODIAC. Significance is based on number symmetries of the circle. Most generally considered are the conjunction, opposition (180 degrees), sextile (60 degrees), square (90 degrees), trine (120 degrees), quintile (based on a fivefold division of the circle), and septile (based on a sevenfold division of the circle).

Astrological Mandala/Metaphysical Chart The Astrological Mandala, or Metaphysical Chart, is the heart of Anthony's method of philosophic symbolization and synthesis. Using a mandala schema, Anthony creatively applied and combined the Pythagorean number meanings, geometrical symbolism, astrological symbols, and concepts from a variety of philosophic and religious traditions. Through this process, he discovered and developed a language increasingly more adequate to his vision—one that sustains deep thought on the nature of reality and the complex interrelationships of its metaphysical, cosmological, psychological, and physical levels.

While he used astrological symbolism to pictorialize philosophical principles, at the same time he showed that astrological principles have a philosophical basis. The astrological mandala also refers to a series of diagrams that deploy astrological symbolism to explore a range of meanings connecting metaphysical principles with the life of an individual person. (See also MANDALA. The Astrological Mandala is discussed in detail in Anthony's Invocatory Prelude, pp. 6 ff. His paper "Astrology: The Topography of the Subtle World," p. 263, sets forth his thinking on the astrological mandala as "a fragment of an archaic wisdom.")

Astronoesis Literally, "star wisdom." The term was coined in juxtaposition to *astronomia* (star arranging or ordering) and *astrologia* (telling about the stars, or star-speaking). *Astro* means star or other heavenly body, and also some kind of light, shining intelligence. NOESIS means Wisdom or "seeing into." Star-wisdom means not only the wisdom *of* the stars, but also the wisdom which *is* the stars—what a star-soul sees or en-visions. Star-wisdom is also the radiating intelligence of the stars as portals to *noesis.*

Augoiedes* The CELESTIAL or perpetual vehicle of the soul—its "robe of glory."

Authentic/Authentic Existence That which has/is Real Being; often MacKenna's translation of *ousia.* (See also BEING and REALITY.)

Becoming As a technical term, refers to the world and experience as moment-to-moment process that is never completed. The SYSTEM OF NATURE is the world of appearance or "becoming." (In Platonism, becoming and BEING are often viewed as dualistic opposites. Plotinus discusses becoming and Being in VI.2.1.)

Being For Plotinus, Being always means Authentic Being (which MacKenna translates as "AUTHENTIC EXISTENCE"). In V.5.5, Plotinus says: "Pressing . . . on the word for Being (*on*) we have the word for One (*hen*)," indicating by this derivation that "Being . . . is that which proceeds from . . . The One." Being is not a static principle for Plotinus: It is void of specific forms but is supremely powerful and Knowing, inseparable from NOESIS. In a narrow sense, Plotinus uses Being (*on*) to refer to the NOUS only, and in this sense the ONE is beyond Being. But sometimes Plotinus does refer to the Being of the One in an analogical sense, as in VI.8.16: "That His being is constituted by this self-originating self-tendance . . ."; this super-essence Anthony in one place calls "be-ness." (Being is discussed in further detail on p. 304 in Note 6D.)

Buddhi A Sanskrit word meaning intellect at the level of spiritual subjectivity. The power of cognition. Applies at the level of valid or "public" knowledge, i.e., knowledge that does not vary according to the subjective constitution of the knower.

Celestial/Celestial Body* This immaterial "body" arises where the unity of the soul first associates with the cosmos; the world of light or AUGOIEDES; also referred to as the "radiant body" or the "robe of glory." (See G.R.S. Mead's *The Subtle Body* for information on this radiant body.)

Chaldean Planets/Chaldeans The seven planets ordinarily visible from Earth to the ancient naked-eye astronomers (Chaldeans or Magi): Sun ☉, Moon ☽, Mercury ☿, Venus ♀, Mars ♂, Jupiter ♃, Saturn ♄. (See Editors Extension IV "Seven Soul Powers" (p. 191), pp. 291–296 in our Epilogue, and Note 54 (p. 351) "Planets as Rational and Instinctual Functions.")

Citta-Vrittis Modifications of the mind (*citta*), or thoughts in the widest sense of the term.

Cosmic Circuit The solar system and planetary spheres, considered as a living organism that expresses the will and reason of Soul. (See Plotinus II.3.9 and III.8.6, and also Plato's *Timaeus.)*

Cosmic Soul This term refers to a variety of levels of "individual" souls, but usually means the Sun soul. "Cosmic" means that it functions jointly with the Universal Soul to manifest a cosmos, providing the habitation for other individual UNIT SOULS. There is an uncountable nested hierarchy of cosmic souls, when considered in terms of their manifestations. However, all cosmic souls are also of the nature of INDIVIDUAL SOUL in general, having an individuality connected with the cosmos, and an inviolable essence.

Cosmocrators Literally, world-lords. These are the PLANETS considered as the powers of the COSMIC SOUL.

Daimon (or "Daemon") Daimons are divine powers, intermediary between the gods and human beings. Note 35, p. 331, provides references on the concept of the daimon in Neoplatonic and other traditions.

Degrees / 360 Degrees See also DRAGON, REASON-PRINCIPLES, and TROPICAL ZODIAC. The tropical zodiac, the Earth's view of the zodiac, has 360 equal intervals called degrees. (These degrees and degree symbolism are discussed in Anthony's papers "The Third Phase of Soul," (p. 256–258) and "Astrology: The Topography of the Subtle World," (p. 264–269) as well as on pp. 288 ff. in our Epilogue.) Degrees are counted from zero. The first degree of Aries, for example, is from 0° plus minutes to exactly 1°, the second from 1° plus minutes to exactly 2°, etc. Many symbolic approaches

---

Terms that appear in SMALL CAPS *indicate cross references within this glossary. An asterisk (*) indicates a term related to soul's "modes" with body.*

to the meaning/intelligences of these degrees are available. Anthony most commonly uses the "Sabian" symbols provided by M.E. Jones.

Demiurge A term Plato and Proclus use for Universal Soul, the creative principle that underlies every and any manifestation. Plato discusses the Demiurge most completely in his *Timaeus.*

Dignities Dignities designate planetary positions in astrological SIGNS with which the functions of the PLANETS are specially connected. Anthony uses Rulership, Exaltation, Detriment, and Fall dignities to associate planetary powers with specific Ideas in his metaphysical chart. (For more on dignities, see Note 16, p. 314, "Placing Dignities in Rings: Monads," and Note 17, p. 317, "Deities in Their Series of Henads.")

Double Act Refers to the double nature of the One's power, which both remains in the One *and* EMANATES. Plotinus says in v.4.2: "There is in everything the Act of the Essence and the Act going out from the Essence: the first Act is the thing itself in its realized identity, the second Act is an inevitably following outgo from the first, an emanation distinct from the thing itself." Double Act is one meaning of DYAD.

Dragon An ancient term for the totality of traces of life-activity deposited in the subtle memory of the Earth. It is the totality of TROPES or VASANAS, subtle memory impressions in the imagination of the Earth. It is parallel to the psychological idea of the unconscious, and is a powerful force at the basis of the EGO's instincts. The terrestrial dragon is another term for the TROPICAL ZODIAC, so called because it is the life-traces of entities on Earth. The celestial dragon refers to the zodiac of the Sun or SIDEREAL ZODIAC, the archetypal energy radiating from the stars, as relative to our solar system.

Dyad Dyad, or indeterminate Dyad, is a term that tries to capture a sense of the two-ness or paradox of the One's power. *Astronoesis* offers a variety of slightly different interpretations of the Dyad: a) the relation of the ONE and POWER; b) the complementary nature of the One's power as active and passive; c) the nature of power as abiding and emanating. (Note 8, p. 306, "The Dyad," discusses these in more detail.)

Ego* The PSYCHOSOMATIC organism, roughly equivalent to the ANIMATE of Plotinus: a conjunction of the cosmic elements and the light of the soul. It usually means the total matrix of thoughts, body, and emotions which constitutes one life. In a deeper sense, it is an expression of the soul's ongoing individuality. The ego's status is best understood in the context of the wider issue of understanding the mental nature of existence itself. Paul Brunton writes: "When it is declared that the ego is a fictitious entity, what is meant is that it does not exist as a real entity. Nevertheless, it does exist as a thought." (*Notebooks,* v. 6, pt. 1, ch. 2, no. 32)

Elements In the ancient sense, the subtle constituents of the individual and cosmos. In *Astronoesis,* elements have two contexts. One context is the 12th house Idea of MAHABHUTAS or Five Great Elements of Pisces. The other context is the "earthly" elements of Fire, Earth, Air, Water.

Emanation This term means overflowing or flowing out from. Plotinus says that all "AUTHENTIC EXISTENCES" emanate, or overflow. In v.1.6, he describes this ESSENTIAL PRODUCTION: "All existences, as long as they retain their character, produce . . . in virtue of the power which must be in them—some necessary, outward-facing hypostasis continuously attached to them and representing in image the engendering archetypes." In v.2.1, he says: "Seeking nothing, possessing nothing, lacking nothing, the One is perfect and, in our metaphor, has overflowed"—even though there is nothing in it, and nowhere to flow out to. Emanation is possible precisely because there is nothing in the One, and nowhere else to go.

Epistemology The study of *how* we know (theories of perception, subject-object, and so forth).

Eros The feeling-correlate of LOGOS. Platonism speaks of it as "a mighty DAIMON," the energy of the divine which informs the individual, leads us to the divine, and pressures us to unfold our true nature. It is also that which unites the INDIVIDUAL SOUL with its true IDEA.

Essential Production A view of causality that says that as long as anything is AUTHENTIC, it EMANATES or produces an outward-facing HYPOSTASIS.

Geocentric/Heliocentric Viewpoints The geocentric (Earth-centered) point of view is the framework established by the Earth soul; the heliocentric (Sun-centered) point of view is that of the COSMIC SOUL. (For more on the appropriateness of the geocentric viewpoint for us as psychological beings, see the margin note on page 4. For more on cosmic soul, see "Cosmic Soul," p. 200, and Note 42, p. 336, "Notes on the Solar Logos/Cosmic Soul." See also figure 106, p. 248, and the accompanying text for more on the intersection of the Earth's and Sun's points of view.)

Glimpse A mystical "glimpse," as the word is used by Paul Brunton, means a direct, immediate experience of reality. Plotinus refers succinctly to one form of such experience in VI.5.7: "To know without image is to be." It may last for a moment or for months, but its eventual passing is what distinguishes it from permanent enlightenment.

Gods, Orders of Proclus (*The Six Books of Proclus on the Theology of Plato* and *Proclus' Elements of Theology*) delineates various orders of gods: intelligible, super-mundane, liberated, mundane, and so forth. G.R.S. Mead (*Orpheus*, pp. 58–59) has a diagram of the various orders of gods. The intelligence of the Universal Soul or Demiurgic Mind (World-Mind) is said to include twelve deific aspects, called the liberated order of Gods in the Orphic theology. The mundane gods, also called cosmic gods (which Anthony usually placed in the context of the cosmology), are associated with the realm of manifestation of the individual cosmic souls. In order, they are Jupiter, Vesta, Mercury, Ceres, Neptune, Minerva, Venus, Juno, Vulcan, Mars, Apollo, Diana. Also, Anthony sometimes speaks of each of the planets (COSMOCRATORS) as a "mundane god."

Head Astronomically, the North Node or Head of the Dragon (☊) represents one direction on the line where the planes of the Sun (☉) and Moon (☽) intersect. In a Hindu myth that Anthony retold in his introduction to *True Lunar Nodes: 1850–2000*, it refers to the immortal head of a demon who swallows the elixir of immortality as far down as his throat before being beheaded. The myth concerns the principle that draws out the relation of NUMBER and INTELLECTION.

Henads The word is Greek for "Ones." Plotinus says that every AUTHENTIC BEING has its unity in the ONE. Anthony uses the word to refer to the super-essential principles of all things in the One, beyond BEING, Knowing, or any other characterization. He writes: "We will conceive of the One as having indivisible aspects that we can examine." (p. 29) Henads are also called "the transcendent paradigms" (p. 31) and each henad includes a whole series of divine principles which will unfold from it. (Note 17, p. 317, "Deities in Their Series of Henads," uses the symbology of the DIGNITIES to develop the concept of henads.)

Horoscope Literally, hour-observer. See NATAL CHART.

House One of the twelve principal sectors of the individualized astrological circle. In astrology, each house marks out one realm of our experience or circumstances. Anthony often uses the words house and SIGN interchangeably when referring to the sectors of the METAPHYSICAL CHART.

Hyparxis Neoplatonic term for summit or root.

Hypostasis A self-determining level of reality. Hypostasis indicates that something has the status of standing in its own right. Literally, hypostasis means "standing under," as the ground or foundation, the support, the underlying principle or substance. Plotinus uses this term when describing how something other than the One can be real. In V.1.8–10, he describes the three pure hypostases—ONE, INTELLECTUAL-PRINCIPLE, and SOUL. (Refer also to Note 6E, p. 305, for more on this term.)

I AM* Refers to the INDIVIDUAL SOUL as an EMANANT of the absolute, the highest aspect of the soul as individual. (See also MERCURIUS, UNIT SOUL, and "WE.")

Idea From the meaning of the word *Eidos*, Greek for Idea, as "the form or figure, the characteristic look that a thing presents to vision . . ." (Findlay, pp. 29–30), we get a sense that Idea is the way the One is seen by, or looks to, contemplation or vision. Anthony interprets the Plotinus V.3.11 "successive impressions" of the NOUS, and its description as an "eye full of determination," as referring to the Ideas. Ideas are powerful facets of Divine Intelligence, not mere concepts or finite thoughts. In Neoplatonic thought, Ideas provide a bridge between a view of REALITY as an absolute unity, and a view of reality as a complete relativism of appearances. (Anthony discusses Ideas in the chapter on the INTELLECTUAL-PRINCIPLE. See also Editors' Extension I, "What Is an Idea?" on p. 121, and Editors' Extension II, "The Twelve Idea-Principles," p. 131.)

Idea of Man Anthony's term for one of the divine Idea-intelligences, the 5th IDEA in the METAPHYSICAL CHART. It is associated with the Idea that the One-Reality can be present in living consciousness, or that Divine Intelligence can contract itself and reproduce something of itself through unique centers of intelligence. (Editors' Extension III, "The Fifth Idea" (p. 147), elaborates on this Idea as a special form of the NOUS.)

Individual Soul* "Individual soul" refers to a wide variety of souls that emanate from ABSOLUTE SOUL, and is interchangeable with the term UNIT SOUL. Individual or unit souls are not *parts* of

---

Terms that appear in SMALL CAPS *indicate cross references within this glossary. An asterisk (*) indicates a term related to soul's "modes" with body.*

SOUL, but ways of looking at the whole of Soul as a principle of individuation. They are distinct from the Absolute by their relation to manifestation. We may think of individual souls as rays continually pouring out from the Absolute Soul, but never ending in their manifestation. We cannot completely separate the individual unit soul from the Absolute Soul, nor can we separate the notion of individual from the unfolding SYSTEM OF NATURE.

Inerratic Means moving without error. It is a term for the sphere of fixed stars, the intelligence of the cosmos.

Insight See VISION.

Instantiation A word for the "instant-to-instant" activity of the Mind actualizing IDEAS in thought.

Intellection MacKenna's translation of *NOESIS*.

Intellectual-Principle MacKenna's translation of NOUS. In the ASTROLOGICAL MANDALA, the Intellectual-Principle can be viewed as the second quadrant or as the two outer rings of the whole chart. It can also be viewed as the whole circle when viewed from an ontological perspective. (The Intellectual-Principle is discussed in the second chapter of *Astronoesis*. See also Note 6C, p. 304.)

Jiva* A Hindu term for INDIVIDUAL SOUL. Literally, a "unit of life."

Logos A term Plotinus uses to mean an IDEA *in the mode of* SOUL, a union of intelligence and life. MacKenna translates Logos as REASON-PRINCIPLE. It has a sense both of REASONING and of revelation of Ideas. It also can mean and be translated as Word. Anthony uses it for the guiding vision of the INDIVIDUAL SOUL seeking actualization, the intelligence in the soul. Logos is both individual and cosmic. In the cosmic sense, the Logos is the Idea of the world, or WORLD-IDEA held by the universal or COSMIC SOUL. Editors' Extension III, p. 147, is about the 5th house Idea, the IDEA OF MAN, as a logos.

Mahabhutas/Great Elements, Five The 12th house IDEA as designated by T.S. Row. Anthony sometimes refers to these elements as "directionalized space."

Mandala A symbol that represents philosophical and deific principles and their dynamic interplay through spatial and sequential arrangements. The Sanskrit word *mandala* is translated in the Tibetan language by words that mean literally center-circumference (*dkyil-khor*), but also have the additional interpreted meaning of "to grasp the essence." This indicates a circular process of churning the waters of experience to distill the core or essence (center). Mandala is a spiritual process as well as a sacred space. (See Herbert V. Guenther, *Matrix of Mystery*, pp. 42–45, and Giuseppe Tucci's *The Theory and Practice of the Mandala*. See also ASTROLOGICAL MANDALA.)

Matter Matter means something quite different for Plotinus than it does in contemporary thought, and has different meanings for him in different contexts. For Plotinus, Matter is not some kind of stuff, but the being of NON-BEING. His view of Matter has three principal meanings: Matter as separateness or otherness, Matter as receptivity and passivity, Matter as underlie of appearance or BECOMING. In the context of the ONE, where there is nothing else but the One, the POWER of the One is metaphorically called "Matter," through which the One is able to appear as infinity, and Ones. (See pp. 72–75 on Saturn in "The One: Symbolization" and Note 47, p. 341, on "Applying Three Contexts to Saturn as the Principle of Matter.")

Mentalism "By mentalism we mean more precisely this: that *all* things in human experience without any exception are wholly and entirely mental things and are not merely mental copies of material things; that this entire panorama of universal existence is nothing but a mental experience and not merely a mental representation of a separate material existence." (Paul Brunton, *Wisdom of the Overself*, p. 27) "The entire world that you perceive comes from your mind and is the experience of your mind by your mind." (Anthony Damiani, *Looking into Mind*, pp. 21–22) Mentalism has both cosmic and individual levels. The individual (human) soul's experience of the world is nested within the COSMIC SOUL'S WORLD-IDEA.

Mercurius* A term equivalent to UNIT SOUL, coined from the alchemical meaning of the Mercurius principle (the creative imagination capable of taking on any forms). In the METAPHYSICAL CHART, the symbol that represents this individual soul is the Mercury in Pisces—hence the appropriateness of the name.

Metaphysical Chart See ASTROLOGICAL MANDALA.

Monad An ultimate unit of being, from the root "to cut off." Anthony uses this term for the unific threads of Divinity which are continuous from the ONE to manifestation. Each divine principle is a monad series, or a "many" taken as a One. So the term is meant to include a unity *per se* and its many participating EMANANTS. In the METAPHYSICAL CHART, monads are the planetary groupings in Being, Life, Intellect, Nature, Soul, Body. (See also TRIAD, and Note 16, "Placing Dignities in Rings: Monads," p. 314.)

Monism The reduction of everything to one homogeneous unific substance. The Greek Parmenides, for example (on whom Plato's dialogue of that name is based), equated BEING with the ONE, and thus denied the possibility of multiplicity in Being. Anthony juxtaposed Monism to NONDUALITY, which he preferred as closer to his view of ultimate REALITY. For Plotinus, the One is infinite in "fathomless depths of power"—it is not to be belittled to a "point" (VI.9.6), as it is in monism's view of ultimate One-ness.

Mulaprakriti Primordial Matter, represented by Saturn in Aries. T.S. Row describes *Mulaprakriti* as the "veil" thrown over the Absolute (Brahman). His term *Mulaprakriti,* peculiar to Theosophy, means the root source which is beyond the made. (See his *Notes on the Bhagavad Gita.*)

Mysterium Magnum Literally, "Great Mystery." Paracelsus' term for Primordial Matter.

Natal Chart/Horoscope/Nativity The ASTROLOGICAL MANDALA, with a particular selection of PLANETS, SIGNS, HOUSES, DEGREES; the chart of the heavens (from the Earth view). This particular configuration of the moment of birth "stamps" the life of an INDIVIDUAL SOUL coming into earthly manifestation. (See also "The Third Phase of Soul" (p. 261–262), "Astrology: The Topography of the Subtle World" (pp. 264–270), and p. 292 ff. in our Epilogue.)

Terms that appear in SMALL CAPS *indicate cross references within this glossary. An asterisk (*) indicates a term related to soul's "modes" with body.*

Nature The standard translation of the Greek word *physis* or *phusis,* meaning the manifest cosmos. Plotinus writes in III.8 that even Nature contemplates. (John Deck's *Nature, Contemplation, and the One* explores this concept in detail.) See also SYSTEM OF NATURE.

Nirguna Brahman *Nirguna* means "without attributes." Nirguna Brahman means the transcendent unmanifest nature of the Absolute Reality (Brahman).

Noesis Knowingness—not knowing in the usual relative sense, but Wisdom. Awareness, Knowing, INTELLECTION, wisdom, contemplating, illuminating, all these terms help to capture something of the meaning of *Noesis. Noesis* is the state of self-knowing, and also the presence that makes Soul "insightful." The *Enneads* often employ illumination metaphors for *Noesis.* (Many translators use luminosity and knowingness to translate the Tibetan *rig-pa,* which is similar to *noesis.*) Paul Brunton and Anthony use the term INSIGHT.

Non-Being Plotinus says that universe has the "Being of non-Being." This is his technical description of the cosmos as becoming to be. It is also the ultimate meaning of MATTER. Non-Being does not stand outside of REALITY, but is a description of the status of appearances as contrasted with that of the PRIMALS.

Nonduality A term honoring the inviolability of REALITY as inclusive of multiplicity. It describes the nature of reality as "not-two," rather than as one homogeneous featureless substance. (See MONISM.)

Nous The Greek word for Mind, Awareness, Divine Intelligence, the Knower as immediately united with its content. MacKenna translates *Nous* as "INTELLECTUAL-PRINCIPLE," as it is the state or stasis of being self-knowing. Translating Nous by the terms Divine Mind, or Infinite Mind or Infinite Intelligence is also useful as well. Nous is from the root *Noeo,* which means both the contact of the mind and its object, and also the intent or tendance of the mind toward something. Nous is both awareness and that of which it is aware; it is inseparable knowing and that which is known. This identity of BEING and INTELLECTION has two sides: It *is* that which it knows, and it *knows* completely what it is. (See Note 6C, p. 304, on Intellectual-Principle or Nous.)

Number Platonism regards numbers as AUTHENTIC principles. They are qualitative rather than quantitative structuring principles of the Divine Mind and its vision, and they extend their power into the ordering of manifestation. Each number is a lens for viewing REALITY, and has a specific power of knowing. (See the Editors' Introduction to Number on page 78.)

Omnipresence Plotinus writes about the omnipresence of any AUTHENTIC EXISTENT to the whole of manifestation in tractates VI.4 and VI.5. Anthony discusses the Soul's omnipresence *vis á vis* Plotinus in the first pages of the Soul chapter. He sums up this concept as "everywhere is present to mind." (*Looking into Mind,* p. 58)

One, The Plotinus' term for ultimate or supreme REALITY, or for the ultimate nature of Reality. Plotinus equates The One with The Good. The One does not mean the first term in a series, but refers to the inviolability of Reality as the utter self-sufficiency and source of all POWER. Synonyms for the One are Absolute and Unity. Anthony discusses the One in the first chapter of *Astronoesis* and represents it by the first quadrant/first HOUSE. (For more on the term ONE see Note 6B, p. 303. See also MONISM.)

Ontology The study and science of BEING, derived from the Greek *on* (Being).

Overself* A term Paul Brunton coined for the higher self or soul. Overself is the meeting point of the human and the divine. It is a fragment or particle of the divine, a "ray shining from" the divine, but not the Sun itself. In one sense the Overself is the core of an individual, and in another sense the individual is held by the Overself.

Paradox This term comes from the words *para* and *doxa*. It may mean a way of understanding which is beyond (*para*) the opinion and belief (*doxa*) of the conventional mind. Opinion always sees things one way or another, but the REASONING of the Soul works with complementarities, or even contradictories, and sees things as combining two contraries—as paradox.

Participation This term means the way that any AUTHENTIC being, soul, and so on, shares in its principle. Think of this sharing in terms of immaterial Mind, not as everything getting a piece of something like a cake. For example, everyone in love is participating in the divine quality Love. Literally, they are *in* the *DAIMON* Eros, and EROS is in them. The entire dialogue *Parmenides* describes how individuals can participate in divine IDEAS, and then ultimately, how everything is enabled to participate in REALITY because the nature of Reality is Oneness. Participation is the obverse side of EMANATION.

Planets Anthony views planets both as souls in their own right and as functions at the metaphysical, cosmological, and individual (NATAL CHART) levels. (Pages 291 ff. in our Epilogue discuss the planets in astrology. See also CHALDEAN PLANETS, DIGNITIES, and TRANS-SATURNIAN PLANETS.)

Poiesis Means a "making" or a "production." For Plotinus (III.8.5–7), all production is through contemplation and for the purposes of contemplation; at every level production (*poiesis*) is for contemplation (*NOESIS*).

Power Power is an inseparable aspect of REALITY—not only something that Reality does, but something that it IS. Power not only radiates from the ONE, but also in-tenses in the One, is the One's self-reference. Power is related to Intelligence turning in on itself prior to radiating. (See the discussion of power on pp. 30–34. See also DYAD.)

Prajna "Wisdom." Prajna consciousness is often referred to as the causal or deep-sleep state of awareness, in which noetic awareness occurs without an appearance of objective content. Anthony sometimes calls this "spiritual subjectivity." See also TURIYA.

Prakriti Literally, "Nature" in Sanskrit; also means "matter" in the SAMKHYA Yoga school. Prakriti is the ever-active basis of all manifestation.

Primals Quoting Plato for his source in V.1.8, 10, Plotinus refers to the ONE, NOUS, and SOUL as "Primals."

Primordial Image The lucid state of world-image as held by the Earth mind and COSMIC SOUL. See also UNDIVIDED MIND (OF THE EARTH).

Providence What the Divine "provides" for the cosmos and the INDIVIDUAL SOULS within it. Almost equivalent to the grace of the divine, it is distinct from fate, which is the Greek idea of karma.

Psychosomatic* Literally, soul (*psyche*) and body (*soma*). Refers to the individual living being as a union of the light of the soul and the body, including subtle body. This term is equivalent to the Plotinian ANIMATE.

Purusha The SAMKHYA Yoga term for the unattached pure soul or consciousness. It means "person."

Reality That which is AUTHENTIC, which never ceases to be itself, and which is not able to be negated. It is not a term that translates exactly any Greek term in Plotinus, though often "ousia" is translated as Reality by MacKenna. For Anthony, reality is *both* transcendent and immanent, and not to be conceived as limited to either phase. In the transcendent view, Reality is equated to the One-Only. In the immanent view, there are degrees or levels of reality, so that in this way reality can include NOUS, SOUL, and NATURE. (Reality is discussed in Note 6A on page 303.)

Reason/Reasoning Anthony uses these terms for *dianoia*, the Greek word meaning "divided Nous." This power is much greater than what is meant by the modern "reason" (that equated with logic or opinion). It is deeper and also includes a component of EROS. Reason for Anthony means the function of the rational soul, that which forms a ratio between the divine and manifestation. In its Plotinian sense, reason is the power of the soul to reconcile the outer experience of the cosmos with its own innate vision and intelligence. (Anthony equates cosmic reason with the activity of the planetary SPHERES.)

Reason-Principles Distinct intelligences, the presence of the IDEAS in SOUL and NATURE. The Greek word *logoi* is used by Anthony to mean the Idea-intelligences present in soul, "unrolled and separate." (Plotinus, I.1.8) These units of meaning constitute the WORLD-IDEA, and the soul assimilates them to itself through experience. Anthony uses many terms—TROPES, *VASANAS,* embodied reason-principles, memory traces, residues, and tendencies—to evoke a feeling for the traces of life-activity that constitute the Earth's memory. The totality of these images, which are both repositories of habitual activity (tropic or memory traces) and instances of reason-principles or Idea-intelligence, is called the TROPICAL ZODIAC or the terrestrial DRAGON. (The sidereal dragon, or SIDEREAL ZODIAC, in contrast, is the realm of the stars and represents the Ideas in themselves.) (See II.3.13, 16; III.3.1–3; III.2.12–16; III.5.9; III.8.1–3; III.8.8. for Plotinus on reason-principles.)

Sage The fully developed and balanced human being who has permanent INSIGHT into the nature of REALITY. MacKenna uses the term "Adept."

Sahaja The SAGE's permanent realization of the natural, intrinsic nature of Mind. Anthony contrasts *sahaja samadhi* with *nirvikalpa samadhi:* In *sahaja,* the world continues to appear within the experience of REALITY, while *nirvikalpa* is the experience of reality without thought.

Samadhi Rapt contemplation. When the individual mind becomes so thoroughly absorbed in the object or subject of its contemplation that what is being contemplated shines forth as it is in its own light, uncolored by EGO-subjectivity, one has attained the state of *samadhi.* In yoga philosophy, this state is the culmination of a three-step process that begins with concentration (*dharana*) and continues with an intermediate state called meditation (*dhyana*). If the object is a form of or within the WORLD-IDEA, then this absorption is called *savikalpa* (with appearance) or *sabija* (with seed). If the absorption is into the formless Mind or OVERSELF, without the appearance of any form of the World-Idea, then it is called *nirvikalpa* (without appearance) or *nirbija* (without seed).

Samkhya A dualistic school of Indian philosophy that proposes a pure unattached *PURUSHA* (soul/consciousness) and grades of *PRAKRITI* (matter—everything else). This system of thought provides the theoretical basis for Patanjali's Yoga, in which the seeker's aim is to see what is not-oneself and isolate the pure *Purusha. Parinama* (transformation) is Samkhya's term for the unfolding of *Prakriti* as a whole sequence of layers constituting the universe. See K.C. Bhattacharyya's essay "Studies in Samkhya Philosophy" in *Studies in Philosophy.*

Sephiroth Attributes of the Divine, as taught in the Jewish mystical tradition of Kabbalah. For the Kabbalists, the tree of the ten *sephiroth* is a sacred MANDALA. Leo Schaya's book, *The Universal Meaning of the Kabbalah,* is an excellent introduction to the philosophy of the Kabbalah. His second chapter explains the meanings of the individual *sephiroth.* For a concise overview of the *sephiroth,* see Manly Palmer Hall, *The Secret Teachings of All Ages,* p. 121.

Sheaths/Bodies* Body in the wider sense means whatever the soul has identified with or made its own of the WORLD-IDEA through experience. Plotinus' term "organized" (as in "organized body" in I.1.7) means that which has been made into an organ of experience. At various levels, body means the organs of experience provided by the cosmos for the experience of the world. These levels of gross, subtle, and causal "body" are called *sheaths* or *vehicles.* Different levels of soul have different relations to their body, the entire cosmos being a vast body for the COSMIC SOUL. Body here is not physical/material in the usual sense, but is a vast thought-form projected by the soul. Each individual body is a vast thought-form in the cosmic ideation; we know that body only in a limited way. (Note 52, p. 346, "The Sheaths/Subtle Bodies and Plato's Divided Line" gives Anthony's overview of the sheaths in a cosmological context. See also G.R.S. Mead's *The Doctrine of the Subtle Body in the Western Tradition.*)

Sidereal Zodiac Sidereal means "referring to stars." The sidereal zodiac is a coordinate system on the circle of "living starry intelligence" as present in the Sun soul's (or SOLAR LOGOS') field of vision. It constitutes the cosmic intelligence as grasped by or in the Sun soul's contemplation (one form of the WORLD-MIND'S WORLD-IDEA in Paul Brunton's terminology). In *Astronoesis* the sidereal and TROPICAL ZODIACS are distinguished as representing the intelligence of the Solar mind (or soul) and the Earth mind, respectively.

Signs Twelve principal divisions of the ZODIAC. In astrology, each sign marks out a sector of cosmo-

---

Terms that appear in SMALL CAPS *indicate cross references within this glossary. An asterisk (*) indicates a term related to soul's "modes" with body.*

logical or transpersonal experience. T. Subba Row correlates metaphysical meanings to each of the signs: Aries = Self-Existent Brahman, Taurus = Pranava Aum, etc. (See Editors' Extension II, p. 131, "The Twelve Idea-Principles.") Anthony often uses the HOUSES and signs interchangeably in his discussion of the metaphysical chart, with Aries corresponding to the first house, Taurus to the second house, and so on. The glyphs for the signs are Aries ♈, Taurus ♉, Gemini ♊, Cancer ♋, Leo ♌, Virgo ♍, Libra ♎, Scorpio ♏, Sagittarius ♐, Capricorn ♑, Aquarius ♒, Pisces ♓.

Solar Logos The Sun as a transmitter of intelligence into the universe. The soul of the Sun unfolds the IDEAS through its life. (See Note 42, p. 336, "Notes on the Solar Logos/Cosmic Soul.")

Soul A translation of the Greek word *psyche.* It is the third Plotinian hypostasis and is represented in the third quadrant of the METAPHYSICAL CHART. It is discussed in the third chapter of *Astronoesis,* although the conjunction of living soul with body takes place in the SYSTEM OF NATURE.

Spheres: Inerratic, Planetary, Sublunary This threefold division of the cosmos corresponds to INTELLECTION, REASONING, and Imagination.

Sublunary In the system of planetary spheres that represents cosmic REASONING, "sublunar" means that which is under the Moon. The Moon rules the imagination and memory. Sublunary refers to the aspect of the cosmos and of the individual which is inseparable from the elemental forces and the DRAGON: the totality of *VASANAS* or memorized traces of life energy which constitute the organized body, emotions, and mentality.

Substance Means "stands-under." For Anthony, the true substance is Intelligence or Mind. (See HYPOSTASIS.)

Synesthesis A term meaning "with perception." In Anthony's and Plotinus' use of it in discussing the ONE, it means more precisely self-perception or self-contained perception.

System of Nature The entirety of universal manifestation. Chapter IV discusses this last of the Plotinian metaphysical stadia, which Anthony represents with the fourth quadrant, or alternatively, the last four houses of the METAPHYSICAL CHART. Through the system of Nature, the soul enters the realm of BECOMING, existence, and appearance in which to gain experience.

Taijasa Consciousness The state of consciousness usually designated the "dream" state; also correlated by Anthony with the REASONING phase of the soul (psychical subjectivity).

Theophany Literally this means the showing forth (*phanes*) of deity (*theos*), i.e., the Deities' showing forth. It may also be translated as "Divine Ideation," meaning that the cosmos is a WORLD-IDEA. It is very close to the meaning of EMANATION and effulgence. Henry Corbin's book *Creative Imagination in the Sufism of Ibn 'Arabi* has much valuable discussion of this term.

Trans-Saturnian Planets: These three planets (beyond Saturn)—Uranus ♅, Neptune ♆, Pluto ♇—represent the higher or cosmic modalizations of Knowing, Feeling, Willing. (For more on Uranus, Neptune, and Pluto, see the margin note on p. 271 and also pp. 296–297 in our Epilogue.)

Transits Usually mean significant angular relations (ASPECTS) made to "fixed" points in a natal chart by the PLANETS in their continual movement in the sky.

Triad A word meaning a view of the ONE as Ones, following the view of One as Unity, and One as POWER (DYAD). Anthony symbolizes the Triad by the DIGNITIES in the first quadrant of the METAPHYSICAL CHART.

Trope Trace of life experience, habit formation, structuralized form of energy = Sanskrit *VASANA.* The totality of the tropes is called the terrestrial DRAGON.

Tropical Zodiac The tropical zodiac provides a coordinate system in which the Earth "orientation" is crucial. One meaning of the term TROPE is turning. In this sense, the "tropical" zodiac refers to the "turning" points of the Earth orientation we call the solstices (and the balance points of the turning called equinoxes). "Tropical" also is derived from the word tropism which means a kind of instinctual "turning" (like helio-trope, a turning toward the Sun). In *Astronoesis,* the tropical zodiac is the repository of memory, the traces of all life-activity on Earth, symbolically/mathematically organized into the 360-fold symmetry of the tropical DEGREES. Throughout history there have been various attempts to symbolize in image the 360 degrees of the tropical zodiac. By contemplating the degree symbols, we may get a glimpse of the IDEAS and their meanings in our lives.

Turiya Sanskrit word meaning "fourth"; it is the fourth, the deepest and ultimate nature of consciousness. Turiya is distinguished from and includes the three "states" of consciousness called *VAISHVANARA* (wakeful), *TAIJASA* (dream), and *PRAJNA* (sleep). (See margin note on page 70, Note 13 (p. 312) on "The Maitri Upanishad," and the Mandukya Upanishad.)

Undivided Mind (of the Earth) Anthony uses this term to mean the deeper mind, the Mind of the World or the Earth-Mind, which is the most immediate source of the WORLD-IDEA for individuals here. He describes the world-thought or "world-ideogram" of the Undivided Mind as "transparently lucid and subtle." (p. 233)

Unit Soul* The INDIVIDUAL SOUL as a ray or intimate EMANANT of the PRIMAL principles.

Unities These are the "Ones" in the ONE, which Anthony symbolizes by the DIGNITIES of the first quadrant of the METAPHYSICAL CHART. (See HENADS and TRIAD. Figures 20–25 (pp. 61–66) and their accompanying text describe the unities in the One.)

Universal (or World) Constructor Reference is to the entire mind-realm as intermediate between *sunyata* (emptiness) and appearances. Maitreya (*Madhyanta-Vibhanga,* chapters 2 and 3) describes this activity as "productive imagination."

Vaishvanara Consciousness The cosmic state of wakeful consciousness, or that of embodiment (physical subjectivity).

Vasana Energic habit pattern, usually subconscious memorized impression of life-activity. (See TROPE. See also Mircea Eliade, *Yoga: Immortality and Freedom,* p. 42.)

Vedanta, Advaita Vedanta A nondual school of Indian thought based on the Vedas, sutras, and commentaries.

Vision The vision or Contemplation (Plotinus' term) of the SAGES is the source of Anthony's exposition. Contemplation is, in the words of Plotinus, a "Seeing that lives." (III.8.8) The ancient Vedas, which contain the Upanishads, are themselves "seeings." According to T.M.P. Mahadevan, the root word for *Veda* is "vid," a cognate of "wis"—wisdom, also meaning to see—and thus *veda* is the vision of REALITY delivered by the seers. *Buddha* has its root in *Budh,* also meaning seeing, or one who sees. Paul Brunton uses the term *insight,* which might be summarized as the direct immediate experience of Mind in itself by the mind in ourselves: intrinsic self-cognition. Insight does not imply a blankness nor only a single static fact, but is living intelligence, continually overflowing wisdom and activity.

Vivarta Hindu theory of illusory manifestation in which (to use the famous example) the trinkets (world) are only illusory manifestions of gold (Brahman or REALITY).

"We"* A term Plotinus uses to refer to the double nature of the soul, usually the rational soul, standing behind or beyond the individual PSYCHOSOMATIC EGO. That organism is a "We" in the sense of being a union of soul and IDEAS, or as being the place where souls are interrelated. (See Plotinus, I.1.7, 10. Plotinus also discusses the "We" in I.1.7–13, II.3.9, IV.4.18, II.1.5, and V.3.3.)

Witness-I* The soul in relation to the WORLD-IDEA at any level. It consists of four aspects or functions: a) the witness of the empirical subject and object, b) the creative power of the mind to transform the World-Idea into our experience, c) the accumulated wisdom of experience, d) the presence of the soul's life-current as our awareness.

World-Idea Paul Brunton's term for the divine order of the universe, the nature of the world as a vast thought held by the WORLD-MIND. The World-Idea is behind "scientific laws" and gives coherence to the experience of individuals participating in it. Our task to consciously align our actions with the evolutionary direction of the World-Idea.

World-Mind Paul Brunton's term for the active power of Mind. Often he uses it synonymously with God, for the higher creating mind (or soul). Our soul as OVERSELF is a particle of this World-Mind. Anthony sometimes uses this term to mean the Universal Soul or DEMIURGE, leader of souls, but sometimes to mean the SOLAR LOGOS or COSMIC SOUL. Of WORLD-IDEA and World-Mind Paul Brunton writes: "Whatever we call it, most people feel—whether vaguely or strongly—that there must be a God and that there must be something which God has in view in letting the universe come into existence. This purpose I call the World-Idea, because to me God is the World's Mind . . ." (*Perspectives,* p. 360)

Zodiac "Zodiac" comes from the Greek *zoidion* meaning a "place for life" or "living place." Zodiac means the presence of the placeless eternal Divine Intelligence or WORLD-IDEAS in a living place. Since cosmos is manifest in soul (life), Zodiac is the presence of the (World) IDEAS in the soul. On the other hand, as the living activity of soul takes place within the infinite Divine Mind, we can also say that the Zodiac represents that "place" in the Divine Mind where the soul lives. Astrology is based on this grand vision: the cosmos as living wisdom, and every soul or mind from the cosmic to the individual human actualizing a form of wisdom—represented by a Zodiac. (See also SIDEREAL ZODIAC, TROPICAL ZODIAC.)

Bibliography

This bibliography includes only works cited or used directly in the making of *Astronoesis*. It is by no means a complete list of the books Anthony read or considered to be essential philosophical works.

Affifi, A.E. *The Mystical Philosophy of Muhyid Din-Ibnul Arabi.* Lahore, Pakistan: Sh. Muhammad Ashraf, 1964.

Apuleius. *The God of Socrates.* Foreword by Daniel Driscoll. Translated by T. Taylor. Gillette, NJ: Heptangle Books, 1993.

Aranya, Hariharananda. *Yoga Philosophy of Patanjali.* Albany, NY: State University of New York Press, 1983.

Aristotle. *Metaphysics.* Cambridge, MA: Loeb Classical Library, Harvard Univ. Press, 1933.

Avens, Roberts. *Imagination Is Reality: Western Nirvana in Jung, Hillman, Barfield, and Cassirer.* Irving, TX: Spring Publications, 1980.

Bhagavad Gita. Translated by S. Nikhilananda. New York: Ramakrishna Vivekenanda Mission, 1964.

Bhaduri, Sadananda. *Studies in Nyaya-Vaishesika Metaphysics.* Poona: Bhandarkar Oriental Research Institute, 1947.

Bhattacharyya, Krishna Chandra. *Studies in Philosophy.* 2 vols. Calcutta: Motilal Banarsidass, 1958. Reprinted in 1 volume: Delhi: Motilal Banarsidass, 1983.

Blavatsky, H.P.B. *The Secret Doctrine* (with index). 6 vols. Adyar: Theosophical Publishing House, 1971.

Brahma, Nalini K. *Philosophy of Hindu Sadhana.* London: Kegan Paul, Trench, Trubner & Co., 1932.

Brahma-Sutras. Translated by Swami Vireswarananda. Calcutta: Advaita Ashrama, 1936.

Brumbaugh, Robert. *Plato's Mathematical Imagination.* Bloomington, IN: Indiana University Press, 1977.

Brunton, Paul. *Essays on the Quest.* York Beach, ME: Samuel Weiser, Inc., 1985.

——. *The Hidden Teaching Beyond Yoga.* New York: Samuel Weiser, Inc. 1971.

——. *The Notebooks of Paul Brunton.* 16 vols. Burdett, NY: Larson Publications, 1984–1988.

——. *Perspectives.* Vol. 1 of *The Notebooks of Paul Brunton.* Burdett, NY: Larson Publications, 1984.

——. *The Quest of the Overself.* London: Rider & Co., 1937. Reprint, New York: Samuel Weiser, 1975.

——. *The Wisdom of the Overself.* York Beach, ME: Samuel Weiser, Inc., 1984.

Burckhardt, Titus. *Alchemy.* Translated by William Stoddart. London: John M. Watkins, 1967.

Chan, Wing-Tsit, ed. *A Source Book in Chinese Philosophy.* Princeton, NJ: Princeton University Press, 1963.

Chang, Garma C. C. *The Buddhist Teaching of Totality: The Philosophy of Hwa Yen Buddhism.* University Park, PA: Pennsylvania State University Press, 1971.

Charubel and Sepharial. *The Degrees of the Zodiac Symbolized by Charubel to which is added Sepharial.* Chicago: The Aries Press, 1943.

Chatterji, J.C. *Kashmir Shaivism.* 1914. Reprint, Chandigarh, India: Galav Pubs., 1974.

Conze, Edward, ed., trans. "Hridaya Sutra" (Heart Sutra) from *Selected Sayings from the Perfection of Wisdom.* Boulder, CO: Prajna Press, 1978, pp. 74–75.

Coomaraswamy, Ananda K. "The One and Only Transmigrant." In Coomaraswamy, *Selected Papers: Metaphysics.* Edited by Roger Lipsky. Princeton, NJ: Princeton University Press, 1972.

Corbin, Henry. *Creative Imagination in the Sufism of Ibn 'Arabi.* Translated by Ralph Manheim. Princeton, NJ: Princeton University Press, 1969.

——. *The Man of Light in Iranian Sufism.* Translated by Nancy Pearson. Boulder, CO: Shambhala, 1978.

Dalai Lama XIV Tenzin Gyatso. *The Buddhism of Tibet and the Key to the Middle Way.* New York: Harper & Row, 1975.

——. *Kindness, Clarity, and Insight.* Ithaca, NY: Snow Lion, 1984.

Dalai Lama XIV, Tsong-ka-pa, and Jeffrey Hopkins. *Tantra in Tibet.* Ithaca, NY: Snow Lion, 1977.

Damiani, Anthony. *Living Wisdom.* Burdett, NY: Larson Publications, 1997.

——. *Looking into Mind.* Burdett, NY: Larson Publications, 1990.

——. Notes and Papers. Unpublished papers and class transcripts. Wisdom's Goldenrod Library, Valois, NY.

——. *Standing in Your Own Way.* Burdett, NY: Larson Publications, 1993.

——. Introduction to *True Lunar Nodes: 1850–2000.* Ithaca, NY: Digicomp Research Corporation, 1975.

Damiani, Paul. "An American Philosopher." Valois, NY: Wisdom's Goldenrod Library, 1984.

Danielou, Alain. *Hindu Polytheism.* New York: Bollingen Series LXXIII, Pantheon Books, 1964.

Deck, John. *Nature, Contemplation, and the One.* Toronto: University of Toronto Press, 1967. Reprint, Burdett, NY: Larson Publications, 1991 (page references are to reprint edition).

Desjardins, Arnaud. *The Message of the Tibetans: Second Part: Tantrism.* 52 minutes. AD Editions Alize Diffusion, 1994. Videocassette.

Dionysius the Areopagite. *Mystical Theology and the Celestial Hierarchies.* Translated by the Editors of the Shrine of Wisdom. Nr. Godalming, Surrey, England: The Shrine of Wisdom, 1965.

Dyczkowski, Mark. *The Doctrine of Vibration: An Analysis of the Doctrine and Practices of Kashmir Shaivism.* Albany, NY: State University of New York Press, 1987.

Eckhart, Meister. *Meister Eckhart: Sermons and Tractates.* Translated by C. deB. Evans. Edited by Franz Pfeiffer (Leipzig, 1857). London: John M. Watkins, 1956.

Eliade, Mircea. *Myths, Dreams, and Mysteries.* Translated by Philip Mairet. New York: Harper & Row, 1960.

——. *Yoga: Immortality and Freedom.* Translated by Willard R. Trask. New York: Pantheon Books (Bollingen), 1958.

Findlay, J.N. *Plato: The Written and Unwritten Doctrines.* London: Routledge & Kegan Paul, 1974.

Fox, Matthew, and Rupert Sheldrake. *The Physics of Angels.* San Francisco: HarperSanFrancisco, 1996.

Gilson, Etienne. *Being and Some Philosophers.* 2nd ed. Toronto: Pontifical Institute of Mediaeval Studies, 1952.

Govinda, Anagarika. *Foundations of Tibetan Mysticism.* New York: Samuel Weiser, 1973.

Guénon, René. *The Multiple States of Being.* Translated by Joscelyn Godwin. Burdett, NY: Larson Publications, 1984.

——. *The Reign of Quantity and the Signs of the Times.* Translated by Lord Northbourne. London: Luzac & Co., 1953.

——. *Man and His Becoming According to the Vedanta.* London: Luzac & Co., 1945.

Guenther, Herbert V. *Matrix of Mystery.* Boulder, CO: Shambhala, 1984.

——. *Tibetan Buddhism in Western Perspective.* Emeryville, CA: Dharma Publishing, 1977.

Hadot, Pierre. *Plotinus: Or The Simplicity of Vision.* Translated by Michael Chase. Chicago: University of Chicago Press, 1993.

Halevi, Zev ben Shimon. *Adam and the Kabbalistic Tree.* New York: Samuel Weiser, 1974.

——. *A Kabbalistic Universe.* New York: Samuel Weiser, Inc. 1977.

Hall, Manly Palmer. *The Adepts in the Eastern Esoteric Tradition, Part One, the Light of the Vedas.* Los Angeles: The Philosophical Research Society, Inc., 1952.

——. *Lectures on Ancient Philosophy.* Los Angeles: The Hall Publishing Co., 1929.

——. *Man, Grand Symbol of the Mysteries.* Los Angeles: The Philosophical Research Society, 1972.

——. *The Secret Teachings of All Ages.* Los Angeles: The Philosophical Research Society, 1977.

Harris, T. Lake. *Arcana of Christianity.* Vol. 1. New York: New Church Publishing Association, 1858. Reprint, New York, AMS Press, 1976 (page reference is to reprint edition).

Hartmann, Franz. *The Life and Doctrines of Philippus Theophrastus Bombast of Hohenheim Known by the Name of Paracelsus.* New York: Macoy Publishing and Masonic Supply Company, 1932.

Hegel, Georg W.F. *Philosophy of Subjective Spirit.* Vol. 1. Edited and translated by M.J. Petry. Boston: D. Reidel Publishing Co., 1978.

Hopkins, Jeffrey. *Meditation on Emptiness.* London: Wisdom Publications, 1983.

I Ching: The Classic Chinese Oracle of Change. Translated by Rudolf Ritsema and Stephen Karcher. Rockport, MA: Element, Inc., 1994.

Iamblichus. *On the Mysteries of the Egyptians, Chaldeans and Assyrians.* 3rd ed. Translated by Thomas Taylor. London: Stuart and Watkins, 1968.

Jones, Marc Edmund. *Lecture Notes on Pythagorean Astrology.* n.p.: 1928–1934.

——. *The Sabian Symbols in Astrology.* 3rd ed. Stanwood, WA: Sabian Publishing Society, 1969. Reprint, Santa Fe, NM: Aurora Press, 1993.

——. "Symbolical Astrology" [Sabian Symbols]. In *Lecture Notes,* 1931.

Julian. *Two Orations of the Emperor Julian.* Translated by Thomas Taylor. 1793. Reprint, Chicago: Hermetic Publishing Co., 1932 (page references are to reprint edition).

Jung, Carl G. *The Collected Works.* Volumes 3, 6, 8, 9 parts I and II. Translated by R.F.C. Hull. Princeton, NJ: Princeton University Press, 1971–1974.

——. *On the Nature of the Psyche.* Translated by R.F.C. Hull. Princeton, NJ: Princeton University Press, Bollingen, 1960.

——. "On the Nature of the Psyche." In *Structure and Dynamics of the Psyche,* vol. 8 of *The Collected Works of C.G. Jung.* 2nd ed. Translated by R.F.C. Hull. Princeton, NJ: Princeton University Press, Bollingen, 1969.

——. "Psychological Commentary on Kundalini Yoga—1932" in *Spring Magazine, 1975–1976.*

Kozminsky, Isidore. *Zodiacal Symbology and Its Planetary Power.* Tempe, AZ: American Federation of Astrologers, Inc., n.d.

Long-ch'en Rb-jam-pa. *The Four-Themed Precious Garland.* Translated by Alexander Berzin, with commentary by Dudjom Rimpoche and Beru Khyentse Rimpoche. Dharmsala, India: Library of Tibetan Works and Archives, 1979.

Lawlor, Robert. Introduction to *Symbol and the Symbolic* by Schwaller de Lubicz. New York: Inner Traditions, 1978.

Hans Levy. *Chaldean Oracles and Theurgy.* Cairo: de l'Institut Francais D'Archeologie Orientale, 1956.

Mahadevan, T.M.P. *The Philosophy of Advaita.* Madras: Ganesh & Co., 1969.

Maitreya. *Madhyanta-Vibhanga: Discourse on Discrimination between Middle and Extremes.* With commentary by Vasubandhu and Sthiramati. Translated from the Sanskrit by Th. Stcherbatsky. Moscow: Academy of Sciences of USSR Press, 1936.

——. *The Sublime Science of the Great Vehicle to Salvation.* Leningrad: E. Obermiller, 1931.

Mayo, Jeff. *The Astrologer's Astronomical Handbook.* London: L.N. Fowler, 1972.

Mandukya Upanishad with Shankara's Commentary. Translated by S. Nikhilananda. Mysore: Sri Ramakrishna Ashrama, 1965.

Mead, G.R.S. *The Doctrine of the Subtle Body in the Western Tradition.* 3rd ed. Shaftesbury, U.K.: Solos Press, n.d.

——. *Orpheus.* London: John M. Watkins, 1965.

——. *The World-Mystery: Four Comparative Studies in General Theosophy.* London: The Theosophical Publishing Company, 1907. Reprint, Mokelumne Hill, CA: Health Research, 1974.

Mukhopadyaya, G. *Studies in the Upanishads.* Calcutta: Sanskrit College, 1960.

The New English Bible. Cambridge: Oxford University Press, Cambridge University Press, 1970.

Papus (Gerard Encaussse). *The Qabalah.* New York: Samuel Weiser, Inc. 1977.

Pausanias. *The Description of Greece.* 3 vols. Translated by Thomas Taylor. London: Richard Priestly, 1824.

Penny, A.J. *Studies in Jacob Boehme.* London: John M. Watkins, 1912.

Plato. *The Collected Dialogues.* Edited by Edith Hamilton and Huntington Cairns. New York: Bollingen Foundation, 1961.

——. *The Works of Plato.* 5 vols. Translated by Thomas Taylor and Floyer Sydenham. London: R. Wilke, 1804.

Plotinus. *The Enneads.* 4th ed. Translated by Stephen MacKenna. With Appendix, "A Suggestive Outlline of Plotinian Metaphysics." Burdett, NY: Larson Publications, 1992.

——. *The Enneads.* 7 vols. Translated by A.H. Armstrong. Cambridge, MA: Harvard University Press (Loeb Classical Library), 1967–1988.

___. *Five Books of Plotinus.* Translated and with introduction by Thomas Taylor. London: 1794. Reprinted as *Selected Works of Plotinus.* New edition with preface and bibliography by G.R.S. Mead. London: G. Bell and Sons, Ltd., 1912. Frome, Somerset, UK: The Prometheus Trust, 1994. Reprinted in *Collected Writings of Plotinus.*

——. *Plotinos: Complete Works.* Translated by K.S. Guthrie. Yonkers, NY: The Platonist Press, 1918.

——. *The Six Enneads.* 1st ed. Translated by Stephen MacKenna. Chicago: Encyclopedia Britannica, Inc., William Benton Pub., 1952.

Proclus. *Commentaries on the Timaeus of Plato.* Translated by Thomas Taylor. London: 1820.

——. *A Commentary on the First Book of Euclid's Elements.* Translated by Glenn R. Morrow. Princeton, NJ: Princeton University Press, 1970.

——. *Elements of Theology.* Translated by Thomas Taylor. (1792) Frome, Somerset, England: The Prometheus Trust, 1994.

——. *The Philosophical and Mathematical Commentaries of Plato on the First Book of Euclid's Elements.* 2 vols. Translated by Thomas Taylor. London, 1792.

——. *Proclus' Commentary on Plato's Parmenides.* Translated by Glenn R. Morrow and John M. Dillon, with introduction and notes by John M. Dillon. Princeton, NJ: Princeton University Press. 1987.

——. *The Six Books of Proclus on the Theology of Plato* and *Proclus' Elements of Theology.* London: A.J. Valpry, 1816.

Ptolemy, Claudius. *Tetrabiblos.* Books I (with an Appendix "Ptolemy's Hypotheses of the Planets: Book II"), III, IV (Book II not yet published). Translated and annotated by Robert H. Schmidt. Berkeley Springs, WV; Cumberland, MD: The Golden Hind Press, 1994–1998.

Raine, Kathleen, and George Mills, eds. *Thomas Taylor: The Platonist: Selected Writings.* Princeton, NJ: Princeton University Press, 1969.

Radhakrishnan, Sarvepalli, ed. and trans. *The Principal Upanishads.* New York: Humanities Press, Inc., 1953.

Ramamurthi, S.V. "Advaita as the Culmination of Mathematics" in *Kumbhakonam Advaita Sabha Golden Jubilee Commemoration Volume.* Kancheepuram: Sri Sankara Bhaktha Jana Sabda, 1978.

Row, T. Subba. *Esoteric Writings.* (1895): Adyar, Madras: The Theosophical Publishing House, 1980.

——. *Notes on the Bhagavad Gita.* Pasadena, CA: Theosophical University Press, 1978.

——. *The Twelve Signs of the Zodiac.* Minneapolis: Wizards Bookshelf, 1977.

Sallust. *On the Gods and the World.* Translated by Thomas Taylor. (1795) Los Angeles: Philosophical Research Society, 1976.

Schaya, Leo. Translated by Nancy Pearson. *The Universal Meaning of the Kabbalah.* Secaucus, NJ: University Books, 1971.

Sharma, I.C. "The Plotinian One and the Concept of Parampurusha." In *Neoplatonism and Indian Thought,* edited by R. Baine Harris. Albany, NY: State University of New York Press, 1982.

Shrine of Wisdom, eds. "The Chaldean Oracles." *The Shrine of Wisdom* (1925–1926).

——. "Extracts from Thomas Taylor's Dissertation on the Platonic Theory of Ideas." *The Shrine of Wisdom,* ix, nos. 41, 42 (1929).

——. "The Simple Way of Lao Tsze," *The Shrine of Wisdom,* v, no. 18 (1923).

——. *The Simple Way of Lao Tsze: An Analysis of the Tao Teh Canon.* Nr. Godalming, Surrey, England: The Shrine of Wisdom, 1974 (1924).

——. "Symbols: Numbers." *The Shrine of Wisdom,* iv, no. 16 (1923).

Taimni, I.K. *Man, God and the Universe.* Adyar, India: The Theosophical Publishing House, 1969.

——. *The Science of Yoga.* Wheaton, IL: The Theosophical Publishing House, 1975.

Taylor, Thomas. *Hymns and Initiations.* Reprint of *The Mystical Hymns of Orpheus* [2nd ed., 1824] and extracts from other works. Frome, Somerset, England: The Prometheus Trust, 1994.

——. *The Hymns of Orpheus.* Introductory preface by Manly P. Hall. Facsimile reprint of 1792 edition. Los Angeles: The Philosophical Research Society, 1981.

——. Introduction and Notes to *The Parmenides* in *The Works of Plato,* Vol. III. London: R. Wilke, 1804.

——. *The Theoretic Arithmetic of the Pythagoreans.* New York: Samuel Weiser, 1978.

Tucci, Giuseppe. *The Theory and Practice of the Mandala.* Translated by A.H. Brodrick. New York: Samuel Weiser, 1978.

The Upanishads. 3 vols. Translated by Swami Nikhilananda. New York: Harper & Brothers, 1952.

von Franz, Marie-Louise. *Number and Time.* Evanston: Northwestern University Press, 1974.

Wachsmuth, Guenther. *The Etheric Formative Forces in Cosmos, Earth and Man.* Vol. 1. translated by O. Wannamaker. New York: Anthroposophic Press, 1932.

Woodroffe, John (Arthur Avalon). *The Garland of Letters.* Madras: Ganesh & Co., 1974.

——. *The Serpent Power.* 7th ed. Madras: Ganesh & Co., 1964.

Index

THIS INDEX TREATS REFERENCES, topics, authors, titles, and (where appropriate) figures. An asterisk (*) indicates where a term is defined or used in a context that helps explain its meaning (or is otherwise significant).

An asterisk (*) indicates where a term is defined or used in a context that helps explain its meaning (or is otherwise significant).

An asterisk (*) indicates where a term is defined or used in a context that helps explain its meaning (or is otherwise significant).

An asterisk (*) indicates where a term is defined or used in a context that helps explain its meaning (or is otherwise significant).

An asterisk (*) indicates where a term is defined or used in a context that helps explain its meaning (or is otherwise significant).

An asterisk (*) indicates where a term is defined or used in a context that helps explain its meaning (or is otherwise significant).

An asterisk (*) indicates where a term is defined or used in a context that helps explain its meaning (or is otherwise significant).

An asterisk (*) indicates where a term is defined or used in a context that helps explain its meaning (or is otherwise significant).

An asterisk (*) indicates where a term is defined or used in a context that helps explain its meaning (or is otherwise significant).

An asterisk (*) indicates where a term is defined or used in a context that helps explain its meaning (or is otherwise significant).

An asterisk (*) indicates where a term is defined or used in a context that helps explain its meaning (or is otherwise significant).

Index of Passages Cited from Plotinus' Enneads

Unless otherwise indicated, all passages are from the fourth edition of Stephen MacKenna's translation of *Plotinus: The Enneads* published by Larson Publications, 1992.

Corrections and Additions since the First Printing

MANY CORRECTIONS ARE INDICATED simply as "x should be y." Changes made in the text are also noted here, with reference to the printing in which they were made. The number of the printing in your hands is indicated by the last number from the right on the bottom line of the copyright page (p. *iv*). Readers may also visit www.larsonpublications.org for the most up-to-date errata.

P. 11, FIG. 3: From information made available for the second printing, figure 3 and its caption now differ slightly from what appeared in the first printing. The current chart is the one Anthony Damiani used for himself, from which he accurately predicted several significant events. His birth time was given as "between 10 and 11 AM."

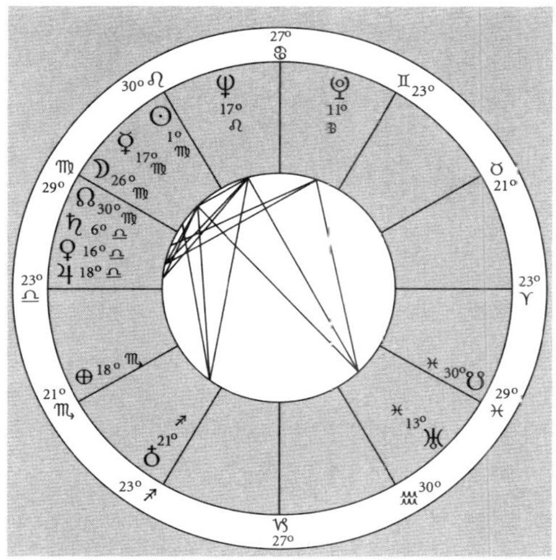

FIG. 3: ANTHONY DAMIANI, NATAL CHART
AUGUST 24, 1922, 10:42 AM EDT
NEW YORK, NY, USA
LATITUDE 40:44 N, LONGITUDE 73:59 W
PLACIDUS HOUSES

From that starting point, Anthony arrived at this chart by combining his insights into his own soul with his knowledge of the degree symbols—especially those of the Ascendant, Midheaven, and Part of Fortune. In addition, he carefully observed the ingress of various planets over the cusps and other sensitive points—balancing the ideal with the practical and testing his own self-knowledge against the impersonal judge of circumstance. *Source rating: A/C. Source: Timothy J. Smith.*

P. 17: ▼ changed to ⊞ and moved up to end of first sentence in paragraph ("... of astrology.") in second printing.

PP. 50–51: margin notes moved to p. 51 in second printing.

P. 68, LINE 3: [13, P. 314] should read [13, P. 312]

P. 111, FIG. 45: In the second printing, ☽♉ in "Being" row (dark orange) changed to ☽♋; also, ☽♋ in "Intellect" row (mid-gray) changed to ☽♉.

P. 126, MARGIN NOTE: Should begin PLATO, *REPUBLIC* (T. Taylor trans., V. 1 PP. 345–352):

P. 146, MARGIN NOTE, LINE 3: "thirds, quarters, for example..." should read "quarters, for example ..."

P. 228, LINE 11: "... amplification is ..." should read "... amplification are ..."

P. 241, SECOND MARGIN NOTE, LINE 10: "... deploying to the ..." should be "... deploying the ..."

P. 248, FIG. 105: ⛢ is repositioned one ring closer to ☉, no longer in the same ring with ♃. Changed in second printing.

P. 260, FIG. 112: Misprints in Emerson's natal chart are corrected for the second printing:

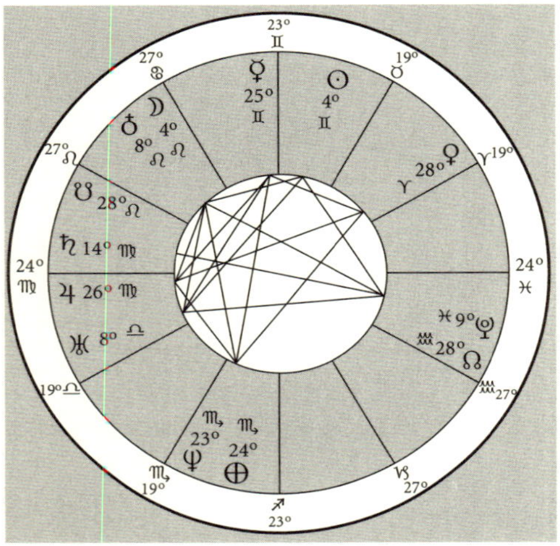

FIG. 112: RALPH WALDO EMERSON
MAY 25, 1803, 1:01 PM EST
CONCORD, MA, USA
LATITUDE 42:28 N, LONGITUDE 71:21 W
PLACIDUS HOUSES

Note that ♅ is at 8°♎ rather than 9°♎, and that ♍ replaces ♓ on the Descendant. From the variety of options available for Emerson's birth chart, Anthony used this one in his astrology class, based on information in M.E. Jones, *The Sabian Symbols in Astrology,* pp. 350, 376. These corrections are also made as needed in Emerson charts on pages 262, 264, 268, 271, and 275.

P. 262, FIG. 113: ⛢ at 8°♌ rather than 7°♌; ♅ at 8°♎ rather than 9°♎.

P. 264, FIG. 114: same changes as in fig. 112, p. 260.

P. 268, FIG. 116, middle circles (Emerson natal chart): same changes as p. 260; extra ♃ removed at bottom of column of Chaldean planets in left circle.

P. 271, FIG. 117: Emerson natal chart, same changes as p. 260.

P. 275, FIG. 119: Emerson natal chart at "Imagination" level, same changes as P. 260.

P. 334, COLUMN 3, LINE 6: "$♃_{♐}$ is the . . ." should be "$♃_{♊}$ is the . . ." Changed in second printing.

P. 343, FIG. 164: caption should read "FIG. 164: BODY MONAD, LED BY $♄_{♈}$"

P. 346, COLUMN 2, EDITORS' NOTE, LINE 7: correct title is: *The Doctrine of the Subtle Body in the Western Tradition.*

P. 357, COLUMN 2, ASMITA ENTRY, LINE 7: "I AM" should appear as "I AM".

P. 365, COLUMN 2, LAST TWO LINES: ". . . on p. 271 and also . . ." should be ". . . on pp. 270–271 and also . . ."

P. 366, COLUMN 3, LINE 4: "Our task to . . ." should be "Our task is to . . ."

P. 371, COLUMN 1, LINE 1: "This index treats references, topics, authors . . ." should be "This index references topics, authors . . ."

PP. 371-390 (THROUGHOUT INDEX): References to 51 N.1 and 51 N.2 apply to notes mislocated on page 50 in the first printing, and corrected to page 51 in the second printing.

Text set in Adobe Minion, supplemented with Monotype Centaur and with AstroLarson, designed for this book.

Imageset, printed, and bound in the United States of America by Syracuse Lithography Company.

Printed on Finch Opaque Vanilla Vellum.

manuscript preparation	June Fritchman
design & layout	Phil Wilson Design
revisions & production	Paperwork